STRENGTH OF METALS AND ALLOYS
(ICSMA 6)

Proceedings of the 6th International Conference
Melbourne, Australia, 16-20 August 1982

Volume 3

INTERNATIONAL SERIES ON THE
STRENGTH AND FRACTURE OF MATERIALS AND STRUCTURES
General Editor: D. M. R. TAPLIN, D.Sc., D. Phil., F.I.M.

Other Titles of Interest

BELY	Friction and Wear in Polymer-Based Materials
BUNSELL	Advances in Composite Materials (ICCM 3) - 2 volumes
EASTERLING	Mechanisms of Deformation and Fracture
FRANCOIS	Advances in Fracture Research (ICF 5) - 6 volumes
GARRETT	Engineering Applications of Fracture Analysis
HAASEN et al	Strength of Metals and Alloys (ICSMA 5) - 3 volumes
KRAGELSKY	Friction, Wear and Lubrication - 3 volumes
KRAGELSKY et al	Friction and Wear - Calculation Methods
MILLER & SMITH	Mechanical Behaviour of Materials (ICM 3) - 3 volumes
OSGOOD	Fatigue Design, 2nd Edition
RADON	Fracture and Fatigue (ECF 3)
SIMPSON	Fracture Problems and Solutions in the Energy Industry
SMITH	Fracture Mechanics: Current Status, Future Prospects
TAPLIN	Advances in Research on the Strength and Fracture of Materials (ICF 4) - 6 volumes

Related Pergamon Journals *(Free Specimen Copy Gladly Sent on Request)*

Acta Metallurgica

Canadian Metallurgical Quarterly

Corrosion Science

Engineering Fracture Mechanics

Fatigue of Engineering Materials and Structures

Journal of Mechanics and Physics of Solids

Journal of the Physics and Chemistry of Solids

Materials Research Bulletin

Metals Forum

Physics of Metals and Metallography

Scripta Metallurgica

Solid State Communications

STRENGTH OF METALS AND ALLOYS

(ICSMA 6)

Proceedings of the 6th International Conference
Melbourne, Australia, 16-20 August 1982

In Three Volumes

Edited by

R C GIFKINS

Department of Materials Engineering, Monash University, Melbourne, Australia

Volume 3

PERGAMON PRESS

OXFORD · NEW YORK · TORONTO · SYDNEY · PARIS · FRANKFURT

U.K.	Pergamon Press Ltd., Headington Hill Hall, Oxford OX3 0BW, England
U.S.A.	Pergamon Press Inc., Maxwell House, Fairview Park, Elmsford, New York 10523, U.S.A.
CANADA	Pergamon Press Canada Ltd., Suite 104, 150 Consumers Rd., Willowdale, Ontario M2J 1P9, Canada
AUSTRALIA	Pergamon Press (Aust.) Pty. Ltd., P.O. Box 544, Potts Point, N.S.W. 2011, Australia
FRANCE	Pergamon Press SARL, 24 rue des Ecoles, 75240 Paris, Cedex 05, France
FEDERAL REPUBLIC OF GERMANY	Pergamon Press GmbH, Hammerweg 6, D-6242 Kronberg-Taunus, Federal Republic of Germany

Copyright © 1983 Pergamon Press Ltd

All Rights Reserved. No part of this publication may be reproduced, stored in a retrieval system or transmitted in any form or by any means: electronic, electrostatic, magnetic tape, mechanical, photocopying, recording or otherwise, without permission in writing from the publishers.

First edition 1983

Library of Congress Cataloging in Publication Data
International Conference on the Strength of Metals and Alloys (6th: 1982: Melbourne, Vic.)
Strength of metals and alloys (ICSMA 6)
(International series on the strength and fracture of materials and structures)
Includes bibliographies and index.
1. Metals—Congresses. 2. Alloys—Congresses.
3. Physical metallurgy—Congresses. I. Gifkins, R. C., 1918-   II. Title. III. Series.
TA460.I532   1982   620.1'63   82-9851

British Library Cataloguing in Publication Data
Strength of metals and alloys.—(International series on the strength and fracture of materials and structures)
1. Physical metallurgy
I. Gifkins, R.C.   II. Series
620.1'6'3   TA405
ISBN 0-08-029325-5

In order to make this volume available as economically and as rapidly as possible the author's typescript has been reproduced in its original form. This method unfortunately has its typographical limitations but it is hoped that they in no way distract the reader.

Printed in Great Britain by A. Wheaton & Co. Ltd., Exeter

Organizing Committee

I J POLMEAR, *Chairman*
D W BORLAND
I S BRAMMAR
G G BROWN
L M CLAREBROUGH
M E De MORTON
J A EADY
R C GIFKINS
E O HALL
M HATHERLY
B E HOBBS
R HOBBS, *Secretary*
P M KELLY
D R MILLER
R B NETHERCOTT
B A PARKER, *Deputy Chairman*
P M ROBINSON
R L SEGALL

Contents of Volume 3

KEYNOTE PAPERS

Fundamentals of strengthening mechanisms
    H Gleiter   1009

The application of the fundamentals of strengthening to the design of new aluminum alloys
    E A Starke, Jr.   1025

Reflections on the industrial application of fundamental research
    G F Bolling   1045

Microstructure and mechanisms of fracture
    E Hornbogen   1059

Routes to higher strength and ductility of steels
    T Gréday   1075

Ductile fracture
    J D Embury   1089

Deformation at high temperatures
    T G Langdon   1105

Strengthening from the melt: castings and weldments
    G J Davies   1121

Deformation problems in minerals and rocks
  M S Paterson   1137

Design against variable amplitude fatigue - an approach through cyclic
stress-strain response
  C Laird, F Lorenzo and A S Cheng   1147

Deformation for manufacture: forming and shaping
  T Furukawa   1165

Deformation at high strains
  M Hatherly   1181

Doing more with less materials: perspectives for research and development
in the Eighties
  D G Altenpohl   1197

POST-CONFERENCE PAPERS

Cyclic response of the directionally solidified superalloy 73C
  M A Abdellatif and A Lawley   1213

The role of deformation character on fatigue crack growth in titanium
alloys
  J E Allison and J C Williams   1219

The development of a high strength manganese steel
  R D Jones, G R Palmer, V Jerath, S·Kapoor and R J Yeldham   1225

The effects of composition and temperature on the dislocation structure
of cyclically deformed Ti-Al Alloys
  H M Kim and J C Williams   1231

Relationship between microstructure and strength properties of copper-
tin-nickel alloys prepared by powder metallurgy
  N C Kothari   1237

Thermomechanical treatment of dual-phase low carbon steels
  T C Lei, D Z Yang and H P Shen   1245

AUTHOR INDEX   1251

SUBJECT INDEX   1255

Contents of Volume 1

FUNDAMENTALS OF STRENGTHENING MECHANISMS

Temperature and strain rate dependence of the flow stress of fatigued niobium single crystals
    M Anglada and F Guiu   3

Fundamental plastic behaviors in high-purity BCC metals (Nb, Mo and Fe)
    Y Aono, E Kuramoto and K Kitajima   9

The effect of solute clusters on screw dislocations in BCC alloys
    R J Arsenault and D M Esterling   15

The effect of non-glide stress on the Peierls stress in BCC metals
    Z S Basinski and M S Duesbery   21

A mechanism for slip transfer across high angle grain boundaries
    C T Forwood and L M Clarebrough   27

A strain hardening law for FCC and BCC crystals derived from latent hardening and multislip tests
    P Franciosi   33

Multislip in FCC and BCC crystals: A theoretical approach compared with experimental data
    P Franciosi and A Zaoui   39

The behaviour of dislocations near a free surface
    P M Hazzledine and S J Shaibani   45

Cross-slip and work hardening
    P J Jackson   51

Low temperature plasticity of molybdenum single crystals of high purity
    H-J Kaufmann   57

Creep of LiF crystals in the temperature range 1.6 - 300 K
    H-J Kaufmann, S V Lubenets and V V Abraimov   63

Understandings of slip behaviors in BCC metals by computer analyses
    E Kuramoto, Y Aono and T Tsutsumi   69

The strength of polycrystals
    T Leffers and O B Pedersen   75

A new way to derive activation enthalpies of serrated flow?
    E Pink and A Grinberg   83

On the high temperature internal friction background
    F Povolo and B J Molinas   89

Dislocation motion in FCC metal crystals
    K Shinohara, S Kitajima and H Kurishita   95

Moving dislocations in aluminium and in aluminium-copper alloys
    H Tamler, O Kanert and J Th M De Hosson   101

Stoneley waves at grain boundaries in copper and aluminium
    A R Thölén   107

ALLOY DESIGN FOR STRENGTH

    (A) FERROUS ALLOYS

Void damage in dual phase steels during plastic deformation
    P S Baburamani, R A Jago and R M Hobbs   115

The effect of inclusions on the structure and properties of HSLA steel
weld metals
    G S Barritte, R A Ricks and P R Howell   121

Micromechanical concepts for the deformation and fracture behaviour of
coarse ferrite-martensite steels
    J Becker and E Hornbogen   127

Transformation behaviour and mechanical properties of rapidly quenched steels
    H W Bergmann and B L Mordike   135

Microstructure and fatigue behaviour of nickel-base eutectic composites
    K Dannemann, T Ishii, D J Duquette and N S Stoloff   141

Thermomechanical treatment of a plain low carbon steel
    P Deb and M C Chaturvedi   147

Property - structure relations in quenched and tempered 2%Mn steel
    M E de Morton   153

A study of the effect of low temperatures on the strength and ductility of commercial HSLA steels
    L R Cutler and M R Krishnadev   161

Internal friction measurements, hardness and tensile tests during the tempering of 1100°C as quenched 17% chromium steels
    B Dubois, F Hernandez and M Bouhafs   167

Precision measurements of pre-yield microstrain in amorphous alloys
    T Hashimoto, S Kobayashi, K Maeda and S Takeuchi   173

A thermodynamic model to predict the precipitation behaviour of both titanium nitride and sulfide during steel solidification. Effects of the globularisation of the sulfides on the isotropy of the mechanical properties
    J-C Herman, P Messien and T Gréday   179

The contribution of cobalt to the tempering resistance and hot hardness of tool steels and cobalt replacement
    J P Hirth, E J Dulis and V K Chandhok   185

Effects of Co, Mo and Ti contents on age-hardening in 15%Ni maraging steels
    K Hosomi, Y Ashida, H Hato and H Nakamura   193

Dynamic strain aging of cast iron
    D Löhe, O Vöhringer and E Macherauch   199

Athermal, stress and strain induced transformation strengthening in multiphase stainless steels
    F D S Marques and N N Thadhani   205

Comparison of the efficiency of various rolling schedules to obtain dual-phase steel plates
    H Mathy and T Greday   211

The influence of plate gauge on the strength and impact properties of steels
    B Mintz   217

Deformation behaviour of austenite-ferrite structures
    J Ruzzante, G Carfi, J Tormo and A M Hey   223

The influence of cementite content and cementite shape on the Bauschinger effect of plain carbon steels
    B Scholtes, O Vöhringer and E Macherauch   229

Ferrite grain refinement in HSLA steels
    R M Smith, T Chandra and D P Dunne   235

Control of strain age hardening in Ti bearing hot rolled strip steels
    J G Williams   241

Coated sheet steel viewed as a composite material
    D J Willis   247

(B) NON-FERROUS ALLOYS

Designing of an optimum aluminium alloy for de-salination applications
    Z Ahmed   255

Effect of microstructure on the mechanical properties of Transage 134
    R W Coade, T W Duerig and G H Gessinger   263

The tensile deformation of aluminium foil
    I R Dover, J A Eady and R C Gifkins   269

Effect of the phase transformation on the mechanical properties of initially cold-worked cobalt and cobalt alloys
    G Bouquet and B Dubois   275

Mechanical properties during the tempering of martensitic copper-aluminium alloys
    B Dubois, G Ocampo, G Demiraj and G Bouquet   281

The influence of post-extrusion thermomechanical treatments on the tensile properties of Zr-2.5 wt% Nb alloy
    R G Fleck and G K Shek   289

Strengthening mechanisms in dispersion hardened copper polycrystals
    N Hansen and B Ralph   295

Deformation behaviour of the ordering alloy Cu$_2$NiZn
    J Th M De Hosson, G J L Van Der Wegen and P M Bronsveld   301

Analysis of strengthening factors in thermomechanical treatment of an
Al-Cu-Mg-Mn alloy
    T-C Lei and D-Z Wang   307

Dispersion strengthened titanium alloys by rapid solidification
processing
    B C Muddle, D G Konitzer and H L Fraser   313

Alloy softening in molybdenum-carbon single crystals
    M I Ripley and J W Christian   319

Solution hardening of FCC alloys by the chemical interaction between
solute atoms and a dislocation
    H Suzuki   327

An investigation of mechanical thermal treated Al-Al$_3$Ni eutectic alloy
    C M Wan, S M Kuo, Y C Chen, T H Wang, M T Jahn, C T Hu and J Heh   333

The flow stress-grain size relationship in a precipitation-hardening
Al alloy
    J Wert   339

(C) CERAMICS AND GLASSES

The hardening reaction in dental amalgam
    J R Abbott, D R Miller and D J Netherway   347

Strengthening of alumina by heat treatment
    D G Brandon, E Y Gutmanas, Z Nissenholz, J Ozeri, D Shechtman,
    N Travitzki and Y Yeshurun   353

The shear band deformation process in metallic glasses
    P E Donovan and W M Stobbs   361

Metal glass composites as engineering materials
    K U Kainer, H W Bergmann and B L Mordike   367

Strength/ductility-microstructure relationships in two-phase materials
    K Tangri and A H Yegneswaran   373

YIELD, FLOW AND STRENGTH OF POLYCRYSTALLINE MATERIAL

Origin of the recrystallization texture in rolled low zinc brass
 C Carmichael, A S Malin and M Hatherly   381

Yield of mild steel at very high strain rates
 N P Fitzpatrick and P L Pratt   387

Luders band fronts in mild steel
 E O Hall, R J Carter and G Vitullo   393

Microstructures and deformation mechanisms in polycrystalline aluminium
 N Hansen and B Bay   401

Yield points in micro-alloyed titanium steels
 R W K Honeycombe and G M Smith   407

Heterogeneous formation of slip bands in neutron-irradiated copper crystals
 S Kitajima and K Shinohara   413

Plastic instability in an Al-Mg-Si alloy
 G J Lopriore and P G McCormick   419

Substructure strengthening of cold rolled aluminium alloys
 E Nes, A L Dons and N Ryum   425

On the nature of the Bauschinger effect in copper and copper aluminium
 H Ono, D J H Corderoy and H Muir   431

The effect of strain rate and geometry on development of microstructure during the forming of an aluminium-magnesium alloy
 B A Parker and R G O'Donnell   437

Analysis of local plastic heterogeneities by use of the Kossel technique
 C Rey, M Berveiller and A Zaoui   443

Control of the anisotropy of mechanical properties in steel plate
 I D Simpson and G J M MacDonald   449

Role of internal stress for the initial yielding of dual-phase steels
 T Sakaki, K Sugimoto and T Fukuzato   455

Tensile deformation of a fine-grained Al-alloy
 H Westengen   461

The effect of penetrator geometry on the deformation of ductile metal targets
    C J Osborn and R L Woodward   467

HIGH STRAIN AND HIGH STRAIN RATE PROCESSES

Cyclic cracking and plastic flow in metal cutting under built-up-edge conditions
    R L Aghan and G R Wilms   475

Plastic stress-strain relationships of some polycrystalline metals tested in biaxial tension to high strain
    M Atkinson   481

The effect of a dispersed phase upon the deformation structure of rolled copper crystals
    I Baker and J W Martin   487

The tensile properties of cold twisted steel bars
    G G Brown and J D Watson   493

Effect of manganese on the dynamic recrystallization behavior of titanium-bearing microalloyed steels
    T Chandra, M G Akben and J J Jonas   499

Dynamic recovery and recrystallization in a duplex stainless steel
    T Chandra, D Bendeich and D P Dunne   505

The significance of high strain deformation in machining
    E D Doyle and D M Turley   511

Dynamic recrystallization of copper observed by SEM channeling contrast
    H J McQueen   517

Microstructure and texture of rolled CPH metals
    A S Malin, M Hatherly and V Piegerova   523

Dynamic recrystallization in ferritic stainless steel
    T Maki, S Okaguchi and I Tamura   529

Microstructure and mechanical properties of heavily deformed copper
    R B Nethercott, J A Retchford and R A Coyle   535

Microstructure and texture of heavily deformed copper
    R G Solomon, A S Malin and M Hatherley   541

"Microbands" in cold worked metals
    F J Torrealdea and J Gil Sevillano   547

Substructural size effect in metal cutting
    D M Turley and E D Doyle   553

Formability of Ni-Fe alloy sheets
    Y C Yoo and D N Lee   559

Contents of Volume 2

HIGH TEMPERATURE PROCESSES, CREEP, FATIGUE, SUPERPLASTICITY

Experimental analysis of the relationship between dip test techniques and the plastic equation of state
  J Birocheau and C Oytana   569

Biaxial dip tests measurements of internal stresses during high temperature plastic flow
  P Delobelle, A Mermet and C Oytana   575

Verification of a microstructurally-based equation for elevated-temperature transient isotropic hardening
  M E Kassner, K A Rubin and A K Miller   581

The interpretation of the scaling relationship shown by the log σ - log $\dot{\varepsilon}$ creep and stress-relaxation curves
  F Povolo and G H Rubiolo   589

Stress and microstructure dependence of the creep resistance of Mo-5% W alloy
  L Bendersky, A Rosen and A K Mukherjee   595

Stress rupture behaviour of copper-chromium weather resistant steels
  R H Edwards, T Payne and K W Gunn   601

Multiple strengthening in creep of wrought Ni-Cr-base superalloys
  F Gabrielli and V Lupinc   607

Anomalous and constant substructure creep transients in pure aluminum
  J C Gibeling and W D Nix   613

Creep rupture properties of a niobium bearing pressure vessel steel containing small amounts of molybdenum
  K Gunn and T Payne   619

Inhomogeneous deformation of some aluminium alloys at elevated temperature
    F J Humphreys   625

Creep and fatigue interactions in a nickel-base superalloy
    H J Kolkman and R J H Wanhill   631

The influence of grain size on elevated temperature deformation behaviour of a type 316 stainless steel
    S L Mannan, K G Samuel and P Rodriguez   637

Development of creep resistant magnesium rare earth alloys
    J E Morgan and B L Mordike   643

The influence of traces of Sb and Zr on creep and creep fracture of Ni-20% Cr
    J H Schneibel, C L White and R A Padgett   649

The effect of pre-aging on the creep behaviour of type 321 stainless steel
    K U Snowden, P A Stathers and D S Hughes   655

High temperature creep and fatigue of Cu-Cr alloys
    N Y Tang, D M R Taplin and G L Dunlop   665

High temperature intergranular fracture enhanced by grain boundary migration in alpha iron-tin alloy
    T Watanabe, M Obata and S Karashima   671

Small angle neutron scattering study of creep deformation and fracture of type 304 stainless steel
    M H Yoo, J C Ogle, J H Schneibel and R W Swindeman   677

Cyclic deformation of superplastic zinc-aluminum alloys
    J W Bowden and B Ramaswami   683

A metallographic study of cavitation during superplastic deformation of CDA 638 alloy
    W J Clegg, J A Rooum and A K Mukherjee   689

The role of grain boundary dislocations in superplastic flow
    P R Howell, L K L Falk and G L Dunlop   695

Mechanisms for the low-stress regime I of superplasticity
    R C Gifkins   701

Strain effect on stress-strain rate relations of superplastic IN744 steel
  B P Kashyap and A K Mukherjee   707

Elevated temperature deformation and fracture of ingot and powder processed aluminum alloys
  N E Paton, C C Bampton and A K Ghosh   713

Anelasticity in superplastic metals
  S H Vale and P M Hazzledine   721

Enhanced superplasticity in modified Ti-6%Al-4%V alloys
  J A Wert and N E Paton   727

STRENGTH OF ROCKS AND MINERALS

Deformation mechanisms in dunite - the results of high temperature testing
  J D Fitz Gerald and P N Chopra   735

Measurements of strain rate continuity in LiF crystals
  T H Alden and J C Gibeling   741

The deformation of polycrystalline sodium nitrate (calcite structure)
  P D Tungatt and F J Humphreys   747

Recovery and recrystallization in olivine
  S Karato   753

Deformation of olivine, and the application to lunar and planetary interiors
  T G Langdon, A Dehghan and C G Sammis   757

STRENGTH RELATED TO FRACTURE - FRACTURE TOUGHNESS

Factors influencing the fracture toughness of high strength aluminum alloys
  K R Brown   765

Fracture toughness and critical crack length parameters in fatigue failure
  G Clark   773

The Bauschinger effect
  R Hsu and R J Arsenault   781

The influence of the addition of boron on the grain-boundary segregation of phosphorus and fracture behavior of medium-carbon ultra-high strength steel
    T Inoue and Y Namba   787

The effects of thermomechanical treatment on the notch toughness of a 4340 Ni-Cr-Mo alloy steel
    M T Jahn, C K Huang and C M Wan   793

The crack-tip ductility of structural steels
    J F Knott   799

Deformation and fracture behaviour of single fibre coated with ductile or brittle layer
    S Ochiai, Y Murakami and K Osamura   805

A comparison of spall fractures with fractures in tensile tests
    J M Yellup   811

CYCLIC DEFORMATION

Low amplitude fatigue of copper crystals at room temperature
    Z S Basinski and S J Basinski   819

Back stress variation along the hysteresis loop
    Y S Chung and A Abel   825

Fatigue crack initiation on slip bands
    M E Fine   833

Detection of crack propagation in fatigue with acoustic emission
    F Hamel and M N Bassim   839

Fatigue crack growth in the near-threshold region
    A J McEvily and K Minakawa   845

Fatigue behaviour of nitrided steel
    J M Cowling   851

Corrosion fatigue of SiC/Al metal matrix composites in salt ladened moist air
    C R Crowe and D F Hasson   859

The effect of overloads on fatigue crack propagation in offshore structural steels
M Drew, K R L Thompson and L H Keys   867

The influence of dispersoids on the low cycle fatigue properties of Al-Mg-Si alloys
L Edwards and J W Martin   873

The effects of frequency and hold times on fatigue crack propagation rates in a nickel base superalloy
S Golwalkar, N S Stoloff and D J Duquette   879

Fatigue crack propagation characterization of 18-8 austenitic stainless steel under repeated impact loading
J-H Zhu, M-J Tu and H-J Zhou   887

The fatigue studies of steel with bainitic structure
T F Liu, M H Yang, C M Wan, M T Jahn, C T Hu and J Heh   895

High temperature fatigue of an austenitic stainless steel
A J Pacey and A Plumtree   901

Dislocation structures in fatigued Cu polycrystals
R Pascual and L C Rolim   907

Influence of cyclic stress on substructure in aluminum and iron
A Plumtree and E S Kayali   913

Influence of overloads on the subsequent crack growth of a fatigue crack in a E 36 steel
C Robin, C Chehimi and G Pluvinage   919

The effect of strain rate in air and vacuum on the high-temperature low-cycle fatigue behaviour of cast IN100 alloy
D Ranucci, M Marchionni, E Picco, F Gherardi and O Caciorgna   927

Fatigue of spinodally decomposing alloys
H R Sinning and P Haasen   933

PAPERS RECEIVED LATE OR DELAYED BY REVISION

The work hardening of copper-aluminium alloys
D J H Corderoy and N Ono   941

A model for plane strain ductile fracture toughness
    D Firrao and R Roberti   947

In situ plastic deformation of metals and alloys in the 200 kV electron microscope
    L P Kubin, J Lépinoux, J Rabier, P Veyssière and A Fourdeux   953

Control of discontinuous yielding by temper rolling
    J S H Lake   959

Deformation mechanisms, microstructure development and grain orientation in rolled copper-alloys
    K Lücke and J Hirsch   965

The constancy of density of deformation energy at macro- and microlevels in case of automodelle elastic-plastic behaviour of materials
    L Maslov   971

The coarsening of precipitates
    M B McGirr and S Y Lee   977

Fabrication of a Cu-Al composite wire and its mechanical properties (part 1)
    Y Mitani and H Balmori   983

Fracture toughness of soft soldering composite laminates made from 4340 steel in high vacuum atmosphere
    M R Sabayo   989

The calculations of weight functions for certain configurations
    K-R Wang and G Pluvinage   995

Influence of microarea chemical composition and microstructure on fracture properties of a medium carbon ultrahigh strength steel
    S-L Xu and X-F Feng   1001

Keynote Papers

Fundamentals of Strengthening Mechanisms

H. Gleiter

University of Saarbrücken, D-6600 Saarbrücken, Federal Republic of Germany

ABSTRACT

The present understanding of mechanical strengthening of crystals by one of the following mechanisms' is outlined. Lattice friction strengthening in pure metals, strengthening due to grain size effects, solid solution strengthening, order strengthening, strengthening by precipitates and by dynamical effects. An attempt is made to point out how the flow stress and work-hardening can be understood on the basis of the interaction between dislocations and obstacles.

KEYWORDS

Lattice friction strengthening; dislocation core structure; grain size effects; solid solution strengthening; order strengthening; precipitation hardening; dispersion hardening; dislocation dynamics.

INTRODUCTION

In this brief review, an attempt was made to expound the present understanding of the most prominent strengthening mechanisms of crystalline solids in outline. Details may be found in the reviews mentioned. By mechanical strengthening we refer mainly to the stress/ strain relationship measured at constant strain rate and/or temperature. Following the framework of the conference, polymers, composits, glasses, intermetallic compounds, molecular crystals and ceramic materials will not be included. The same applies to deformation at high temperatures, effects based on radiation damage, and on work hardening processes in pure metals. The various types of strengthening mechanisms are discussed separately. Strengthening of real crystals represents inevitably a superposition of different strengthening processes. The problem of superposition of strengthening acting simultaneously in an alloy has been considered in excellent reviews recently (Kocks et al. 1975, Hornbogen 1980, Kocks 1980).

Investigations on crystals of different lattice structure and different atomic bonding have evidenced the significance of crystal structure for the strength of materials. In pure metals, the factors of prime importance are the interaction, dissociation and the core structure of dislocations. In metals with f.c.c. and h.c.p. structure, dislocations are dissociated into partials. The results of atomic and continuum calculations (Peierls-Nabarro model) indicate that the atomic structure of the cores of the partial dislocations varies little when they move through the lattice. Hence, the stress required to move dislocations in f.c.c. or h.c.p. metals is small. In metals with b.c.c. structure strengthening effects due to the dislocation core structure seem to be significant. Atomic calculations for the core structure of screw dislocations indicate (Vitek, 1974; Basinski et al., 1971) that the total Burgers vector ($b = 1/2 <111>$) is removed from the center of the dislocation and shared among three fractional dislocations in the three $\{110\}$ planes intersecting along a <111> direction. All the cores of non-screw dislocations seem to be planar and rather narrow (Vitek, 1974). If a dislocation lies parallel to close packed directions it encounters a Peierls barrier and glides (at finite temperatures) by stress-assisted, thermally activated formation of pairs of kinks which subsequently move sidewise in opposite direction. The critical resolved shear stress (CRSS) of b.c.c. metals rapidly deceases at high temperatures due to the thermally-assisted kink formation on migration, and hence b.c.c. close packed metals become alike as far as the stress required for dislocation motion is concerned. However, at low temperature the CRSS differs by orders of magnitude from the f.c.c. and h.c.p. metals primarily due to the core structure of the screw dislocations. Of course, the different dislocation structures also affects other parameters of deformation like glide geometry, mobility of screw or edge dislocations, cross-slip processes etc. For details we refer to the review by Sestak (1980). The modifications of the above effects in single phase or multiphase alloys are considered in the corresponding subsequent sections.

GRAIN SIZE EFFECTS

In polycrystalline materials, grain refinement is well known to represent a method of improving strength and fracture resistance (for a review see Armstrong 1969). The effect was first treated quantitatively by Hall (1951) and Petch (1953) and is now generally referred to as Hall-Petch relationship. It seems to apply to single-phase materials of various lattice structure (f.c.c., h.c.p., and b.c.c.) as well as two phase materials (e.g. pearlite and eutectic structures) and formulates the resistance of interfaces to the spread of slip bands. The stress necessary to propagate slip from one grain to an other may be determined by
(i)   activating a dislocation source in the undeformed crystal
(ii) unpinning of dislocations in the undeformed crystal
(iii) generation of lattice dislocations from the interface
(iv) absorption of lattice dislocations at the interface.
In the original papers by Hall and Petch the propagation of slip was assumed to occur by the nucleation of slip in undeformed neighbouring grains due to the stress concentration at the head of slip bands stopped at the grain boundaries. In this treatment the length of the dislocation pile-up was assumed to be proportional to the grain size. This has been subject of much debate. At sufficiently small grains (which cannot contain pile-ups) this assumption should not hold, resulting in appreciable departure from the Hall-Petch relationship at small grain sizes. This has indeed been observed (Haessner and Müller, 1981). If dislocation unpinning

is the critical step of slip propagation, the flow stress is expected to show a marked temperature dependence. Measurements of the temperature dependence of the Hall-Petch relationship (Cottrell, 1963; Wilson, 1967) as well as the kinetic observations during annealing treatments (Wilson, 1967) are consistent with this idea. However, the observed temperature dependence and the kinetic observations may equally well be interpreted in terms of
(i) the segregation of solute atoms to grain boundary dislocations and the unpinning of those defects (Bäro and Hornbogen, 1969)
(ii) the incorporation of lattice dislocations in grain boundaries (Pumphrey and Gleiter, 1974; Lojkowski and Grabski, 1980).
The attractive feature of the latter interpretations is the fact that they do not require the presence of dislocation pile-ups for which little evidence exists in several materials although the Hall-Petch relationship applies. During plastic deformation of polycrystals, the grains are forced to retain contiguity which is equivalent to establishing a gradient of dislocation density within the grain (geometrically necessary dislocations). If these dislocations dominate work hardening, a Hall-Petch type behaviour of the material is obtained (Embury, 1971). The model becomes invalid if statistically stored dislocations or dynamic recovery processes dominate work hardening. The ease of propagating slip into a neighbouring grain and its relation to the contiguity condition was considered by several workers in terms of Taylors theory (Armstrong et al., 1962; Mecknig, 1980). All factors affecting the critical shear stress, the slip band distribution, and the orientation factor of the grains (which may affect the number of available slip systems) are found to increase the Hall-Petch slope. Indeed, strong dislocation pinning effects were observed (Embury, 1971) to give rise to large slopes of the Hall-Petch relation. The same applies to materials exhibiting long range or short range order (cf. subsequent paragraph), and to materials having a strong texture, so that a large orientation factor results. Investigations of the grain size on the flow stress of ordered alloys have shown that one of the most potent methods to strengthen ordered alloys is to refine the grain size.

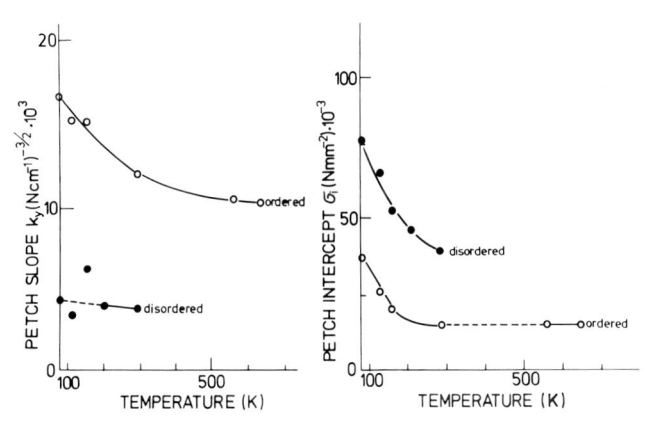

Fig. 1. The effect of long range order on the slope (k_y) and the intercept (σ_i^-) of the Hall-Petch relationship for FeCo-2% V (Jordan and Stoloff, 1968).

Fig. 1 shows a comparison between the slope and the intercept of the Hall-Petch relation for FeCo-2% V alloys in the ordered and disordered state.

Similar effects have been noted for FeCo and Ni$_3$Mn (Marcinkowski and Fisher, 1965; Johnston et al., 1965). The reason for the enhanced grain boundary strengthening is believed to be the restriction of cross-slip at grain boundaries (Armstrong et al., 1962).

By analogy to the regions I, II, and III of the stress-strain curve of f.c.c. or h.c.p. single crystals, the deformation curve of polycrystals is also divided in stage II and III which are characterized by a high athermal (stage II) and a thermally activated (stage III) hardening rate.

Polycrystals and single crystals are assumed to obey the same flow stress equation. Only the relationship between the deformation and the dislocation density is modified (Ashby, 1971; Thompson, 1977) to account for compatibility effects at the boundaries. The main effect of grain size on stress results from the enhanced work hardening in the boundary area which becomes more dominant with decreasing grain size. The grain size effect in stage III has not been investigated extensively so far. However, the development of the substructure in stage III reduces the influence of the boundaries. The inhomogenious distribution of dislocations in the vicinity of grain boundaries (geometrically necessary dislocations) in the early stages of deformation is expected to convert into a homogeneous cell structure as deformation proceeds. Therefore, all effect related to the presence of grain boundaries should be reduced when the dislocation density has become high enough for dynamic recovery processes to occur. As dynamic recovery starts from the boundary regions, the stress strain curves of fine grained materials flatten at smaller strains than the curves for materials with coarse grains.

It has been shown that low angle grain boundaries may have a similar effect on the flow stress as high angle boundaries (e.g. Rezek and Craig, 1961; Warrington, 1963). This effect was interpreted in terms of pushing a glide dislocation through a low angle boundary (dislocation wall) or unpinning a dislocation of the wall. The mode of behaviour depends on the amount of segregation and the dislocation structure of the low angle boundaries.

High strength wires represent a special application of grain size (substructural) strengthening. The high strength is achieved by extensive cold working mostly combined with segregation and/or precipitation processes. The method has been successfully applied to materials with f.c.c. structure (Ni, Ni - C, Ni-C-Cr-V-Mo-Co, stainless steel, Inconel X, Rene 41) h.c.p. structure (Ti-V-Cr-Al, Be) and b.c.c. structure (W, Mo, Mo-Ti-Zr-C, eutectoid Fe-C) (Embury, 1971). The major contribution to the high strength of these wires (up to 6000 Nmm^{-2}) comes from the development of a dislocation cell structure. The dimensions of the cells are progressively reduced during drawing. In iron-carbon alloys and nickel based alloys an additional strengthening was shown to originate from finely dispersed particles (Wilcox and Gilbert, 1967; Kindin et al., 1966) and dislocation locking by interstitials (Chandok, 1966). Microcrystalline (sub-micron) structures of high strength have recently been investigated in several systems. Polycrystalline iron whiskers produced by a chemical vapour deposition technique showed extraordinary strength up to 8 GPa. Microstructural investigations (FIM, TEM) (Wilsdorf et al., 1978) revealed the whiskers to consist of grains of α-iron between 5 and 30 nm diameter containing a microdispersion of iron oxides, carbides, and carbon dissolved in the iron lattice. The grain boundaries between neighbouring α-iron crystals and the phase boundaries between the α-iron and the precipitates seem to be several interatomic spacings wide and seem to contain a mixed metallic and covalent bonding. Microcrystalline structures of high strength were also produced in recent years by rolling or co-extruding powder mixtures or conventionally prepared multiphase alloys. The compositions of the powder mixtures and alloys studied were Fe-Cu, Cu-glass, Al-AgCl and Al-polytetrafluoroethylene (Bergmann, 1976; Frommeyer and Wassermann, 1976; Wassermann, 1976). Apparently, large plastic deformations of all components of the microcrystalline material are the essential prerequisites for good bonding between the partners (e.g. metals and ionic or metals and covalent crystals). The maximum tensile strength (Fig. 2) achieved by this method approaches the theoretical values. Remarkable strengthening effects (Fig. 3) due to submicron structures were obtained by vapour deposition of Au and Ni layers so that a sandwich structure resulted (Yang et al., 1977). The enhanced modulus for layer thicknesses below 2 nm was rationalized in terms of the interaction of the Fermi surface with the additional Brillouin zone created by the modulated structure. It may be mentioned at this point that microcrystalline structures in the sub-micron range show also striking electric, magnetic and optical properties some examples of which are documented in the literature (Gleiter, 1980).

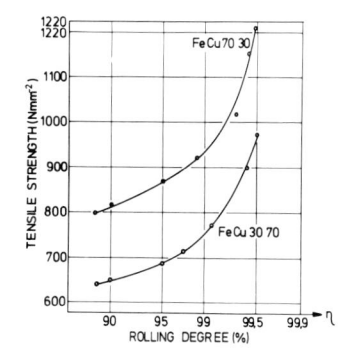

Fig. 2. Tensile strength of non-copper composite foils as a function of the degree of rolling (Wassermann, 1976).

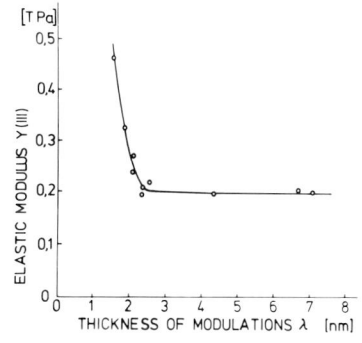

Fig. 3. Variation of the elastic modulus Y (111) with the thickness λ of the compositional modulation for gold-nickel films of Au : Ni = 1 : 1. The constant value of $Y(111)$ at $\lambda > 3$ nm corresponds to the elastic modulus of bulk crystals (Yang et al., 1977).

SOLID SOLUTION STRENGTHENING

a) Mechanisms of interaction between dislocations and solute atoms

The increase of strength associated with the presence of solute atoms may be divided in the collective and the direct interactions between solute atoms and dislocations. The collective interactions in general affect the dislocation width through a variation in stacking fault energy, the interaction, density and arrangement of the dislocations. The most significant effect is the change of the dislocation width as it affects cross-slip and mechanical twinning.

The direct interactions can be divided in two parts:
(i) dislocation locking due to solute segregation at dislocations at rest

(ii) dislocation friction due to the forces excerted by the solute atoms on the moving dislocations. As all of the mechanisms involved have been reviewed in the literature (e.g. Haasen, 1970,1979), we shall limit ourself on summarizing the most important mechanisms. The locking mechanisms are either based on elastic or on electronic interaction effects. Elastic locking (Cottrell, 1958) occurs in a crystal to compensate for the respective distorsions in the vicinity of dislocations and solute atoms. Solute atoms differing in size from the matrix atoms find energetically favourable sites near dislocations and, hence, segregate in the vicinity of the dislocation.
Order locking (Schoeck and Seeger, 1959) is based on stress induced local ordering in the vicinity of a dislocation. The elastic dipole moments associated with certain solute atoms are aligned by the strain field of the dislocation, thus locking the dislocation.
Electrostatic locking (Cottrell et al., 1953; Friedel, 1956) occurs by the electrostatic interaction of the expanded and, hence, negatively charged dislocation core with the charge located predominantly in the vicinity of a solute atom differing in valancy from the host metal. Chemical locking (Suzuki, 1957) is due to the different chemical potential of solute atoms in the perfect lattice and in the vicinity of stacking faults. Several friction mechanisms have been proposed. Again, all of them are either based on elastic or electronic effects. If solute atoms differ in size from the solvent atoms or if the solute/solvent banding strength is different from the solute/solute interaction, then the interaction with a moving dislocation is of the paraelastic or dielastic type (Mott and Nabarro, 1948; Fleischer, 1961,1963; Kröner, 1964). The paraelastic interaction is due to the permanent internal stress field of the misfitting solute atoms felt by the moving dislocation. The dielastic interaction is induced by the strain field of the dislocation and is based on the different interatomic coupling between solvent/solvent and solute/solvent atoms. In other words, the solute atoms act as soft or hard spots of the matrix. In alloys with short range order, order friction effects result from the replacement of energetically favourable bonds by less favourable ones across the dislocation ship plane (Fisher, 1954). In solid solutions of atoms of different valency, dislocations assume a zig-zag shape compromising between the line energy and the electrostatic interaction of the dislocation core with the solute atoms (Friedel, 1956).

b) Solute effects on the macroscopic deformation behaviour

The absence of yield points in most solid solutions with f.c.c. structure and the same temperature dependence of the flow stress during easy glide in alloys and pure metals suggests friction mechanisms to be the dominant strengthening processes. The approximately linear increase of the CRSS with temperatures observed in the low temperature regime (Hendrickson and Fine, 1961) is interpreted in terms of the dislocation core interaction with solute atoms. The temperature independent part of the CRSS at higher temperatures (e.g. Haasen, 1970,1979) was found to be controlled by paraelastic, dielastic or order friction effects. The yield points observed in some carefully treated noble metal alloys seem to be due to chemical locking as they appear in alloys with and without misfitting solute atoms (Kopenaal and Fine, 1961; Hendrickson and Fine, 1961). The work hardening rate in stage II of f.c.c. alloys is little affected by solute atoms. This may be understood in terms of the same elastic interaction between the dislocations accumulated in the slip bands of pure metals and solid solutions. The variation in stacking fault energy increases the stress required to initiate cross slip (beginning of stage III).
In the case of solid solutions with h.c.p. structure, the CRSS increases

nearly linearly at low temperatures and exhibits a temperature independent part (plateau) above about 300 K just as in the case of f.c.c. metals. The observed plateau hardening for Mg-, Cd- and Zn-alloys seems to be due to paraelastic and/or dielastic interaction between dislocations and solute atoms (Akhtar and Teghtsoonian, 1968; Boček et al., 1961; Haessner and Schreiber, 1957; Wielke, 1979). The low temperature strengthening agrees with the core interaction mechanism. In addition to the features of the f.c.c. structure, the c/a ratio of h.c.p. crystals varies with alloy content. This variation affects the ratio of basal and non-basal slip and thus the shape of the stress-strain curve.

If solute atoms are added to b.c.c. solid solutions, the temperature dependence of the CRSS is modified. At intermediate temperatures, the CRSS always increases with increasing solute content, whereas at low temperatures solute softening is observed by many alloys (Fig. 4)

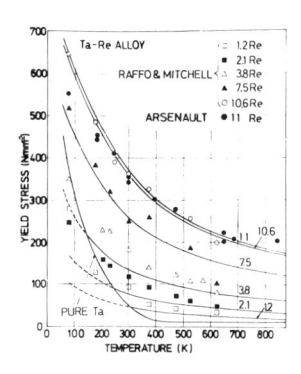

Fig. 4. Yield stress of tantalum-rhenium alloys. The numbers indicate the rhenium content of the alloy (Raffo and Mitchell, 1968; Arsenault, 1969).

In fact, the Peierls stress seems to control the motion of screw dislocations at low temperatures in b.c.c. metals. Therefore, the present theories of strengthening of b.c.c. alloys assume that solute atoms affect the formation of kink pairs on screw dislocations and impede the sidewise motion of kinks or act as weak obstacles and shorten the effective length of screw segments (Kubin and Louchet, 1978) or change the activation enthalpy for the nucleation of kink pairs (Kubin and Louchet, 1979; Sato and Meshii, 1973). The quantitative confirmation of these models is still difficult because of the adjustable parameters. The decrease in the transition temperature upon alloying may be understood in terms of the interaction between solute atoms and dislocations of different type (Novak et al., 1976). At low temperatures, the relative change of the mobility of screw dislocations by solute atoms is weaker than for edge dislocations. Hence, edge dislocations have a relative high mobility and are, thus, affected by any obstacle more than the immobile screws. This view agrees with experiments on dislocation mobilities in pure Fe and Fe-6 at % Si alloys (Stein and Low, 1960; Turner and Vreeland, 1970). The significance of dislocation locking for solid solution hardening in b.c.c. alloys has recently be questioned on the following grounds (Suzuki, 1980). Low strain experiments have shown a considerable fraction of the grown in dislocations to start motion at stresses below the yield point. If the majority of dislocations were locked, prestaining should result in crystal softening. This was not observed.

A special case of strengthening of b.c.c. metals is the strength of ferrous martensites which is based on the superposition of several strengthening mechanisms. Much of the strength is due to carbon in

solution (Winchell and Cohen, 1963). Yield stress measurements as a function of carbon content (Winchell and Cohen, 1963; Roberts and Owen, 1968) showed that carbon atoms inhibit dislocation motion primarily by dielastic interaction. However, the observed temperature dependence of the hardening effect is inconsistent with this idea and favours paraelastic interaction or stress induced ordering. Furthermore, dynamic or static dislocation locking by elastic interaction forces has been reported for martensites (Owen and Roberts, 1968). In addition to solid solution hardening processes, the following effects seem to contribute to the strength of ferrous martensite (Leslie, 1966; Speich and Warlimont, 1968; Hornbogen, 1970; Christian, 1971). (i) Grain size strengthening: it was shown (Grange,1966) that smaller martensite plates result in higher yield stress of martensite. This contribution becomes significant if the grain sizes are in the order of 1 μ or less. (ii) precipitation strengthening: precipitates have been formed in martensites after low temperature ageing of freshly quenched martensites and in quenched carbon steels in which they formed during quenching (Leslie and Sober,1967; Kelly and Nutting, 1965). The observations so far available seem to indicate that precipitated carbon may be more effective in strengthening the martensite than the carbon in solution. (iii) substructure strengthening: qualitatively the substructure of martensites may account for the increase of strength with carbon content in terms of increasing proportion of twinned plates (Kelly and Nutting, 1961). The same applies to lattice defects inherited from the austenite.
(iv) dislocation core structure effects: although there is presently little information on the dislocation core structures in martensites, the screw dislocations seen in untwinned regions of martensite are quite similar to the structures seen in b.c.c. metals deformed at low temperatures.
(v) elastic transformation stresses which have not been reduced by plastic shear.

STRENGTHENING BY ORDER

As the movement of superlattice dislocations in perfectly ordered alloys (no domain boundaries) produces no disorder, no increment of strength is to be expected in comparison to the fully disordered condition. This is in general confirmed experimentally (Stoloff and Davies, 1966). However, for imperfectly ordered alloys, a peak in the flow stress at or below the critical temperature of ordering (T_c) is observed (Stoloff and Davies,1966). Six models have been advanced to explain this effect.
(i) When the lattice constant changes upon ordering or if the ordered phase has a distorted structure, relative to the disordered one, partially ordered structures contain internal stress fields in the regions where the order parameter varies (Haasen, 1970). At elevated temperatures, the order profile of an ABB of a superdislocation is wider in thermal equilibrium than in rapid motion due to entropy effects. This difference leads to a yield phenomenon as the dislocations have to be pulled away from their equilibriated APBs. The unlocking stress depends on the degree of order, i.e. on temperature and is expected to have a maximum below T_C (Brown, 1959) because at low temperatures entropy effects are unsignificant and if the temperature approaches T_C the APB energy vanishes.
(ii) If dislocations cut through the APBs, the APB area is increased. The increase of the CRSS due to this effect has a maximum when the domain diameter (l) becomes comparable to the domain wall thickness (Cottrell,1954). At smaller values of l, the alloy is disordered before slip, at larger l, the increment in boundary area diminishes.
(iii) Superdislocations moving through an almost perfectly ordered superlattice are restrained by the unfavourable bonds they create (Fisher,1954).

Hence, a maximum strengthening at an intermediate degree of order results
when the motion of single dislocation is impeded by the existing order and
when the motion of superdislocations is impeded by the existing disorder
(Rudman,1962; Stoloff and Davis,1964).
(v) If the minimum energy configuration of the APB between the two partners
of a superdislocation is not the slip plane, the dislocations rearrange at
elevated temperatures into a lower energy configuration. They are then
difficult to move unless the temperature increases further to permit dislocation motion by blimb (Flinn,1960).
Instead of temperature variation, the degree of order may also be changed
by varying the composition off the stoichometry. Compositional variations
may strengthen an ordered crystal by two effects
(i) The surplus of atoms of one sort may result in solid solution hardening
effects.
(ii) If the ordering energy is high, the fully ordered structure may be
obtained and the deficient atoms are replaced by vacant lattice sites.
Experimental evidence supporting the models advanced for the strengthening
of imperfectly ordered alloys (either by high temperatures or off-stoi-
chometric compositions) has been presented by several authors (Marcinkowski
and Miller,1961; Lawley et al.,1961; Savitskii,1960; Syntkina and Yakosleva,
1963; Kornilov,1960).

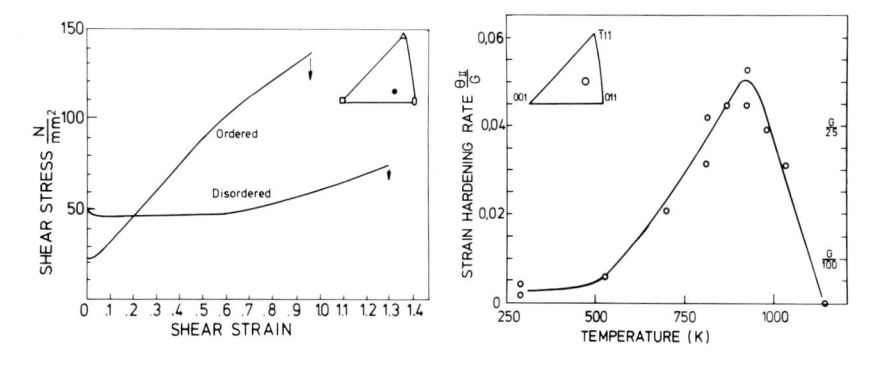

Fig. 5. (a) Effect of long-range order on the stress-strain curve of Cu_3
Au single crystals (Davies and Stoloff,1964) and (b) temperature vs.
hardening rate Θ_{II}/G of Ni_3Al, where G is the shear modulus and
Θ_2 is the slope of the stress-strain curve in stage II (Copley
and Kear,1967).

The effect of order on work hardening is shown in Fig. 5. Stage I is
virtually eliminated and Θ_{II}/G becomes temperature dependent (Fig. 5b)
Two types of models have been advanced to account for these effects. The
first model attributes the strong work hardening to cross slip (Kear,1966).
If the second dislocation of a superdislocation pair cannot follow exactly
in the wake of the first one, both will be pinned, since to move both dislocations as single units requires the formation APBs by both dislocations.
As cross slip is thermally activated, the number of barriers will increase
with increasing temperature. Close to T_m, the barriers can be removed by
climb and Θ_{II}/G will decrease (Davies and Stoloff,1965). Instead of cross-
slip, the onset of cube slip at elevated temperatures has been involved
(Copley and Kear,1967). Alternatively (Vidoz and Brown,1962; Schoeck,1969)
the rapid strain hardening of superlattices has been suggested to be a con-

sequence of creation of tubular anti-phase defects during intersection of primary superlattice dislocations with forest superlattice dislocations. Additional strengthening effects result (Schoeck,1969) from the intersection of glide dislocations with grown-in APBs and from recombination processes between glide and forest dislocations. As all contributions (except of intersection with grown-in APBs) increase with growing dislocation density, ordered materials may be expected to show stronger work hardening than the disordered counterpart. At elevated temperatures, the APB tubes are eliminated. Although this paper is concerned with strengthening effects, it may be mentioned that most ordered alloys, except of Mg$_3$Cd and Cu$_3$Au, exhibit low temperatures brittleness in the polycrystalline state. This brittleness may be caused by an unsufficient number of slip systems, restricted cross slip and/or grain boundary embrittlement due to solute segregation.

STRENGTHENING BY PRECIPITATES

For practical purposes one distinguishes between the hardening by coherent (penetrable) precipitates (usually termed precipitation hardening) and by impenetrable (mostly incoherent) precipitates (usually termed dispersion hardening). For details the reader is referred to existing review articles (Kelly and Nicholson,1963; Gleiter and Hornbogen,1968; Brown and Ham,1971; Haasen,1970; Brown,1980). In this section, we shall consider the physical bases of the strengthening mechanisms only and refer to the papers where they have been discussed first. In all cases, the understanding of the mechanisms has been refined and deepened since by further work, the details of which may be found in the reviews mentioned above.

Precipitation Hardening

The various types of interactions between slip dislocations and coherent precipitates can be divided in two groups. One group is based on elastic interaction forces between dislocations and precipitates. This group is of long range character and does not require the core of the dislocation to intersect the precipitate. The elastic interaction may be paraelastic or dielastic. A paraelastic interaction exists if the precipitate is surrounded by a permanent strain field due to lattice mismatch between the precipitate and the surrounding matrix (Mott and Nabarro,1940). The dielastic interaction dominates in the case of a misfit-free precipitate, the elastic constants of which differ from the surrounding matrix (Fleischer,1960). The second group of interaction mechanisms between precipitates and dislocations involves the generation of additional interfacial energy due to the shear of the precipitate by the dislocation. (i) Order strengthening due to the creation of antiphase boundary energy when a slip dislocations of the (disordered) matrix cuts through an ordered precipitate (Gleiter and Hornbogen,1965). (ii) Stacking fault strengthening (Hirsch and Kelly,1965) results from a difference in the stacking fault energy of the matrix and the precipitate so that the spacing between two Shockley partials differs in the matrix from the one in the precipitate. (iii) Precipitate shear strengthening (Kelly and Fine, 1957) is due to the extra interface created in the form of ledges between the matrix and the precipitate when the dislocation cuts through the precipitate. (iv) Friction stress strengthening is due to a friction stress, for example a Peierls stress, in the precipitate which does not exist in the matrix (Brown and Ham,1971).

The basic physical aspects of the movement of a dislocation through an array of precipitates may be seen by simplifying the real system in two ways (Fig. 6a). Firstly, one considers the dislocation as a string of constant line tension and secondly one regards the obstacles as point obstacles without any long range interaction effects on the dislocation.

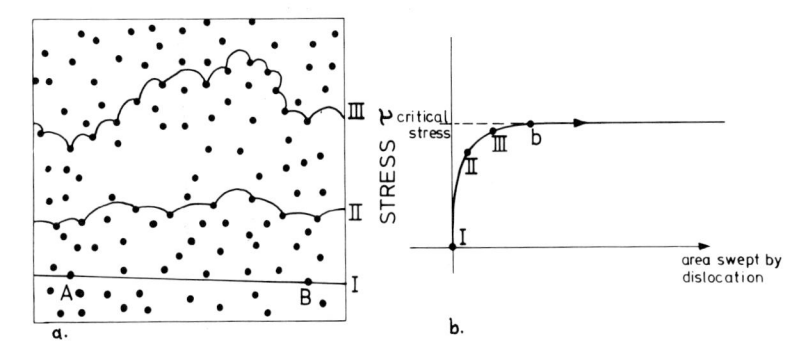

Fig. 6 (a) Movement of a dislocation through a random array of small obstacles. I, II and III show the configuration of the dislocation with increasing stress. (b) Stress/swept area relationship for a dislocation moving through an array of obstacles. The numbers I, II and III refer to the positions of the dislocation shown in Fig. 6a

If the applied stress is zero, the dislocation meets only two obstacles (A,B) of the random array. However, the slightest applied stress bows the dislocation and pushes it forward to meet more obstacles (position II). If the increment of stress is outweighted by additional forces excerted by the increased number of obstacles, the dislocation cannot overcome the obstacles and gets stuck at the position II. A further increment in stress increases the force pushing the dislocation forward (III), but it also bows the dislocation further so that it meets more obstacles than in position II. As the pushing force increases more rapidly with τ, than the sum of the pinning forces, there exists a critical external stress, at which the sum of the forces excerted by all the obstacles cannot counterbalance the force with which the external stress pushes the dislocation through the array of obstacles, and, hence, the dislocation travels large (ideally infinite) distances. This process has been simulated on a computer, and a typical result is shown in Fig. 6b. At the point b, the increment in area for a given increment in stress becomes very large and the stress at this point is identified as the flow stress. In recent years much of this picture has been refined and deepened (Kocks,1977). The quantitative treatment of this process leads to an increase of the flow stress as a function of the precipitate size (r) at constant volume fraction of precipitates as shown in Fig. 7.

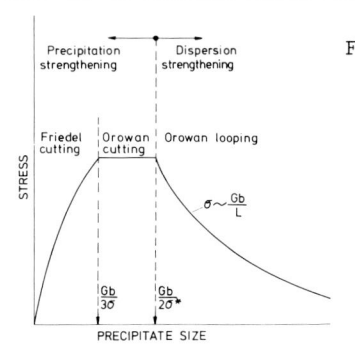

Fig. 7. Schematic variations of the flow stress with precipitate size (Brown,1980). σ^* is the stress excerted by the precipitate on the dislocation, L is the mean planar spacing of the precipitate. G and b are the shear modulus and the Burgers vector of the dislocation, respectively.

The initial increase is due to the superposition of three factors: the larger precipitates excert a larger force on the dislocation. However with increasing precipitate size (r), the number of precipitates met by the dislocation becomes smaller. As r grows, the dislocation bows more and more between the precipitates and, thus, meets (due to the bending) an additional number of precipitates. When the spacing along the precipitates has reached its minimum value, namely L, the mean planar spacing of the precipitates, the alloy has its maximum strength which is controlled by the stress excerted by the precipitates on the dislocation (Orowan cutting). In the regime of Friedel cutting the first and third factor outweighs the second factor, whereas in the regime of Orowan cutting the factors one and three are balanced by the second factor and, hence, the flow stress of the alloy is independent of the size of precipitates. As r increases above $\frac{Gb}{2\sigma^*}$ (Fig. 7), the dislocations by-pass the precipitates by Orowan looping or prismatic cross slip. In principle, precipitation strengthening should be independent of temperature as precipitates are much too large for thermal activation to be significant. However, certain intrinsic properties of the precipitates may depend on temperature and, thus, make precipitate strengthening appear temperature dependent, e.g. the strain surrounding a precipitate may change through differential thermal expansion of matrix and precipitate. Deformation of precipitate hardened alloys may exhibit structural mistabilities that will be treated in the subsequent section.

Dispersion strengthening

Dispersion strengthening in the strengthening of two phase materials by precipitates that cannot be cut by the glide dislocations because they are inherently non-deformable (coherent or incoherent) precipitates. However, also deformable, fully coherent precipitates may become impenetrable if the interaction forces between precipitates and glide dislocations require larger stresses to go through the precipitates than to by-pass them. Impenetrable particles are by-passed by the well known Orowan process.
   As the dislocation does not enter the precipitates, the flow stress (Orowan stress) depends only on the particle spacing and the properties of the matrix and decreases approximately inversely proportional to the precipitate spacing. The Orowan stress has been extensively discussed in the literature (Bacon et al.,1979; Kocks,1977). As long as dislocation climb is negligible, the Orowan stress depends on temperature only through the elastic constants. However, it is unaffected by cross slip (Ashby, 1968). Nevertheless, cross-slip processes in the vicinity of coherent and incoherent precipitates may result in the formation of prismatic loops rather than planar Orowan loops (Hirsch,1957; Gleiter,1967).
Every dislocation by-passing a precipitate generates an Orowan loop or prismatic loops in the vicinity of a precipitate. These loops work harden the material due to the mean stress induced by dislocation loops in the matrix between the precipitates and the effective shortening of the precipitate spacing, as the repulsive stress excerted by a pre-existing loop onto subsequent dislocations makes the subsequent dislocations stand off from the precipitates. Theoretical estimates of both effects (Pedersen and Brown,1977) agree well with the experimental findings. As plastic straining proceeds, the elastic stresses due to the loops may relax by plastic shear of the precipitate (Moan and Embury,1979) prismatic cross slip (Hirsch, 1957; Gleiter,1968) plastic flow by secondary slip systems (Brown and Stobbs,1976; Humphreys and Stuart,1972) and diffusional relaxation (Stobbs, 1973).

Ausformed steels and maraging steels are high strength materials in which strengthening by precipitates plays an important role. For reviews on this subject we refer to the articles by Hornbogen (1970) and Christian (1971). Ausforming is a combined thermal and mechanical process. After an austenitising treatment, the steel is heavily deformed at temperatures in the bainitic bay and is then quenched to produce martensite. The most important factors for the high strength of these materials are believed to be:
(i) the dislocations in the austenite are inherited by the martensite
(ii) coherent precipitates formed in the austenite are sheared into a metastable lattice structure during the martensitic transformation
(iii) the dislocations in the austenite are pinned by the precipitates, thus forming an increased rate of dislocation multiplication. The superposition of all three factors seems crucial for achieving the high strength observed (Leslie et al.,1965). Maraging steels (for a review cf.Hornbogen 1970) are obtained by a heat treatment involving the tempering of martensite. A typical maraging steel contains little carbon, 10-25 % Ni to lower the martensite temperature (M_S) and substitutional elements to produce precipitates. Cooling below M_S leads to dislocation strengthening of the martensite. During subsequent ageing finely dispersed precipitates form and inhibit motion of glide dislocations and prevent recovery of the dislocations introduced by the martensitic transformation. Most effective for maraging strengthening are internally ordered precipitates with large permanent strain fields.

STRENGTHENING BY DYNAMICAL EFFECTS

In general, the stress required to deform crystalline solids increases as the strain rate grows. This dynamical strengthening effect is based on the drag forces exerted on individual moving dislocations and on effects of groups of dislocations. The drag effects on individual moving dislocations (for reviews we refer to Klahn et al.,1970 and Alsitz and Indenbom,1975) are based on the following effects:
(i) phonon dragging by excitons: radiation friction, impurity scattering, Raman scattering and flutter effects (ii) anharmonic phonon mechanisms: scattering, phonon viscosity, thermoelasticity (iii) electron drag (important at low temperatures only) (iv) spin wave drag (limited to ferromagnetic materials).
At high temperatures phonon drag mechanisms dominate (Alsitz and Indenbom, 1975). At very low temperatures electron drag and radiation friction in the Peierls potential are the rate controlling processes (Iida et al., 1979). The understanding of the region of dislocation velocities close to that of sound is still poorly developed. Evidence for break-down of dislocation motion at sound velocity(Weiner et al.,1977) as well as evidence for supersonic dislocation velocities (Weertman,1967) has been presented. In general, groups of dislocations rather than single ones are involved in extensive plastic deformation. Group motion enables slip to proceed over distances exceeding the wavelength of the internal stress fields at stresses smaller than the amplitude of the interal field (Smith,1967). Furthermore, the consideration of dislocation groups has shown that structural and temperature instabilities are important for the resistance of a crystal to deformation. For example, if the first dislocation changes, the obstacle structure in the slip plane, subsequent dislocations can pass more easily. This applies to sweeping of radiation damage (Bapna and Meshii,1974), precipitates (Gleiter and Hornbogen,1965; Klein,1970), break-down of pre-existing dislocation structures in pre-strained (Wessel and Nabarro,1971) or fatigued materials (Neumann,1971). During deformation at very low temperatures, instabilities due to a local rise of temperature and, hence, local increase of the rate of thermally activated processes may

occur (Malygin,1978). At elevated temperatures, solute atom rearrangements resulting in dislocation locking (yield point, Portevin-Le Chatelier effect) have been reported. However, it may be pointed out that the Portevin-Le Chatelier instability may - at least partially - be interpreted in terms of a collective dislocation phenomenon without any long range rearrangement of solute atoms (Kobel et al.,1976).

REFERENCES

Akhtar, A.,and E.Teghtsoonian (1968). Trans Jap. Inst. Metals.
Alshitz, V.A., and V.L. Indenbom (1975). Soc. Phys.-Usp., 18,1.
Armstrong, R.W., I. Codd, R.M. Douthwaite,and N.J. Petch (1962). Phil. Mag. 7, 45.
Armstrong, R.W. (1969). Advances in Materials Research IV, Wiley-Interscience Publ.; p. 1.
Arsenault, R.J. (1969). Acta Met. 17, 1291.
Ashby, M.F. (1968). Unpublished quoted in L.M. Brown, and R.K. Ham in: Strengthening Methods in Crystals; A. Kelly,and R.B. Nicholson eds. Applied Sci. Publ. London, p. 72.
Ashby, M.F. (1971). In: Strengthening Methods in Crystals, Applied Sci. Publishers; A. Kelly,and R.B. Nicholson, eds., p. 137.
Bacon, D.J., D.M. Barnett, and R.O. Scattergood (1979). Prog. Mat. Sci. 23,21.
Bäro, G., and E. Hornbogen (1969). Quant Relation of Properties and Microstructure, Israel University Press, Jerusalem, p. 457.
Bapna, M.S., and M. Meshii (1974). Mat. Sci., and Eng. 16, 181.
Basinski, Z. S., M. S. Duesbery, and R. Taylor (1971). Can. J. Phys., 49,2160.
Bergmann, H.W. (1976). Z. f. Werkstofftechnik, 7, 129.
Bocek, M., P. Kratochvil, and P. Lukáč(1961). Czech. J. Phys. B11, 674.
Brown, N. (1959). Phil. Mag. 4, 693.
Brown, L.M., and W.M. Stobbs (1976). Phil. Mag. 34, 351.
Brown, L.M., and R.K. Ham (1971). In: Strengthening Methods in Crystals; A. Kelly, and R.B. Nicholson eds.; Applied Science Publ., p. 9.
Brown, L.M. (1980). In: Proc. 5th Int. Conf. on the Strength of Metals and Alloys; P. Haasen, V. Gerold, and G. Kostorz eds.; Pergamon Press,p. 1551.
Chandhok, V.K., A. Kasak, and J.P. Hirth (1966). Trans ASM 59, 288.
Christian, J.W. (1971). In: Strengthening Methods in Crystals; A. Kelly, and R.B. Nicholson eds.; Applied Science Publ. London, p. 261.
Copley, S.M., and B.H. Kear (1967) Trans AIME 339, 977.
Cottrell, A.H., C.S. Hunter, and F.R.N. Nabarro (1953). Phil. Mag. 44,1064.
Cottrell, A.H. (1963). The Realtionship between Structure and Mechanical Properties of Metals, p. 456, H.M.S.O. London.
Cottrell, A.H. (1954). In: Relations of Properties to Microstructure, ASM, Metals Park, Cleveland, p. 121.
Davies, R.G., and N.S. Stoloff (1964). Phil. Mag. 9, 349.
Davies, R.G., abd N.S. Stoloff (1965). Phil. Mag. 12, 297.
Embury, J.D. (1971). In: Strengthening Methods in Crystals Applied Science Publ. London, A. Kelly and R.B. Nicholson editors, p. 337 and p. 342.
Fisher, J.C. (1954). Acta Met. 2, 9.
Fleischer, R.L. (1968). Acta Met. 8, 598.
Fleischer, R.L. (1963). Acta Met. 11, 203.
Flinn, D.A. (1960). Trans AIME 218, 145.
Friedel, J. (1956). Dislocations and Mechanical Properties of Crystals, J.C. Fisher ed. J. Wiley and Sons, New York, 361.
Frommeyer, G., and G. Wassermann (1976). Z. f. Werkstofftechnik 2, 136 and 154.
Gleiter, H., and E. Hornbogen (1965). Phys. stat. sol. 12, 251.
Gleiter, H. (1967). Acta Met. 15, 1213 and 1223.
Gleiter, H. (1968). Materials Science and Engineering 2, 285.
Gleiter, H. (1980). In: Proc. 2nd Risø Int. Symp. on Metallurgy and Materials Science. N. Hanse, A. Horsewell, T. Leffers, and H. Lilholt eds.; Risø Natl. Lab. Roskilde/Denmark; 15.

Grange, R.A. (1966). Trans ASM 59, 26.
Haasen, P. (1970). In: Physical Metallurgy. R.W. Cahn ed.; North Holland Publ. Amsterdam, 1011.
Haasen, P. (1979). In: Dislocations in Solids. F.R.N. Nabarro ed.; North Holland Publ. Amsterdam, Vol. IV, 218.
Haessner, F., and D. Schreiber (1957). Z. f. Metallk. 48, 263.
Haessner, F., and H. Müller (1981). Annual Meeting of the German Metal Society, 9-12 June, 1981, Baden-Baden.
Hall, E.O. (1951). Proc. Roy. Soc. B64, 267.
Hendrickson, A.A., and M.E. Fine (1981). Trans AIME 221, 967.
Hirsch, P.B. (1957). J. Inst. Metals 86, 7.
Hirsch, P.B., and A. Kelly (1965). Phil. Mag. 12, 881.
Hornbogen, E. (1970). In: Phyiscal Metallurgy. R.W. Cahn ed., North Holland Publ. Amsterdam, 589.
Hornbogen, E. (1980). In: Proc. 5th Int. Conf. on Strength of Metals and Alloys, Vol. 2, Pergamon Press, New York, 1337.
Humphreys, F.J., and A.T. Stuart (1972). Surf. Sci. 31 , 389.
Iida, F., T. Suzuki, E. Kuramoto, and S. Takenchi (1979). Acta Met. 27, 637.
Johnston, W.G., R.G. Davies, and M.S. Stoloff (1965). Phil. Mag. 12, 305.
Jordan, K.R., and N.S. Stoloff (1968). Trans Jap. Inst. Met. 9, 281.
Kear, B.H. (1966). Acta Met. 14, 659.
Kelly, A., and M.E. Fine (1957). Acta Met. 5, 365.
Kelly, A., and R.B. Nicholson (1963). Progr. Mat. Sci. 10 (3).
Kelly, P.M., and J. Nutting (1961). J. Iron Steel Inst. 197, 199.
Kelly, P.M., and J. Nutting (1965). Iron Steel Inst. Spec. Rep. 93, 166.
Kelly, P.M., and J. Nutting (1960). Proc. Roy. Soc. A259, 45.
Kindin, I.N., A.N. Marshalkin, and Y.N. Mizonoo (1966). Izv. vuz. Chernaya 1, 141.
Klahn, D., A.K. Mukherjee, and J.E. Dorn (1970). In: Proc. 2nd Int. Conf. on Strength of Metals and Alloys, ASM, Metals Park Ohio, 951.
Kobel, A., J. Zasadzinski, and Z. Siekluka (1976). Acta Met. 24, 919.
Kocks, U., A.S. Argon, and M.F. Ashby (1975). Progr. Mat. Sci. 19, 1.
Kocks, U.F. (1977). Mat. Sci. and Eng. 27, 291.
Kocks, U. (1980). In: Proc. 5th Int. Conf. on Strength of Metals and Alloys, Vol. 3, 1661, Pergamon Press, New York.
Kopenaal, T.J., and M.E. Fine (1961). Trans AIME. 221, 1178.
Kornilov, I.I. (1960). Mechanical Properties of Intermetallic Compounds; J.H. Westbrook ed.; Wiley Publ. New York, 344.
Kröner, E. (1964). Phys. kondens. Mat. 2, 262.
Kubin, L.P., and F. Louchet (1978). Phil. Mag. 38, 205.
Kubin, L.P., and F. Louchet (1979), Acta Met. 27, 337.
Lawley, A.E., A. Vidoz, and R.W. Cahn (1961). Acta Met. 9, 287.
Leslie, W.C., D.W. Stevens, and M. Cohen (1965). High Strength Materials; V.F. Zackay ed.; Wiley Publ., 382.
Leslie, W.C., and R.J. Sober (1967). Trans ASM 60, 99.
Leslie, W.C. (1966). In : Strengthening Mechanisms in Metals and Alloys. Syracuse University Press, 43.
Lojkowski, W., and M.W. Grabski (1980). In: Proc. 2nd Int. Symp. on Metallurgy and Materials Science, 1981, Risø National Laboratory, Roskilde/Denmark; N. Hansen, A. Horsewell, T. Leffers, and H. Lilholt eds., 329.
Malygin, G.A. (1977). Sov. Phys. Sol. State 19, 850.
Marcinkowski, M.J., and D.S. Miller (1961). Phil. Mag. 6, 871.
Marcinkowski, M.J., and R.M. Fisher (1965). Trans Met. Soc. AIME 233, 293.
Moan, G.D., and D.J. Embury (1979). Acta Met. 27, 903.
Mott, N.F., and F.R.N. Nabarro (1940). Proc. Roy. Soc. 52, 86.
Mott, N.F., and F.R.N. Nabarro (1948). Rep. Bristol Conf. on Strength of Solids, Phys. Soc. London, 1.
Neumann, P. (1971). Acta Met. 19, 1233.
Novak, V., K.Z. Saleeb, S. Kadečkova, and B. Sestak (1976). Czech. J. Phys. B26, 565.

Owen, W.S., and M.J. Roberts (1968). Int. Conf. on Strength of Alloys, Tokyo, Japan, Inst. of Metals, 911.
Pedersen, O.B., and L.M. Brown (1977). Acta Met. 25, 1303.
Petch, N.J. (1953). J. Iron Steel Inst. 174, 25.
Pumphrey, P.H., and H. Gleiter (1974a). Phil. Mag. 30, 593.
Raffo, P.L., and T.E. Mitchell (1968). Trans Met. Soc. AIME 242, 907.
Rezek, J., and C.R. Craig (1961). Trans. Met. Soc. AIME 221, 715.
Roberts, M.J., and W.S. Owen (1968). J. Iron Steel Inst. 206, 375.
Rudman, P.S. (1962). Acta Met. 10, 253.
Sato, A., and M. Meshii (1973). Acta Met. 21, 753.
Savitskii, M.E. (1960). Mechanical Properties of Intermetallic Compounds; J.H. Westbrook ed.; Wiley Publ. New York, 87.
Schoeck, G., and A. Seeger (1959). Acta Met. 7, 469.
Schoeck, G. (1969). Acta Met. 17, 147.
Sestak, B. (1980). In: Proc. 5th Int. Conf. on Strength of Metals and Alloys, Vol. 3, Pergamon Press New York, 1461.
Smith, E. (1967). Phil. Mag. 16, 1285.
Speich, G,R., and H. Warlimont (1968). JISI 204, 385.
Stein, D.F., and J.R. Low (1960). J. Appl. Phys. 31, 362.
Stobbs, W.M. (1973). Phil. Mag. 30, 1073.
Stoloff, N.S., and R.G. Davies (1966). Prog. Mat. Sci., B. Chalmers ed., 13, 1.
Suzuki, H. (1957). In: Dislocations and Mechanical Properties of Crystals. J.C. Fisher ed., J. Wiley and Sons New York, 361.
Suzuki, H. (1971). Nachr. Akad. Wiss. Gött., Math.-Phys. Klasse 2, Nr. 6, 1.
Suzuki, H. (1980). In: Proc. 5th Int. Conf. on Strength of Solids, P. Haasen, V. Gerold, and. O. Kostorz eds.- Pergamon Press, Vol. 3, 1595.
Syntkina, V.I., and E.S. Yakovleva (1963(. Sov. Phys. (Solid State), 4, 2125.
Thompson, A.W. (1977). In: Work Hardening in Tension and Fatigue, A.W. Thompson ed., ASM, Metals Park, 89.
Turner, A.P.L., and T. Vreeland (1970). Acta Met. 18, 1225.
Vidoz, A.E., and L.M. Brown (1962). Phil.Mag. 7, 1167.
Vitek, V. (1974). Crystal Lattice Defects, 5, 1.
Warrington, D.H. (1963). J. Iron Steel Inst. 201, 610.
Wassermann, G. (1976). In: Proc. 4th Int. Conf. on Strength of Metals and Alloys Vol. 3, Nancy, 1343.
Weertmann, J. (1967). J. Appl. Phys. 38, 5293.
Weiner, J.H., and Y.Y. Amme (1977). J. Appl. Phys. 48, 3317.
Wessel, E.J.H., and F.R.N. Nabarro (1961). Acta Met. 19, 915.
Wielke, B., A. Chalupka, B. Kaufmann, P. Lukáč, and A. Svoboda (1979). Z. f. Metallk. 70, 85.
Wilcox, B.A., and A. Gilbert (1967). Acta Met. 15.
Wilsdorf, H.F.F., O. Inal, and E. Murr (1978). Z. f. Metallk. 69, 701.
Wilson, D.V. (1967). Metals Science Journal 1, 40.
Winchell, P.G., and M. Cohen (1963). Electron Microscopy and Strength of Crystals. Interscience Publ., 995.
Yang, M. T. Tschakalakos, and J.E. Hilliard (1977). J. Appl. Phys. 48, 876.

ACKNOWLEDGMENT

This review was prepared during a stay at the TECHNION - Israel Institute of Technology -, Haifa. The author would like to thank Profs. B.Z. Weiss and Y. Komem from the Department of Materials Engineering of the TECHNION, Haifa for their support and hospitality.

The Application of the Fundamentals of Strengthening to the Design of New Aluminum Alloys

Edgar A. Starke, Jr.

Fracture and Fatigue Research Laboratory, Georgia Institute of Technology, Atlanta, Georgia 30332, USA

ABSTRACT

The strengthening methods in nonferrous metals are briefly reviewed. Recent approaches to alloy design of high strength-high temperature aluminum alloys and high strength-low density aluminum alloys are described in terms of fundamental concepts.

KEYWORDS

Precipitation hardening, dispersion hardening, boundary hardening, high temperature strength, creep, aluminum alloys.

INTRODUCTION

The strengthening of metals and alloys is associated with the obstruction of mobile dislocations by certain components of the microstructure. Our ability to understand the effectiveness of many of these components has led to the development of reliable theories governing strengthening mechanisms. For recent reviews see, for example, Strengthening Methods in Crystals, ed. by Kelly and Nicholson, 1971, and Alloy and Microstructural Design, ed. by Tien and Ansell, 1976. Although we are normally unable to calculate mechanical properties from first principles (Ashby, 1970), the fundamental concepts serve as guidelines for alloy development, increasing the probability for success and minimizing the time spent in evolutionary alloy design based totally on experience. Since the fundamentals of strengthening mechanisms are discussed in detail by Professor Gleiter in the lead-off paper of this conference, I will only cover those points germane to this presentation and then describe how they are being used to develop two new classes of aluminum alloys.

STRENGTHENING MECHANISMS

The strengthening mechanisms most often employed for nonferrous alloys include those associated with solid solutions, precipitates and dispersions, and grain and subgrain boundary effects. These may be used independently, but are most often used in combination. Collectively the contribution of each mechanism is somewhat less than anticipated from an individual basis.

Effective solid solution strengthening requires a high solid solubility of an element having a significant size difference from the solvent which imparts a large lattice distortion (Fleischer, 1961, 1964). Unfortunately, the latter limits solubility and thus restricts solid solution strengthening in most commercial alloys. Effective precipitation and dispersion strengthening depend on a fine distribution of second phase particles. Depending on the size, spacing and degree of coherency, the particles are either looped and bypassed or sheared by dislocations during plastic deformation.

If the particles are bypassed, the strengthening is independent of the properties of the particles and the yield stress (τ_y) can be described simply by the modified Orowan relationship

$$\tau_y = \tau_o + \frac{0.8Gb}{L} \qquad (1)$$

where τ_o is the flow stress of the unstrengthened matrix and may include a component of solid solution strengthening, G is the shear modulus, b the Burgers vector and L the interparticle spacing. The strength decreases with increasing particle size when the volume fraction is constant. One of the problems associated with using this strengthening method results from the difficulty in producing a dispersion of small, hard, closely-spaced, incoherent particles using normal ingot casting procedures. Another is the negative ductility factor that accompanies this, and other strengthening methods.

If the particles are sheared, strengthening depends on the intrinsic properties of the particles and the yield stress may be represented by an equation of the form,

$$\tau_y = cf^m r^p \qquad (2)$$

where f is the volume fraction and r is the particle radius. The exponents m and p are always positive and the strength increases with both volume fraction and particle size. The parameter c is an alloy constant and depends on the misfit between the particles and the matrix, the modulus of the particles relative to the matrix, the degree of order of the particles and the energy associated with the particle-matrix interface (Kelly, 1973). When the particles are sheared their associated strengthening effects are reduced resulting, successively, in a local decrease in resistance to further dislocation motion, planar slip, a tendency toward strain localization, and low ductility (Hornbogen and Zum Gahr, 1975). Concomitant with the uniform precipitation of the coherent metastable phases, heterogeneous precipitation of equilibrium phases can occur along grain boundaries resulting in solute-depleted precipitate free zones (PFZ's) adjacent to the grain boundaries (Starke, 1970). These zones are weaker than the matrix and can be the site of preferential deformation (Geisler, 1949) leading to high stress concentrations at grain boundary triple junctions and low ductility.

Grain and subgrain boundaries generally inhibit dislocation motion and their presence is a source of strength. Grain refinement has also been shown to improve ductility (Lütjering and co-workers, 1977), fracture toughness (Yokobori and co-workers, 1973), stress corrosion resistance (Sarkar and co-workers, 198) and fatigue crack initiation resistance (Starke and Lütjering, 1979). The effect of grain size (d) on the yield stress is normally described by the Hall Petch relationship (Hall, 1951; Petch, 1953),

$$\tau_y = \tau_o + kd^{-\frac{1}{2}} \qquad (3)$$

where τ_o and k are constants. A similar relationship holds for strengthening by subgrains or cells (Thompson, 1977), however, the constant k is usually somewhat smaller than that for grains, and the exponent of d normally has a value of one. The Hall-Petch relationship may be rewritten (Nicholson, 1971) as

$$\tau_y = m\tau_o + (m^2\tau^*r^{\frac{1}{2}})d^{\frac{1}{2}} \qquad (4)$$

where m is an orientation factor related to the number of slip systems available, τ_o is the friction stress on a single dislocation, τ^* is the critical shear stress required to activate a dislocation source and r is a measure of separation of the source from the nearest dislocation pileup. τ_o depends on the various matrix hardening mechanisms, and a large τ_o such as those for precipitation and dispersion strengthened alloys, reduces the grain size contribution to the strength. A sharp texture may also decrease the effectiveness of grain boundaries as slip barriers due to the alignment of slip systems in adjacent grains and hence the ease of propagation of slip (Nicholson, 1971; Sanders and Starke, 1979). Since grain boundary sliding becomes an important deformation mode at high temperatures, alloys designed for that purpose should have their grain boundaries strengthened (Copley and Williams, 1976) or have a large grain size (Ralph, 1980).

APPLICATION OF STRENGTHENING METHODS IN NEW HIGH STRENGTH ALUMINUM ALLOYS

The primary use of high-strength aluminum alloys is in aircraft construction, and today, on an industry-wide basis, they make up 80% of the weight of the airframe. Consequently, property improvements in these materials are most often driven by design criteria for aircraft. Designers currently list in order of importance (1) primary strength to weight ratio, (2) durability, which includes corrosion resistance, damage tolerance, etc., and (3) economic considerations, which include materials costs and their availability, producibility and maintainability. Recently, the higher temperatures generated by supersonic flight and those encountered in high performance gas turbine engines coupled with aluminum's limited high temperature strength has resulted in an increased use of titanium as a replacement. In addition, increasing energy costs and demands for greater payload ratios continue to require minimum weight solutions (Lewis and co-workers, 1978), and projections for future aircraft suggest that many of the aluminum alloy components will be replaced by those manufactured from lighter nonmetallic composites. However, since titanium alloys and nonmetallic composites are costly, difficult to fabricate, and not always readily available, there is a clear demand for new aluminum alloys that combine high strength, fracture toughness and corrosion resistance with improved thermal stability, reduced density, and greater stiffness.

High Temperature Aluminum Alloy Development

In the 1950's attempts were made to produce an aluminum alloy stable at high temperatures by utilizing sintered alloy powder, SAP, which contained a high concentration of fine oxide particles (Bloch, 1961). Although this product was stable to ∿800K, it had limited room temperature strength, poor reproducibility of mechanical properties, and a tendency toward embrittlement during sustained loading at elevated temperatures (Langenbeck and co-workers, 1981). The common aluminum alloys currently being used by the aircraft industry for high temperature applications are based on the Al-Cu system and include 2219 (Al-Cu-Mn) and 2618 (Al-Cu-Mg-Fe-Ni). The transition elements form intermetallic phases which are stable at elevated temperatures and hinder grain boundary sliding. However, both alloys derive their strength from coherent and partially coherent intermediate precipitates which coarsen rapidly and transform to equilibrium precipitates at temperatures above 450K. The rate of coarsening is controlled either by atom transfer across the precipitate matrix interface or by volume or bulk diffusion. Low values of interfacial energy, solid solubility, and solute diffusivity are desirable for low coarsening rates (Lifshitz and Slyosov, 1961; Wagner, 1961).

TABLE 1 Diffusion Parameters and Solid Solubilities of Binary Additions to Aluminum*

Solute	Maximum Solid Solubility Wt.%	Diffusion Coefficient D_o, cm^2/sec	Activation Energy k cal/g atom
Ce	<0.05	1.9×10^{-6}	26.6
Co	<0.02	1.1×10^{-6}	19.19
Cr	0.72	3.01×10^{-7}	15.4
Cu	5.70	0.15	30.2
Fe	0.05	4.1×10^{-9}	13.9
Mg	17.4	0.12	28.6
Mn	1.4	0.22	28.8
Ni	0.05	2.9×10^{-8}	15.7
Ti	0.24	5×10^{-7}	25.8
V	0.37	6.05×10^{-7}	19.6
Zn	82.2	11.6	27.8

*Data from Hansen, 1958 and Bishop and Fletcher, 1972.

The solid solubility and diffusion parameters for a selected number of binary additions to aluminum are given in Table 1. The alloying elements normally used for precipitation hardening, i.e., Cu, Mg, and Zn, have high solubilities and diffusion rates. However, the dispersoid forming transition elements and the rare earth Ce have low solubilities and low diffusion rates. A high volume fraction of small dispersoids formed by these additions can give adequate high temperature strength. The dispersoids can also slow down all three contributions to steady-state creep; grain boundary sliding, diffusional or "Nabarro-Herring" creep; and dislocation creep (Ashby, 1970). Large additions of the transition elements can significantly increase the elastic modulus of aluminum, Fig. 1, making the dispersion-hardened alloys especially attractive for stiffness critical, high temperature applications. Unfortunately, the limited solid solubility makes it impossible to obtain a fine dispersion of closely spaced incoherent particles using normal ingot casting procedures. The inherently slow ingot solidification rates enhance the precipitation of coarse particles which lower ductility and fracture toughness values.

Rapid solidification procedures, based on the concepts developed initially by Duwez and co-workers (1960) for quenching droplets of molten metal, avoids the thermodynamic limitations normally imposed on aluminum-transition metal alloys. The droplets are either solidified in flight at rates somewhat less than 10^5 K/s or during impact on a water-cooled metal hearth at rates exceeding 10^5 K/s. The accompanying supercooling promotes enhanced supersaturation of solute elements and a refinement of the microstructure (Davis, 1978). Large quantities of particulate can be produced using rapid solidification methods and can be consolidated using normal powder metallurgy procedures. Processing parameters must be controlled so that decomposition of the supersaturated alloy produces the desired fine dispersion of precipitates since they cannot be redissolved below the melting temperatures. Although this technique normally results in a small-grained material which is usually undesirable for high temperature applications, the presence of fine aluminum oxides along the particulate boundaries in conjunction with the dispersoids can restrict creep at high temperatures.

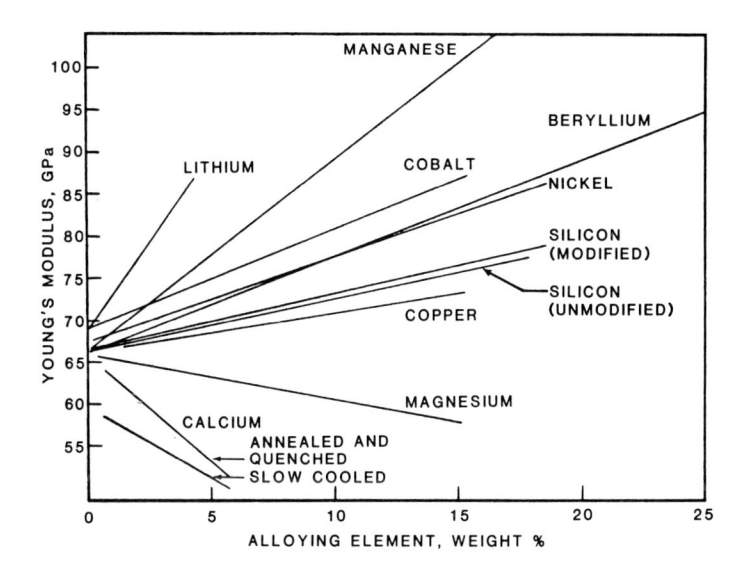

Fig. 1. The effect of solute additions on the modulus of aluminum alloys (complied by George Wald).

Rapid solidification and P/M consolidation is currently being used in aluminum alloy development programs aimed at producing elevated temperature properties competitive with those of titanium (Wald, 1981; Griffith and co-workers, 1982; Paris and co-workers, 1982). Solute element additions having high liquid solubility, low solid solubility and low diffusion rates are being used either individually or in combination. These include Co, Fe, Ni, Ce and Cr. Although Eq. 1 predicts an increase in strength with decreasing interparticle spacing, and therefore with volume fraction for a constant particle size, the negative ductility factor places restrictions on solute concentrations. Argon and Im (1975) have shown that when the volume fraction of particles approaches some critical value, interaction effects come into play and interfacial stresses are enhanced due to overlap of the plastic zones around the particles. This results in interface separation at very low strains (Tien, 1975).

The principal characteristics which dominate the microstructure and properties of dispersion-hardened aluminum alloys fabricated from rapidly solidified particulate arise from the size distribution of the particulate and the quality of the interparticulate bonding (Paris and co-workers, 1982). A typical microstructure of an Al-Fe-Ni-Co splat-quenched flake is shown in Fig. 2 (Mullins, 1982), and contains both coarse and fine precipitates analogous to the microstructure of RS Al-Fe alloys studied by Jones (1969). Jones suggested that the amount of coarse precipitates could be reduced by increasing the solidification rates and/or reducing the solute concentration. Although the bimodal structure normally exists within one particulate particle, the relative amounts of coarse and fine structure depend on the particulate size. A larger amount of the coarse dispersoids appear in the large particulates because they have a lower degree of undercooling during solidification. Since a range of particulate sizes are used for consolidation, the finished product contains a bimodal particle size distribution similar to those shown in Fig. 3 for Al-Fe-Ni-Co extruded plate.

Fig. 2. TEM of rapidly solidified Al-3.3Fe-2.3Ni-4.6Co[1] splat-quenched flake showing regions of coarse and fine dispersoids (from Mullins, 1982).

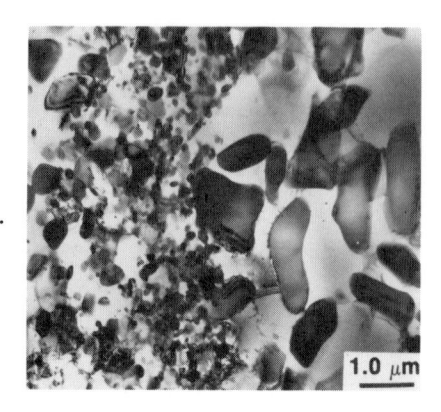

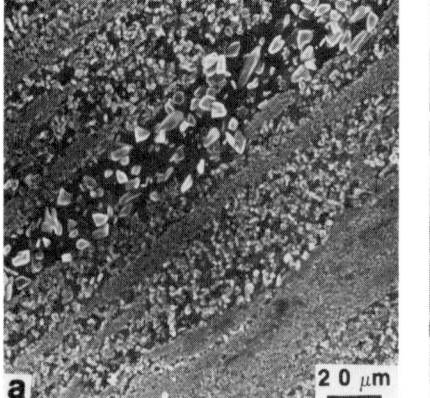

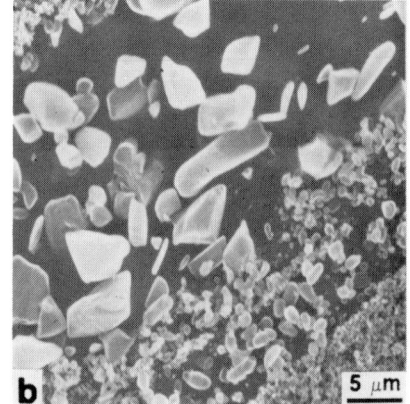

Fig. 3. SEM's of bromine-etched surfaces of extruded Al-3.3Fe-2.3Ni-4.6Co plate made from rapidly solidified flake showing coarse and fine dispersoids associated with large and small flake particulate (from Mullins, 1982).

The fine structure occupied approximately 95% of the volume of the alloys studied by Mullins and appeared to control the flow stress. Although the yield strength could be correlated with the interparticle spacing of the small particle regions, Fig. 4, no one-to-one relationship existed with calculated values using the Orowan equation. This is probably due to the additional strength contribution associated with the high dislocation densities introduced during compaction and primary processing of the wrought product. Nondeformable dispersoids generate a large number of geometrically necessary dislocations

---

[1] All compositions are in weight percent unless otherwise specified.

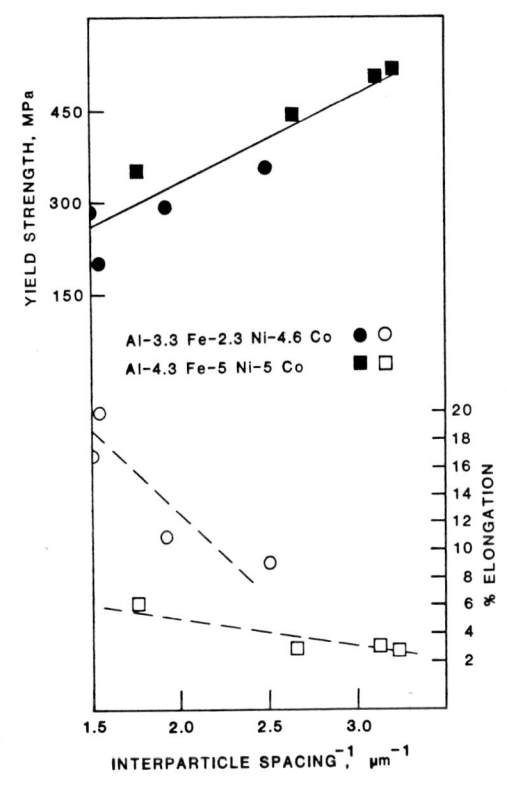

Fig. 4. The effect of interparticle spacing on the yield strength of Al-Fe-Ni-Co plate made from P/M consolidation of rapidly solidified particulate (from Mullins, 1982).

(Ashby, 1968), which contribute to the strength when the final product is unrecrystallized, as for the case described here. The coarse structure occupied less than 5% of the extruded plate but most likely controlled the ductility. Cracks were first formed at the large particles, Fig. 5, and final fracture occurred by the coalescence of microvoids associated with the dispersoids. Other research has also shown that even when small particles far outnumber large ones, the larger particles crack first and at lower strains than the smaller particles (Palmer and Smith, 1968; Gangulee and Gurland, 1967). The difference in strength-ductility relationship of the two alloys of Fig. 4 may be related to more large particles in the higher solute alloy.

As mentioned earlier, Jones suggested that the formation of coarse particles may be suppressed by increasing the degree of undercooling. This is often impractical and other approaches must be taken. It is well known that multiple alloy additions can affect the precipitation in many aluminum systems (Polmear, 1966). In certain cases, compositional modification can increase the nucleation frequency and decrease both the size and coarsening rates of the precipitates, probably by altering interfacial energies and diffusion processes. This has been accomplished for the Al-Fe alloys by additions of Ce. Rapidly solidified Al-Fe binary alloys have microstructures containing both coarse and fine dispersoids (Jones, 1969) similar to those shown in Fig. 2. The coarse precipitates are absent in Al-Fe-Ce alloys, Fig. 6 (Hildeman and Sanders, 1981).

Al-8Fe-3.4Ce is a new aluminum alloy that shows promise for high temperature applications (Sanders and Hildeman, 1981). Figure 7 compares its yield strength

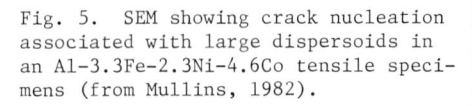

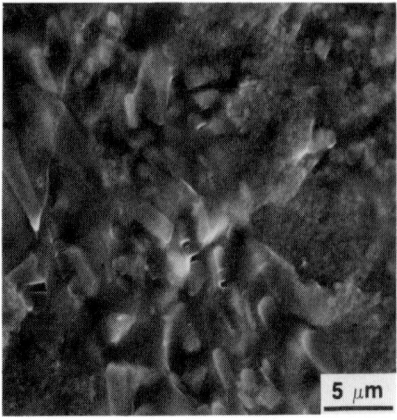

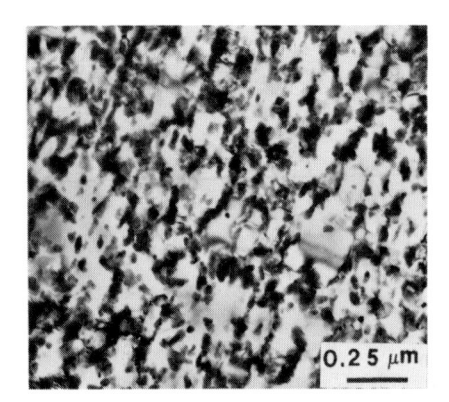

Fig. 5. SEM showing crack nucleation associated with large dispersoids in an Al-3.3Fe-2.3Ni-4.6Co tensile specimens (from Mullins, 1982).

Fig. 6. TEM of a rapidly solidified Al-8Fe-3.4Ce alloy (from Sanders and Hildeman, 1981).

at various temperatures with the program goal and the currently used high temperature alloy 2219-T851. The decrease in strength with temperature is related to both recovery and coarsening processes and occurs at a lower rate than that for 2219-T851 at temperatures up to 505K. The real advantage of the dispersoid-strengthened material, however, is in its creep resistance, Fig. 8 (Sanders and Hildeman, 1981). Note that the creep rate of age-hardened 2219 and 2124 increases rapidly with time while that of the Al-8Fe-3.4Ce alloy seems to level off. The accelerated creep of 2219 and 2124 is associated with extensive overaging that occurs at 505K. However, coarsening of the dis- dispersoid-strengthened Al-8Fe-3.4Ce alloy is unlikely at 505K and the initial creep strain is probably associated with recovery processes. Once these are complete additional creep strain is minimal.

This study clearly demonstrates the feasibility of producing aluminum alloys with usable mechanical properties and creep resistance in the 505-616K range. However, more work needs to be directed toward improving the ductility. The elongation to failure at room temperature was 5% for the Al-Fe-Ce alloy. Argon and Im (1975) suggest that some improvement may be obtained by controlled deformation processing, i.e., keeping the increments of strain between annealing operations below the critical strain for dispersoid fracture and interface separation.

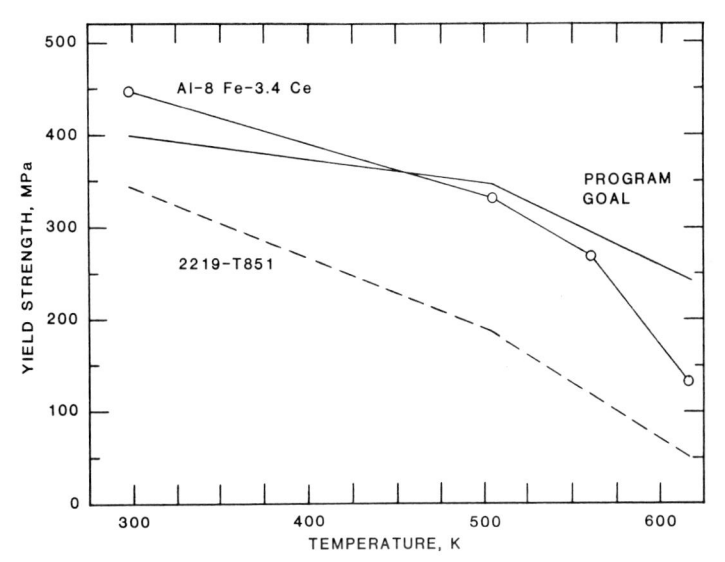

Fig. 7. Comparison of yield-strength-temperature behavior of Al-8Fe-3.4Ce forging made from rapidly solidified particulate with two I/M 2XXX alloys. Tests conducted after 1000h exposure at temperature (from Sanders and Hildeman, 1981).

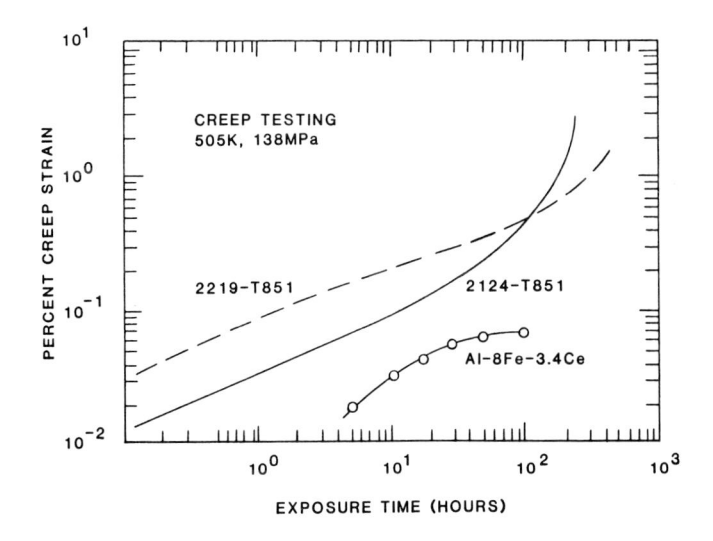

Fig. 8. Comparison of creep strain versus temperature of Al-8Fe-3.4Ce forging made from rapidly solidified particulate with two I/M 2XXX alloys (from Sanders and Hildeman, 1981).

Low Density Aluminum Alloy Development

Lithium is the lightest metallic element and, with the exception of beryllium which has manufacturing and health-related problems, is the only metal that improves both the modulus and density when alloyed with aluminum, Figs. 1 and 9. Each weight percent lithium added to an aluminum alloy reduces the density approximately 3% and increases the elastic modulus approximately 6% for lithium additions up to 4% (Sankaran and Grant, 1981). Alcoa was the first to recognize the effect of lithium on elastic modulus; and the potential for increasing strength and elastic modulus while decreasing density resulted in the development of alloy 2020 (Al-4.5Cu-1.1Li-0.5Mn-0.2Cd) which became available in the late 1950's (Balmuth and Schmidt, 1981). Alloy 2020 was used to form the upper and lower wing skins of the US Navy A-5A and RA-5C Vigilante aircraft. Although this application was successful, the alloy's notch sensitivity and low toughness led to the termination of production in 1969. The current desire to develop high-strength, low density aluminum products has led to renewed interest in Al-Li alloys (Lewis and co-workers, 1978; Starke and co-workers, 1981).

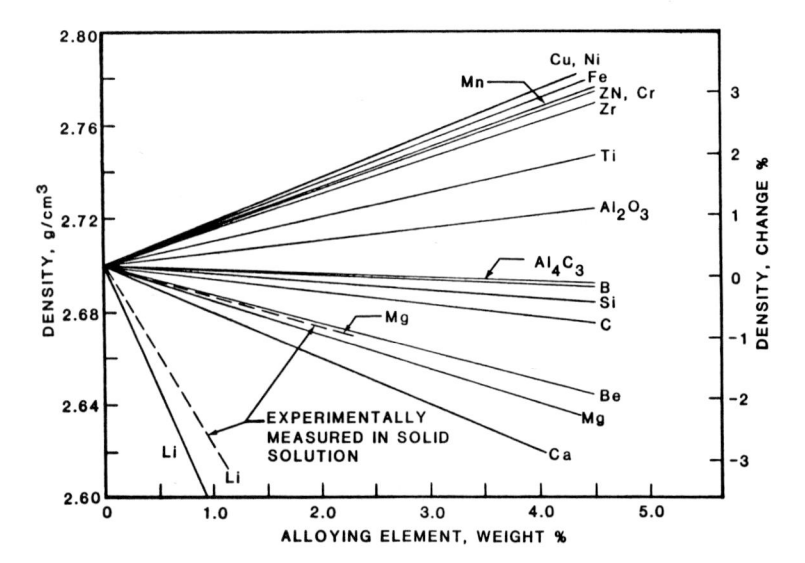

Fig. 9. The effect of solute additions on the density of aluminum alloys (compiled by George Wald).

The Al-Li system is a simple eutectic that contains a metastable miscibility gap (Williams, 1981) and short range order in the solid solution (Cesara and co-workers, 1977). When Al-Li alloys with sufficient solute are quenched from the single phase field and subsequently aged below the critical temperature that defines the metastable miscibility gap, decomposition of the supersaturated solid solution occurs by homogeneous precipitation of the ordered Ll_2 phase Al_3Li ($\delta´$). The similarity in structure and lattice parameter of the $\delta´$ and the fcc matrix results in a small lattice misfit (-0.18%) (Nobel and Thompson, 1971) and spherical precipitates with an interfacial energy between particle and matrix of 180 ergs/cm^2 (Tomura and co-workers, 1970). Concurrently, preferential precipitation and coarsening of the equilibrium AlLi (δ) phase occurs at grain boundaries leading to the development and growth of PFZ's.

Identifying the Ductility Problem. Sanders and Starke (1982) have recently studied how the various microstructures formed in three binary alloys affect the strength, deformation behavior and ductility. One alloy was a solid solution with short range order, one had a small volume fraction of shearable precipitates and one had a large volume fraction of shearable precipitates. The monotonic properties are summarized in Table 2 along with the properties of pure aluminum for comparison. Increasing the amount of lithium and/or aging time progressively increases the elastic modulus, yield strength and tensile strength along with a concomitant reduction in strain to fracture. TEM studies of deformed samples showed that the dislocations were arranged in bundles for the alloys containing short range order, and that well defined planar slip bands were present for alloys containing shearable δ' precipitates. Increasing the aging time resulted in more intense and widely spaced bands.

TABLE 2 Tensile Properties of the Al-Li Binary Alloys*

Alloy	Heat Treatment	E GPa	σ_{02} MPa	σ_{uts} MPa	ε_f %
Pure Aluminum	Annealed	68	11.7	47	60
Al-3.5at.%Li (M-I)	811K CWQ** + ½h @ 473K	73	22	75	43
	811K CWQ + 4h @ 473K	73	24	82	39
Al-5.8at.%Li (M-II)	811K CWQ + ½h @ 473K	76	37	96	42
	811K CWQ + 4h @ 473K	76	46	94	30
Al-8.9at.%Li (M-III)	811K CWQ + ½h @ 473K	80	137	215	21
	811K CWQ + 4h @ 473K	80	195	278	6.6

* Average of two tests.
**Cold water quench.

The observed decrease in macroscopic ductility with increasing lithium content and aging time can be explained qualitatively by the corresponding increased tendency for planarity and inhomogeneity of slip. The intense slip bands produce stress concentrations at grain boundaries (Starke and co-workers, 1981) and in many cases, can shear the boundaries (Gysler and co-workers, 1974) producing grain boundary offsets (ledges). When a small amount of sliding occurs along the grain boundary due to restricted slip and grain rotation, cracks form at the ledges. Once the crack has nucleated, it concentrates the stress, and more intensive localized deformation occurs in its vicinity. The crack then expands transgranularly along the intense slip bands or intergranularly along the grain boundaries as shown in Fig. 10. The PFZ's which form in the late stages of aging are soft with respect to the age-hardened matrix, and plastic deformation can be localized in these regions. Figure 11, from recent HVEM in situ deformation studies on Al-2.3Li by Crooks and co-workers (1982), illustrates this point. Figure 11a shows a wide PFZ adjacent to a high angle grain boundary. and Fig. 11b shows that deformation localized in the PFZ during straining. Cracks then nucleate at the grain boundary triple junctions and propagate intergranularly within these zones.

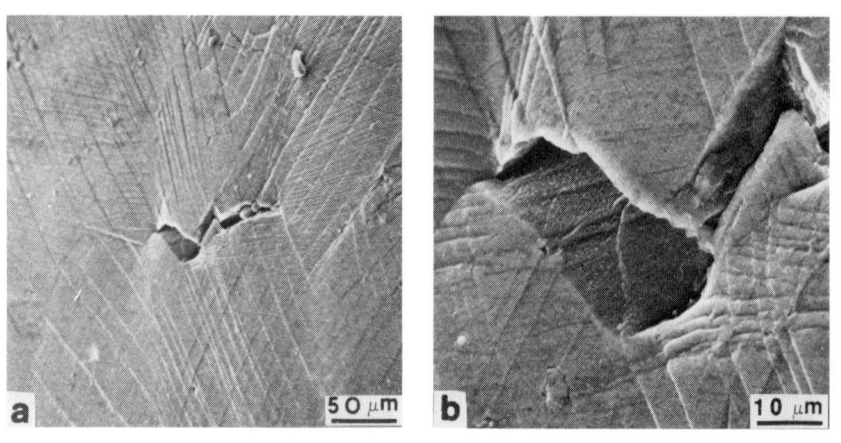

Fig. 10. SEM's of the surface of M-III (Al-8.9at.%Li) aged 4h at 473K showing ledges and dimples along grain boundary (from Sanders and Starke, 1982)

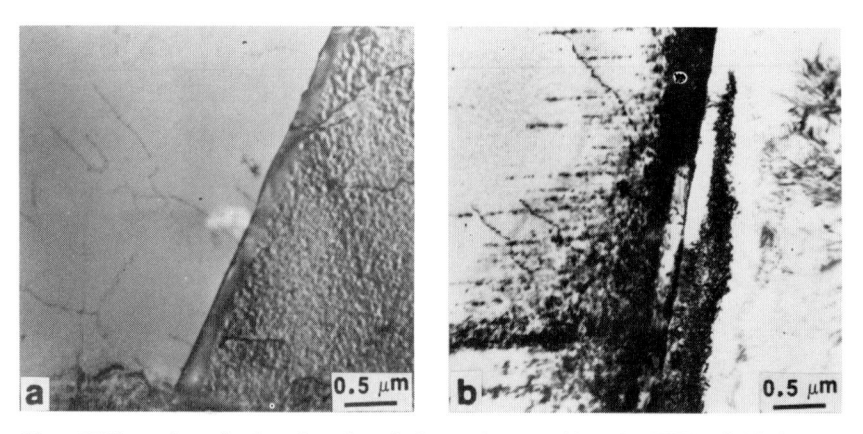

Fig. 11. TEM's taken during in situ deformation studies in HVEM of Al-3Li alloy aged 12h at 473K (a) undeformed showing GB and PFZ (b) deformed showing strain localization in PFZ (from Crooks and co-workers, 1982).

In many cases the intergranular fracture and associated low ductility of Al-Li alloys has been attributed to embrittling tramp elements such as sodium, potassium, sulfur, and hydrogen segregated at grain boundaries. Although such embrittlement may be possible in some cases, our research suggests that the low ductility is most likely associated with the deformation behavior in Al-Li alloys. The extent of strain localization appears to be much worse in Al-Li alloys than in other age-hardenable aluminum alloys. This may be due to very low interfacial energies and lack of coherency strains of the precipitates, and the wide PFZ's that develop during aging.

Improving the Ductility of 2020. The primary strengthening precipitates in 2020, partially coherent θ' (Al_2Cu) and T_B ($Al_{15}Cu_8Li_2$) may also be sheared by moving dislocations resulting in localized planar deformation. In addition, PFZ's exist along high angle grain boundaries (Gysler and co-workers, 1981), and these features, coupled with the very large recrystallized grains present in 2020-T651 plate (Coyne and co-workers, 1981) suggest that the observed low ductility may be associated with strain localization effects similar to those that occur in Al-Li binary alloys. A reduction in grain size, which can be achieved by thermomechanical processing treatments (TMP's) has been shown to be effective in preventing premature failure in other systems by reducing the dislocation pile-up length and therefore the high local stress concentrations which cause early crack nucleation (Peters and Lütjering, 1976; Sanders and Starke, 1976).

Starke and Lin (1982) recently studied the influence of a wide variety of grain structures produced by TMP's on the ductility of 2020 plate. These included a small-grained, completely recrystallized structure R-1, a partially recrystallized structure PR-1, and two unrecrystallized structures UR-1 and UR-2 having a different unrecrystallized grain size. The textures were also significantly different; the recrystallized and partially recrystallized materials had no discernible preferred orientation, while the unrecrystallized materials had a sharp deformation texture. The details of the TMP procedures and resulting microstructures are described in detail elsewhere (Starke and Lin, 1982).

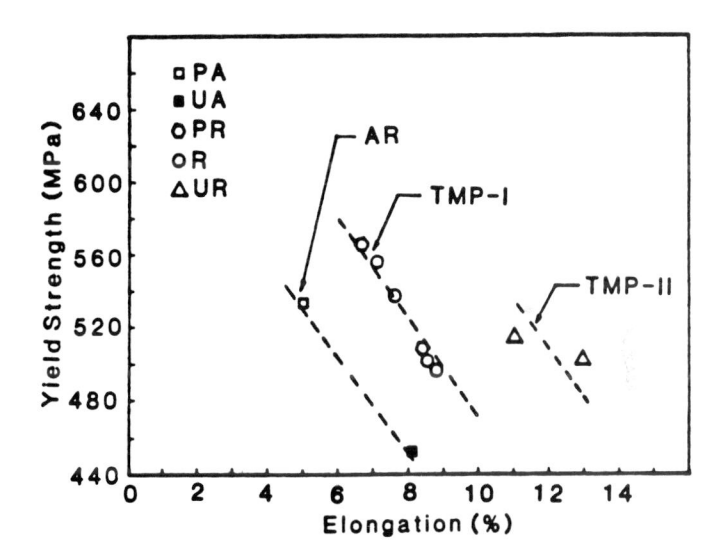

Fig. 12. Yield strength versus elongation from tensile tests of 2020. AR, as-received; PA, peak-aged; UA, underaged, and TMP-II in PA condition (from Starke and Lin, 1982).

The improvements in the strength-ductility relationship obtained by TMP are shown graphically in Fig. 12. Materials processed to produce an unrecrystallized grain structure, UR-1 and UR-2, show a significant improvement over the other TMP-2020, and a strength-ductility relationship comparable to 7075-T6. Using a yield strength of 520 MPa as an example, the as-received-2020 has a strain to failure of 5.4%; the TMP-I processed to produce partially recrystallized and recrystallized structures have a strain to failure of 8.2% (a 43% improvement); the TMP-II processed to produce unrecrystallized structure has a strain to failure of 11.6% (a 115% improvement).

The fracture features of the recrystallized and partially recrystallized TMP materials were very similar and quite different from those of the unrecrystallized materials. On a macroscopic scale, these fracture surfaces were essentially normal to the stress axis and predominately intergranular, although some transgranular features were evident. A high density of very shallow, equiaxed dimples were observed on the intergranular fracture surfaces at higher magnifications. The fractures of the UR samples were completely transgranular and approximately 50 degrees to the stress axis, obviously following a plane of maximum macroscopic shear stress.

The large difference in ductility between the R, PR, and UR samples can be explained in terms of the crack nucleation and fracture. All of the materials have planar slip and a tendency toward strain localization for the aging conditions studied. The slip bands impinge upon grain boundaries and cause stress concentrations across the boundaries; the magnitude of which depends on the slip length (Starke and Lütjering, 1979). In the random-textured R and PR materials there is a high probability that the stress concentrations will not be easily relieved by the transfer of plasticity to adjacent grains. For this case, initiation of voids at constituent phases and dispersoids lying along the grain boundary is easier than initiating slip in an unfavorably oriented adjacent grain. Once the crack is nucleated the plastic zone is localized in the PFZ and propagation occurs quite easily along these regions, aided by coalescence of voids initiated at grain boundary precipitates. The significant improvement in ductility obtained for the UR materials is associated with a change in fracture mode from the predominately intergranular rupture observed for the AR and TMP-I materials, to transgranular dimpled rupture. It is clear that stress concentrations at grain boundaries, resulting from the planar deformation, are relieved in the highly textured UR material by shear in neighboring grains. The low degree of misorientation betwen neighboring grains reduces the effectiveness of the grain boundary as a slip barrier and guarantees nearly parallel slip systems.

New Al-Li-X Alloys. When considering that 2020 has an elastic modulus 10% higher than 7075, superior creep properties (Polmear, 1981), and fatigue crack growth resistance at low ΔK (Coyne and co-workers, 1981; Sanders, 1981), the TMP-II materials look especially attractive. However, the lithium content in 2020 is insufficient to obtain the low density, high specific modulus and speci: strength goals of current alloy development programs (Lewis and co-workers, 1978). Lithium contents higher than 1.5% appear necessary.[2] The strength ductility combinations of Al-Li binary alloys are too low to be commercially useful and other elements must be added to increase the strength, improve the homogeneity of deformation, and the ductility. These additions should not degrade secondary properties or cause large increases in density that would negate the benefits of lithium.

The 2020 research illustrates the benefits of having an unrecrystallized produc since the associated sharp texture seems to reduce the adverse effects of PFZ's In addition, the substructure of unrecrystallized grains obstructs dislocation motion and increases the strength. Zirconium seems to have the most potent effect of all elements added to aluminum for the purpose of preventing recrystallization. Unlike other additions used to control the grain structure, i.e.,

---

[2]For details of alloy composition selection on recent programs see Palmer and co-workers, 1981, 1982; Starke and co-workers, 1981.

manganese and chromium, zirconium precipitates as very small coherent Al$_3$Zr particles which retard subgrain boundary migration and coalescence (Sanders, 1979). The larger incoherent manganese-rich and chromium-rich dispersoids accelerate recrystallization under certain circumstances. Copper has been shown to be very effective in increasing the strength of Al-Li alloys. The main strengthening precipitates that form during aging of Al-Li-Cu alloys containing 2-3% Li and 2-3% Cu are δ' (Al$_3$Li), T$_1$ (Al$_2$CuLi) and Θ' (Al$_2$Cu) (Silcock, 1959/60). The T$_1$ and Θ' precipitates have interfacial strains that may also aid in homogenizing deformation. The equilibrium precipitates T$_2$ (Al$_6$CuLi$_3$) and δ (AlLi) can form additionally (Noble and Thompson).

A composition of Al-3Li-2Cu-0.2Zr was selected on a recent alloy development program (Palmer and co-workers, 1981; Lin and co-workers, 1982). Three percent lithium was chosen to maximize the modulus and density effects without having the fabrication problems associated with higher lithium content alloys. Two percent copper was selected for strength considerations and approaches the solubility that can be obtained in an Al-3Li alloy prior to aging. Zirconium was added to inhibit recrystallization. A comparison of the tensile properties of this alloy produced by three methods is given in Table 3. Although the strength-ductility relationship is considerably better than that for Al-Li binary alloys, it does not meet the program goals and is somewhat below that obtained for TMP-2020. The low ductility was again due to strain localization, but was different from the intense slip bands observed for AR-2020. The Al-3Li-2Cu-0.2Zr alloy was unrecrystallized and PFZ's formed along the subgrain boundaries, Fig. 13a. Similar zones were not observed along subgrain boundaries of 2020, possibly due to the cadmium in that alloy enhancing nucleation of Θ' up to the boundaries. The PFZ's are much weaker than the matrix and many are favorably oriented for slip. Deformation concentrates in these regions, Fig. 13b, and low ductility ensues.

TABLE 3 Comparison of Tensile Properties of I/M, P/M-Splat, and P/M Atomized Al-3Li-2Cu-0.2Zr Alloy Artificially Aged to Peak Strength

Alloy	$\sigma_{0.2\%}$ MPA	σ_{uts}[b] MPA	ε_f[c] %
I/M	503	561	4.0
P/M-splat	490	546	4.2
P/M-atomized	443	562	7.3

[a] Yield strength - 0.2% offset
[b] Ultimate tensile strength
[c] Strain to fracture

Attempts were made to harden the PFZ by solid solution strengthening and additional precipitation and to minimize subgrain boundary precipitation by alloy modification and TMT's (Palmer and co-workers, 1982). Since magnesium is a potent solid-solution strengthener in aluminum, it was selected as an additive and one of the alloys had the nominal composition Al-3Cu-2Li-1Mg-0.2Zr. The age-hardened microstructure of Al-3Cu-2Li-1Mg-0.2Zr is very complex and contains T$_1$ (Al$_2$CuLi), S' (Al$_2$CuMg) and Al$_2$MgLi strengthening precipitates (Crooks, 1982). Some δ' (Al$_3$Li) may also form; however, due to the presence of Al$_3$Zr, it is difficult to detect (Lin and co-workers, 1982). Although the equilibrium T$_2$ and S phases were detected on the boundaries of the elongated unrecrystallized grains, no preferential precipitation and associated PFZ's were observed at subgrain boundaries in TMT processed material, Fig. 14. The strength-ductility

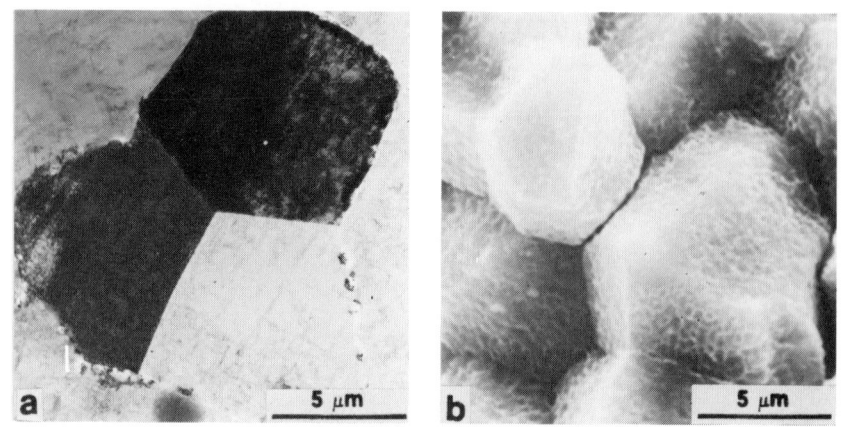

Fig. 13. (a) TEM showing PFZ's along subgrains and (b) intersubgranular fracture of an age-hardened Al-3Li-2Cu-0.2Zr alloy (from Lin and coworkers 1982).

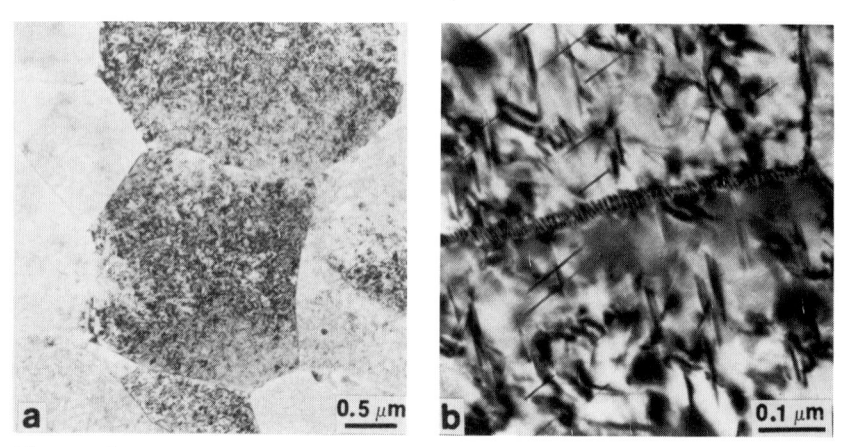

Fig. 14. TEM's of Al-3Cu-2Li-1Mg-0.2Zr showing uniformly distributed precipitates and a lack of PFZ's (from Crooks, 1982).

relationship of this alloy, Fig. 15, meets the program goals, exeeds that of 7075-T76, and compares with that obtained for TMP-II-2020, Fig. 16. The high ductility is associated with homogeneous deformation and a transgranular fracture mode as shown in the TEM's of Fig. 17 from the recent HVEM in situ deformation studies by Crooks and co-workers (1982).

CONCLUSIONS

During the last fifty years or so, there has been a significant increase in our knowledge of strengthening mechanisms and deformation behavior. An understanding of the basic concepts is aiding in the development of new high strength alloys having attractive cominations of properties. Since most of my own

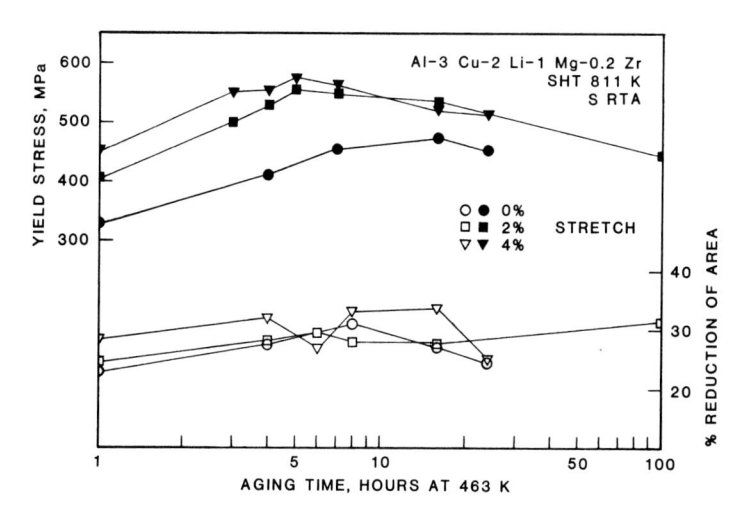

Fig. 15. Yield stress and reduction of area versus aging time after various TMT (stretch) (from Crooks, 1982).

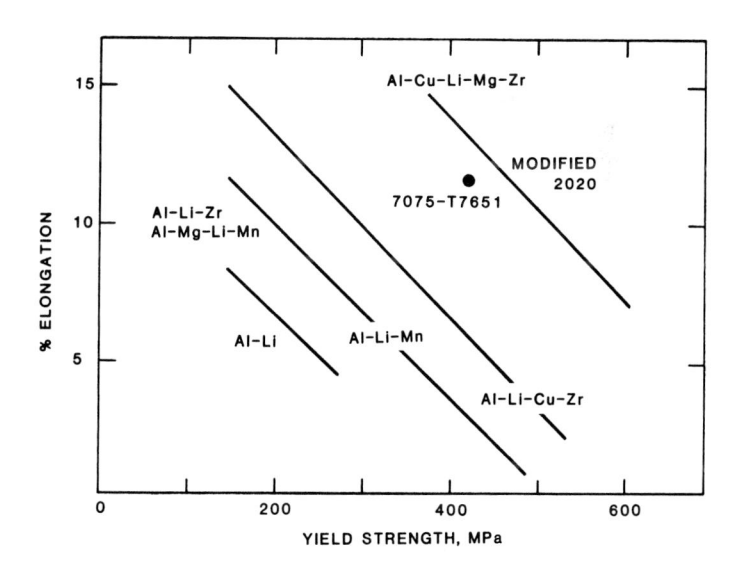

Fig. 16. Schematic of strength/ductility relationships for various Al-Li-X alloys (from Starke and co-workers, 1981).

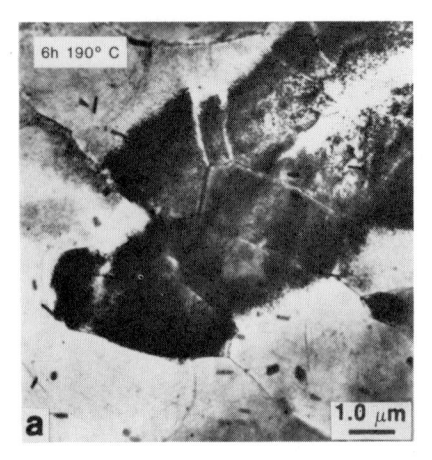

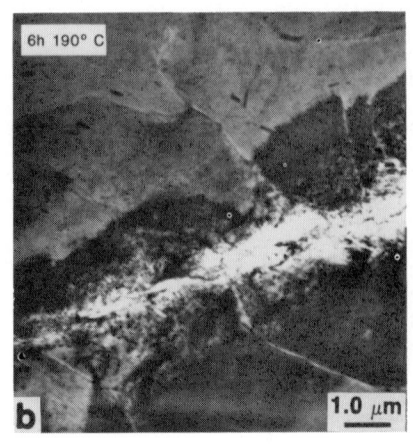

Fig. 17. TEM's taken during in situ deformation studies in HVEM and showing homogeneous deformation and transgranular fracture of aged Al-3Cu-2Li-1Mg-0.2Zr alloy (from Crooks and co-workers, 1982).

research activities revolve around aluminum alloys, I have selected my examples from those materials. However, new alloys are being developed in other systems using similar approaches.

ACKNOWLEDGMENTS

I would like to thank R. E. Crooks, G. Hildeman, J. Mullins, H. Paris, R. E. Sanders and G. Wald for providing information prior to publication; Keith Bresnahan for help in preparing the figures and B. Meadows for typing the manuscript. Financial support by the Army Research Office, Durham, is gratefully acknowledged.

REFERENCES

Argon, A. S. and J. Im (1975). Met. Trans., 6A, 839.
Ashby, M. F. (1968). In G. S. Ansell, T. D. Cooper and F. V. Lenel (Eds.), Oxide Dispersion Strengthening. Gordon & Breach, NY. P. 143.
Ashby, M. F. (1968). In Proceedings Second International Conference on the Strength of Metals and Alloys. Asilomar, California.
Balmuth, E. S. and R. Schmidt (1981). In T. H. Sanders, Jr. and E. A. Starke, Jr. (Eds.), Aluminum-Lithium Alloys. TMS-AIME, Warrendale, PA. P. 69.
Bishop, M. and K. E. Fletcher (1972). Inter. Met. Reviews, 17, No. 168, 203.
Bloch, E. A. (1961). Met. Reviews, 6, No. 22, 193.
Brown, L. M. (1979). In P. Haasen, V. Gerold and G. Kostorz (Eds.), Strength of Metals and Alloys. Pergamon Press, Frankfurt. P. 1551.
Cesara, S., G. Cocco, G. Fagherazzi and L. Schiffini (1977). Phil. Mag., 35, 373.
Copley, S. M. and J. C. Williams (1976). In J. K. Tien and G. S. Ansell (Eds. Alloy and Microstructure Design. Academic Press, NY. P. 3.
Coyne, E. J., Jr., T. H. Sanders, Jr. and E. A. Starke, Jr. (1981). In T. H. Sanders, Jr. and E. A. Starke, Jr. (Eds.), Aluminum Lithium Alloys. TMS-AIME, Warrendale, PA. P. 293.

Crooks, R. E. (1982). The Influence of Microstructure on the Ductility of an Al-Cu-Li-Mg-Zr Alloy, Ph.D. Thesis. Georgia Institute of Technology, Atlanta, GA.
Crooks, R. E., E. Kenik and E. A. Starke, Jr. (1982). Unpublished research.
Davis, H. A. (1978). In B. Cantor (Ed.), Rapidly Quenched Metals III, Vol 1. The Metals Society, London. P. 1.
Duwez, P., R. H. Willens and W. Klement (1960). J. Appl. Phys., 31, 1136.
Fleischer, R. L. (1961). Acta Met., 9, 996.
Fleischer, R. L. (1964). In Strengthening of Metals. Van Nostrand-Rheinhold, Princeton, NJ. P. 63.
Gangulee, A. and J. Gurland (1967). Trans. TMS-AIME, 239, 269.
Geisler, A. H. (1949). Trans. Inst. of Metals Division AIME, 180, 230.
Gleiter, H. and E. Hornbogen (1965). Phys. Status Solidi, 12, 235.
Gleiter, H. and E. Hornbogen (1967-68). Mater. Sci. Engr., 2, 285
Griffith, W. M., R. E. Sanders, Jr., and G. J. Hildeman (in press). In M. J. Koczak and G. J. Hildeman (Eds.), Powder Metallurgy of Aluminum Alloys, Elevated Temperature Aluminum Alloys for Aerospace Applications. TMS-AIME, Warrendale, PA.
Gysler, A. G. Lütjering and V. Gerold (1974). Acta Met., 22, 901.
Gysler, A., R. Crooks and E. A. Starke, Jr. (1981). In T. H. Sanders, Jr. and E. A. Starke, Jr. (Eds.), Aluminum-Lithium Alloys. TMS-AIME, Warrendale, PA. P. 263.
Hall, E. O. (1951). Proc. Phys. Soc., 64, 747.
Hansen, M. (1958). In Constitution of Binary Alloys. McGraw-Hill Book Company, NY.
Hornbogen, E. and K. H. Zum Gahr (1975). Metallography, 8, 181.
Jones, H. (1969-70). Mater. Sci. Engr., 5, 1.
Kelly, A. and R. B. Nicholson (Eds.) (1971). Strengthening Methods in Crystals. Applied Science Publishers, Ltd., London.
Kelly, P. M. (1973). Int. Met. Rev., 18, No. 172-C, 31.
Langenbeck, S. L., G. G. Wald, E. J. Himmel, R. F. Simenz, G. Hildeman, C. M. Adam, D. Hilliard and J. Bjeletich (1981). Elevated Temperature Aluminum Alloy Development, Interim Technical Report for Period April-September, 1981. Lockheed-California Company.
Lewis, R. E., I. G. Palmer, H. G. Paris, E. A. Starke, Jr., and G. Wald (1978). Development of Advanced Aluminum Alloys from Rapidly Solidified Powders for Aerospace Structural Applications, AFWAL Contract F33615-78-C-5203, 1978-1982.
Lewis, R. E., D. Webster and I. G. Palmer (1978). Final Report, Contract F33615-77-C-5186, Technical Report No. AFML-TR-78-102, July, 1978.
Lifshitz, I. M. and V. V. Slyosov (1961). J. Phys. Chem. Solids., 19, 35.
Lin, F. S., S. B. Chakrabortty and E. A. Starke, Jr. (1982). Met. Trans., 13A, 401.
Ludtka, G. M. and D. E. Laughlin (1982). Met. Trans. A.
Lütjering, G. T. Hamajima and A. Gysler (1977). In Fracture 1977, Vol. 2. ICF4, Waterloo, Canada. P. 7.
Mullins, J. W. (1982). The Microstructure and Tensile Properties of RSP Al-Fe-Ni-Co Alloys, M. S. Thesis. Georgia Institute of Technology, Atlanta, GA.
Nicholson, R. B. (1971). In Kelly and R. B. Nicholson (Eds.), Strengthening Methods in Crystals. Applied Science Publishers, Ltd., London. P. 535.
Nobel, B. and G. E. Thompson (1971). Metal Sci. J., 5, 114.
Nobel, B. and G. E. Thompson (1972). Met. Sci. J., 8, 167.
Palmer, I. G. and G. C. Smith (1968). In G. S. Ansell, T. D. Cooper and F. V. Linel (Eds.) Oxide Dispersion Srengthening. Gordon & Breach, NY. P. 253.
Palmer, I. G., R. E. Lewis and D. D. Crooks (1981). In T. H. Sanders, Jr. and E. A. Starke, Jr. (Eds.), Aluminum Lithium Alloys, TMS-AIME Warrendale, PA. P. 241.

Palmer, I. G., R. E. Lewis and D. D. Crooks (in press). In M. J. Koczak and G. J. Hildeman (Eds.), Powder Metallurgy of Aluminum Alloys, Development of Al-Li-X Alloys Using Rapidly Solidified Powder. TMS-AIME, Warrendale, PA.
Petch, N. J. (1953). J. Iron Steel Inst., 197, 25.
Peters, M. and G. Lütjering (1976). Z. Metallkunde, 67, 811.
Polmear, I. J. (1966). J. Aust. Inst. Met., 11, 246.
Polmear, I. J. (1981). Light Metals, Metallurgy of the Light Metals. Edward Arnold Pub., London.
Przystupa, M. A. and T. H. Courtney (1982). Met. Trans., 13A, 873).
Ralph, B. (1980). In R. W. Balluffi (Ed.), Grain Boundary Structure and Kinetics. ASM, Metals Park, OH. P. 181.
Sanders, R. E., Jr. and E. A. Starke, Jr. (1979). In J. G. Morris (Ed.), Thermomechanical Processing of Aluminum Alloys. TMS-AIME, Warrendale, PA. P. 50.
Sanders, R. E., Jr. and G. J. Hildeman (1981). Elevated Temperature Aluminum Alloy Development, Final Report AFWAL-TR-81-4076.
Sanders, T. H., Jr. and E. A. Starke, Jr. (1976). Met. Trans., 7A, 1407.
Sanders, T. H., Jr. (1979). NADC Contract No. N62269-76-C-0271, Final Report, June, 1979.
Sanders, T. H., Jr. and E. A. Starke, Jr. (1982). Acta Met., 30, 927.
Sankaran, K. K. and N. J. Grant (1981). In T. H. Sanders, Jr. and E. A. Starke, Jr. (Eds.), Aluminum Lithium Alloys. TSM-AIME, Warrendale, PA. P. 205.
Sarkar, B., M. Marek and E. A. Starke, Jr. (1981). Met. Trans., 12A, 1939.
Silcock, J. M. (1959/60). J. Inst. Met., 88, 357.
Starke, E. A., Jr. (1970). J. of Metals, 22, 54.
Starke, E. A., Jr. and G. Lutjering (1979). In M. Meshii (Eds), Fatigue and Microstructure. ASM, Metals Park, Ohio. P. 205.
Starke, E. A., Jr., T. H. Sanders, Jr., and I. G. Palmer (1981). J. of Metals, 33, No. 8, 24.
Starke, E. A., Jr. and F. S. Lin (1982). The Influence of Grain Structure on the Ductility of the Al-Cu-Li-Mn-Cd Alloy 2020, submitted to Met. Trans. A.
Thompson, A. W. (1977). Met. Trans., 8A, 833.
Tien, J. K. (1975). In R. I. Jaffee and B. A. Wilcox (Eds.), Fundamental Aspects of Structural Alloy Design. Plenum Press, NY. P. 363.
Tien, J. K. and G. S. Ansell (Eds.) (1976). Alloy and Microstructure Design. Academic Press, NY.
Tomura, M. T. Mori and T. Nakamura (1970). J. Japan Inst. Metals, 34, 919.
Wagner, C. (1961). Z. Electrochem, 65, 581.
Wald, G. G. (1981). NASA Contractor Report 165676, Lockheed-California Company, Burbank, CA. May, 1981.
Williams, D. B. (1981). In T. H. Sanders, Jr. and E. A. Starke, Jr. (Eds.), Aluminum Lithium Alloys. TMS-AIME, Warrendale, PA. P. 89.
Yokobori, T., A. Kamei and T. Kogowa (1973). Fracture 1973. Verein Deutscher Eisenhuttenleute. Dusseldorf, Paper I-431.

Reflections on the Industrial Application of Fundamental Research

G. F. Bolling

North American Research Liaison Office, Ford of Europe Incorporated, Laindon, Basildon, UK

ABSTRACT

Some general characteristics of ICSMA research are described. The question of how we look at the industrial application of fundamental research is then reviewed. The evolving picture of separated science and technology and the concept of a gatekeeper working across these two areas of a modern research group are highlighted. The importance of a recognized need and the construction of "wants lists" as a way of expressing ongoing needs are presented and discussed.

KEYWORDS

Separation of science and technology; technical champion; gatekeeper; importance of need; wants-lists.

INTRODUCTION

The author was asked to discuss the industrial application first of fundamental research and then of ICSMA research ..... "to take a broad look at the range of topics and concepts to be discussed, and to describe the successes and failures in transferring these into practical processes and products."

The presentation does start out with a broad look at the contents of ICSMA 6 and continues with discussion about the industrial application of fundamental research. Background information was drawn from many publications involving the management of research.

At the end of the presentation some few suggestions are made to the conference organization.

A DESCRIPTION OF ICSMA

A Test

At the time of writing there were 212 abstracts available to the author. Several engineering co-workers were approached, handed the abstracts and asked about their industrial relevance, care being taken not to define the word relevance too closely. The returns were criticizing with the most helpful response saying, "I really can't tell, perhaps one in four could be relevant." This exercise was undertaken to show how difficult it is to define industrial relevance and to stress the type of comments to be obtained from engineers working far away from the long-term research front.

Pointedly, the result does not depend on how we define this research since the result was obtained from the abstracts for this conference.

An impersonal test of relevance was then applied as blindly as possible. Those abstracts which noted either that

- the questions asked were generated by industrial or production problems, or that
- the results could be applied in practice

were classed as being industrially relevant. The result was that about one in four showed relevance, the same portion estimated by one of the engineers.

Some Statistics

The content of ICSMA 6 was examined several other ways. It is possible to learn, for example, about

- geographic origin:
    - Based on first authors, papers come from 29 countries.
    - About 50% of the papers are from English-speaking countries.
    - The greatest number of papers is from Australia (about 25%).

- content:
    - 5% of the papers concern nickel-containing alloys
      11% "   "   "   "   copper   "   "
      15% "   "   "   "   aluminium   "   "
      34% "   "   "   "   steels
    - 27% "   "   "   "   steels if all Australian papers are excluded.

- relevance to industry:
    - 23% of the papers are "relevant" by the test just described.
    - 18% "   "   "   "   "   if all Australian papers are excluded.
    - 17% "   "   "   "   "   if Australian papers related to steel are excluded.

Although other questions could be asked about the conference and "answered" beforehand on the basis of abstracts, both the questions and the answers stood on shaky ground. It is perhaps for this reason that <u>keywords</u> "to indicate the main topics discussed and to provide basic terms for indexing ..... which potentially increases the publicity of the paper and recognition for the author" are requested by publishers (Pergamon Press, 1981). These keywords were not available to the author, and he wonders how many of these will relate to product or process and attempt to trigger industrial interest.

HOW WE LOOK AT THE INDUSTRIAL APPLICATION OF RESEARCH

General

Solutions to problems are successes. Applications are successes. Sometimes the early-enough determination that something cannot work should be scored as a success for research. Reducing costs; increasing profits; enabling something to be done that could not be done otherwise; aiding strategic international efforts; aiding humanity; the list of possible successes is endless. Yet there are few examples where fundamental research on phenomenon A produced product or process A'. Most of us know enough about the transistor, the laser, xerography ®, to construct stories about these successes and class them as A → A' examples, but even here there are counter-arguments.

How do we apply fundamental research to (or in) industry? This question is so important there have been hundreds of papers written about it. A good portion of these concern the techniques and skill of management and have therefore appeared for example in Research Management, a journal which started publication in the late 1950's. These papers come from a wide spectrum of experience, and often seem to be based on opinion rather than fact. They make very interesting reading particularly for this reason because it can be seen how opinions taken in the whole have evolved.

Interest in managing research blossomed, let's say from 1950-1965 when it was widely believed that scientific research gave technical development, which led to new industrial products and processes and thus gave economic growth which in turn provided for the new investment in research. Models like the following were in vogue:

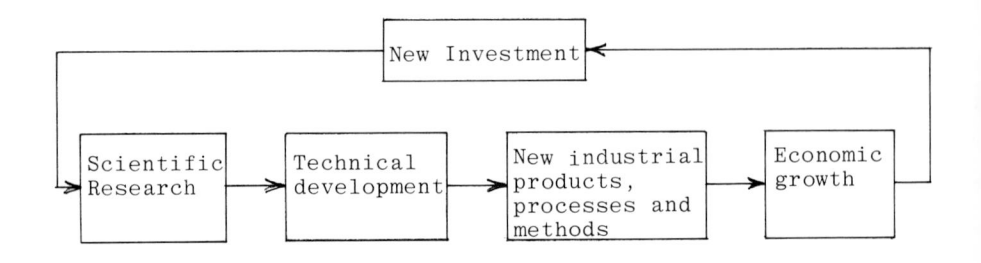

```
                    ┌──────────────────┐
         ┌─────────▶│  New Investment  │◀─────────┐
         │          └──────────────────┘          │
┌────────┴──┐   ┌───────────┐   ┌──────────────┐   ┌──────────┐
│ Scientific│──▶│ Technical │──▶│New industrial│──▶│ Economic │
│ Research  │   │development│   │products,     │   │ growth   │
└───────────┘   └───────────┘   │processes and │   └──────────┘
                                │methods       │
                                └──────────────┘
```

THE 1950-65 IDEAL

Gjostein (1978) discussed this ideal model, called the Innovative Chain, and showed how it failed, personally noting that "....... consolidated research contributes to the education of the next generation technologist. Short of this there is no good mechanism by which science flows directly into technology." Significantly, Gjostein also noted that financial support for fundamental research started to subside after 1965.

This was a time of re-examination and the time of C.P. Snow's two cultures (1959-64) wherein the scientist tends to think from first principles and the technologist rarely finds time to be other than a working storehouse of the state-of-the-art. De Solla Price (1965) added to the distinction between these two members of the development team in a discussion involving the application of research:

> "The technologist's searching of the technical literature is doomed to failure because other technologists are not interested in helping him to their own disadvantage.
>
> If the technologist searches on the other hand through the literature of science it becomes evident that this has not been published for his benefit but for the peer group at the (research) front, in the first instance, and for the eternal archive in the second place. Because of this it must be presumed that science cannot flow into technology from the literature at the research front but rather from some position well behind this front. Basically, the technologist could only monitor the scientific research front if he were in his own right an

From Tanenbaum (1967) (listed in order of importance)	From Gjostein (1978) (unordered; arranged to match)	Observation or Key Element
• Principal investigators were required to make several major changes in directions and goals and had to have the flexibility to do so.	• Many events contributed in a complex manner to a given technical innovation.	Innovation is complex
• Research-engineering interactions had to occur and required close and frequent communications between organizationally independent groups.	• Technological knowledge ("state-of-the art") was a key factor in technical innovation.	A technological knowledge of the State-of-the-art is essential to success
• Recognition of an important need was an essential factor	• An identifiable need was apparent when a technical innovation occurred.	An identifiable need is essential to success.
• Key individuals were necessary to bridge geographic, organizational and functional gaps and to actively stimulate communications: (These could be) - couplers, people especially motivated to see that information flowed between participate groups, or - champions, leading and dominating individuals.	• Technical entrepreneurs, i.e. persons who champion a particular scientific or technical activity, were the most important factor in successful innovations. Often the technical entrepreneurs persisted in the face of many inhibiting factors.	Technical Champions are the key individuals in success.
• In less than half the cases did events require the development of a new solution on a major technical problem .... and few of these solutions could ever be classified as a scientific discovery or even as fundamental research.	• Rarely could one brilliant scientific discovery be identified as playing a major role in the innovation.	It is rare that a scientific discovery plays a major role

TABLE ABOUT INNOVATION

ONE POST-1965 MODEL after Haeffner (1973)

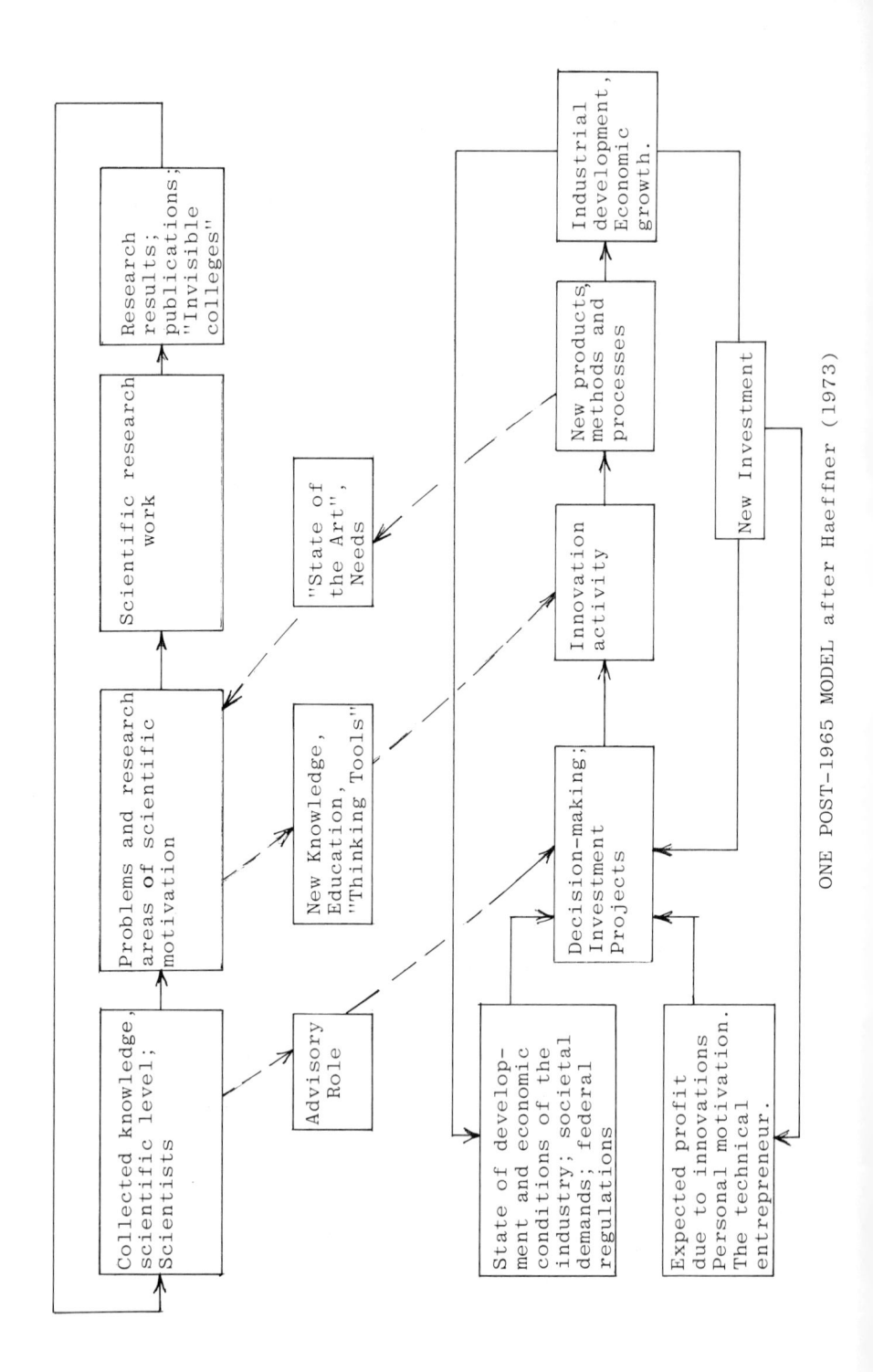

active peer in an Invisible College. Short of this, he must retire to the level where such knowledge has been assimilated as part of the eternal archive."

It seems we reached 1965 with support for fundamental research (however, this was defined) just starting to be reduced, and lines being drawn between technologists and scientists.

* * * * *

Diminishing support for fundamental research was justified (although not on purpose, it seems) by several studies done on innovation. Tanenbaum (1967) had spearheaded a broad study on innovation in the area of interest to ICSMA titled "Research/Engineering Interactions in Materials Science and Technology." More information can be added from Gjostein (1978) who summarized five studies. As shown in the Table the strong parallel results make it difficult not to turn away from the idealism that research leads to all innovation.

There were several other models constructed about the flow of research. Haeffner (1973) for example, constructed a two-feedback-loop system that expands on De Solla Price's words as quoted a few paragraphs back. This figure is reconstructed here because it is a good example of how much we accepted the separation of science and technology.

The problem for the next decade varied by country and by industry but generally the questions asked, involved how much research to support, in what areas and how to manage it.

* * * * *

These studies provided excellent bases for us to understand and then to influence the flow of research into technology, as long as we did not accept all these studies and thinkings too uncritically. One flaw in arguments that can be constructed purely from the information in the table is that technology has no corner on innovation and that the importance of a champion in innovation is common to many other fields ..... business, law, politics, the arts ..... The challenge is to effect change; and one wonders about instances where change has occurred without a champion.

Many of us know how difficult it is to provoke change especially if the present way works or is accepted. Moreover, we know instances of success when the appeal was there and the excitement of the new was communicated, and a champion emerged who eventually succeeded. But the emergence of a champion, and especially his success, may blind us from seeing

other ways of getting the job done. The author believes we can construct cases where need was strong enough ..... let us say urgent ..... and need itself became the champion.
In these instances the system, or company or society appoints and directs task force leaders or managers who act to get the job done. Their success requires that the fundamental tools are there or there will be no solutions. And the task force leaders or managers are only champions in the sense that the system wants them to be; (at the very least these people may not be entrepreneurs who can be out of place when there is an urgent job to be done).

In normal times need is still a directive. For example at Ford we have evolved what is called a "Wants List". This is compiled from requests and needs throughout the operating parts of the company and is used primarily as an input to develop specific advanced projects. Since the same management is also involved with the longer-term view necessary to support fundamental work, the "wants" can also be a guide (but not the only one) to determine areas which should be supported by general research.

A successful research organization is generally not filled with champions and entrepreneurs, but with technical personnel who are expert in the needs of the company paying for the research. A successful research organization (of today) also seems to blend its technologists and scientists and ignore the two cultures and the difficulty of the technologist at the research front. How is this done?

Importance of the Group

Most dramatically the realization of what is needed does not seem to be determined by group size and is applicable on the large scale of national research support as well as on the company scale. Recent personal experience by the author has taught that part of the system for funding research in the United Kingdom is an evolved one that favors industry supported topics. Government sponsored agencies such as the Science and Engineering Research Council co-operate in supporting university located research. Funding occurs easier today when a university project is supported in concept by a private company and this company pays jointly with the government agency to sponsor research in a direction needed by the industrial concern. Often it is not just one company but a transient consortium of companies with related interests that fund the research on a 50:50 basis with the government agency. Professors who seek funding often first look for the appropriate company. There is also the reverse where companies needing extra research tend to seek out specific professors who have been "successful" in responding to industry.

The net result is a tightened network in the U.K. which
expresses a sort of "national" choice in research: the
research is done where the companies (money) want the research
to be done. Recent committee studies are reflected in two
booklets from the Cabinet Office, Advisory Council for
Applied Research and Development, which touch on this effect:

- Industrial Innovation (1978), and
- Technological Change: Threats & Opportunities for the United Kingdom (1979).

In the United States there has also been a contemporary report
from the Advisory Committee on Industrial Innovation (1979)
dealing at length with the national challenge of managing
the environment for innovation. President Jimmy Carter spoke
to Industrial Innovation Initiatives (1979) shortly thereafter
including specific reference to

"..... efforts to assure an adequate investment
in the basic research that will underlie future
technical advances ....."

In addition to this reference to the importance of basic
research, it is interesting for us to note part of President
Carter's concluding statement that says:

"Innovation is a subtle and intricate process,
covering that range of events from the inspiration
of the inventor to the marketing strategy of the
eventual producer."

* * * * *

Today, like these two examples, we have committees helping
in several other countries to advise on technical matters
important in national technical decision-making. Some
Industrial concerns also have advisory panels peopled by
outside/independent scientists and thinkers. At Ford, there
is a research-advisory group (Ford, Annual Report 1981) which
meets in Dearborn several times yearly. Their counsel
naturally relates to the way they believe research should be
used by industry. In this context, Dr.David Saxon (1982)
one of the group analyzed the research array at Ford and
ascribed certain essential functions to the work going on.
This analysis can be summarized as a set of guidelines;
research should provide:

- <u>resident experts</u>, able to answer reasonably expectable questions, expeditiously and correctly.

- <u>the resident capacity to push current state of the art to its limits</u>, (to develop more efficient engines, lighter cars, better pistons, more complete combustion, better rust proofing, "more this and better that in short.")

- <u>the resident capacity for innovation</u>. This function, more difficult and subtle than the others, is for protection from being technologically surprised; also for producing and for understanding new and innovative developments, new products and new ideas, scientific and technological breakthroughs.

The author believes that many research groups in the world are constructed to respond in ways that parallel Saxon's guidelines. This modern research group usually does not separate technology and science nor does it rely on searching for champions; (it should, however, not prohibit them).

Importance of the Individual

The author believes that modern research groups have evolved a new class of individuals who are not champions of specific innovations, nor managers as such, but who have responsibilities for promoting applications of research and are thus closely involved with innovation. In discussions with Horton (1982) this concept was developed and the individual is called a "gatekeeper". This person is an individual who must stand with one foot in science and the other in technology; this person must have worked in both areas and be one who is interested in both areas. More often than not, this individual is a research scientist who has chosen to work in technology, although there are indentifiable engineers/technologists who have done the reverse.

The task of the gatekeeper has not been to be the driving entrepreneur or champion but to be one of the resident experts who knows people and styles in both areas and who can hold the gate open between science and technology. He/she tries to speak both languages; he/she tries to be a member of the two cultures.

*   *   *   *   *

For people who act in the role of gatekeeper it should be most possible to identify successful research and development efforts. This is their job. Several months ago the author asked help from colleagues in different industries (and later supportive help from co-workers at Ford) to detail historical developments for new metals/metal phenomena to serve as illustrations. The analyses chosen were:

- <u>High Strength Steels</u> - a class of similar materials, now used extensively in the automotive industry because of the need for fuel economy/weight reduction which spurred the steel and metal forming industries to new production methods and controls ..... Up to the late 1950's the higher strength hot-rolled gauge steels were basically ferrite-pearlite mixtures with higher strengths obtained by increasing the carbon content and thus the percent pearlite. A key step was taken in the early 1960's when E.R. Morgan and other researchers at J & L Steel started to water-cool the strip as it left the final hot-roll; this led to a finer structure (both ferrite grain size and pearlite spacing) and therefore a higher strength for a given carbon level with concomittant better formability and weldability. Other key contributions followed.

- Aluminium Components/Products - the use of aluminium in LNG tankers through the development of manufacturing and metallurgical practices to produce plate and form it into spherical, trapezoidal sections ..... This advance was keyed to fundamental work in fracture mechanics and fatigue crack growth. New automotive body sheet alloys and the development of 390 alloy for castings, were similarly based on fundamental understandings.

- Modern Fatigue Analysis and Methodology - a departure from the traditional S-N endurance limit approach to one of controlling the onset of significant fatigue by selecting and designing structural elements based on cyclic deformation concepts and finite life design ..... Although there were many champions, JoDean Morrow and his students from the 1960's at the University of Illinois were key in developing use of the closed loop, servo-hydraulic test systems especially coupled with the revolution provided by the computer for quick complex calculations.

These examples survived to be used in this presentation because they relate to classes of questions asked about ICSMA research (but not presented here in "Some Statistics"): high strength steels as a class of materials brought into prominence by a single need; aluminium components as a class of products resulting from the special characteristics of the material; modern fatigue analysis as a system of understanding. The first problem for the author to do more than recognize these examples was to serve as a judge on who or what was the key to any given advance; this was abandoned. The second problem was to define the contribution of fundamental research. In keeping with all the findings of so many others the author has concluded that while these advances all depended substantially on fundamental research, it was in each case an indirect dependence. Need was the greater driving force in eventually leading to success.

APPLICATION TO ICSMA

How then does all this reflection apply to discussing the industrial application of ICSMA research? In a sense the conference organization is like a government on a supranational scale. Concerning the "strength of metals and alloys" it supports fundamental research by giving it publication.
And very recently by inviting inputs from those people in industry who must concern themselves with applied results, the conference organization is interesting itself in transferring research.

The conference organization might have shaped its proceedings by the way it requested or accepted the papers to be presented. But the author believes the spectrum of papers is less a product of direction than a reflection on the work going on throughout the world today, somewhat biased by language, travel and historical associations; (See 'Some Statistics').

What then is ICSMA? Judged by the work to be presented ICSMA is a group that

- is made up of experts
- has the capacity to push the current state of the art to its limits
- has the capacity for innovation

These characteristics are interestingly close to Saxon's guidelines, but while the group has common characteristics, it is not a resident group. It is not managed and does not as a group have common needs announced by others. Moreover, the conference is primarily a scientific group and could be one of those that tends to exclude the technologist (engineer) from the research front ..... and in the long run to turn the opinion of this development team member away from valuing fundamental research.

Possibly it was recognition of characteristics like these that led to the sessions scheduled towards the end of the conference on, Future Directions for ICSMA Research:

I - Fundamental problems still to be tackled, and

II - Research topics arising from industrial practice

It is, in sessions like these that the conference itself will answer questions about the industrial application of ICSMA research, the topic the author was asked to address.
However, the author believes that these sessions would be better if they were not run concurrently. It is not necessary that technologists and scientists attend each other's meetings, but it seems worthwhile. Fundamental scientists are basically very inquisitive people and if they are exposed to the needs of industry some of these needs may "rub-off". In the other direction it would be useful for the technologists to keep up-to-date with the research front, but at the very least the gatekeepers have a responsibility to attend the

scientific sessions and might contribute by expressing their views.

Perhaps our profession requires more gatekeepers who would ask what is needed in the "strength of metals and alloys"; who would examine what has at least the chance of being selected from the "eternal archive" sometime in the future. Perhaps our conferences should be changed so that they are not separated into being part technology and part science. This present conference might consider the construction of some kinds of wants-lists in its concluding discussion on Future Directions for ICSMA Research.

The author believes that success in the industrial application of research involves an active awareness, the process of learning where to look.

ACKNOWLEDGEMENTS

The author is indebted to several people for their generous advice. Dr N.A. Gjostein's 1978 review was a foundation for these thoughts; and discussions with Mr E.J. Horton have been very helpful. Thanks are also due to Mr J.D. Dowd of Alcoa, Dr H.H. Macklin Jr of Reynolds Aluminium and Mr H.A. Gilmore of Kaiser Aluminium for discussions about advances in aluminium; and to Dr R.G. Davies and Dr R.W. Landgraf of Ford, for discussions about high strength steels and fatigue analysis, respectively. Many other colleagues patiently helped the author; they are all equally free from responsibility for his opinions.

REFERENCES

Advisory Committee on Industrial Innovation, Final Report, (September 1979) Under the auspices of the United States Department of Commerce. U.S. Government Printing Office, Washington D.C. 20402 (Stock Number 003-000-00553-4)

Carter, President J.C. (October 31,1979). Fact Sheet and Briefing, Office of the White House Press Secretary, Washington

Cabinet Office, Advisory Council for Applied Research and Development (ACARD), (1978). Industrial Innovation. First published 1978, Her Majesty's Stationery Office,London

Cabinet Office, ACARD (1979). Technological Change: Threats and Opportunities for the United Kingdom. First published 1980, Her Majesty's Stationery Office, London

Ford Annual Report (1981). Ford Motor Company, The American Road, Dearborn, Michigan 48121; pages 13-14

Gjostein, N.A. (1978). The Role of Scientific Research in Industry. Unpublished, Ford Motor Company Ltd, Dearborn, Michigan

Haeffner, E.A. (1973). Technology Rev.March/April, 18.

Horton, E.J. (1982). Private discussions with the author about the need for champions/entrepreneurs. The concept of a "gatekeeper" apparently was mentioned in the U.K. press sometime during late 1981-early 1982, but the exact reference cannot be found.

Pergamon Press (1981). Style Notes for Authors - for the preparation of Camera-Ready Manuscripts. Pergamon Press, Oxford.

Price, D.J. De Solla (1965) *Technology and Culture* 6, 533.

Saxon, D.S. (1982). Private Communication ..... part of an address presented in Dearborn, Michigan.

Snow, Sir Charles Percy (later Baron Snow). (1959-64) Rede Lecture 1959; Godkin Lectures 1961; Two Cultures and a Second Look, 1964, Cambridge University Press

Tanenbaum, M. (1967) Study of Research/Engineering Interactions in Materials Science and Technology, *Coupling Research and Production*, Interscience, Wiley, New York.

Microstructure and Mechanisms of Fracture

E. Hornbogen

Ruhr-Universität Bochum, D-4630 Bochum, Federal Republic of Germany

ABSTRACT

Fracture of alloys can be treated on the level of the atomic bond, of microstructure, and of continuum mechanics. Microstructural aspects of formation, as well as subcritical and critical growth of cracks are discussed, for static and cyclic loading conditions. Four principle considerations are useful to derive models for quantitative relations between microstructure and fracture properties:

1. Homogeneous or localized plastic strain.
2. Partial properties of microstructural components and their morphology in two- or multiphase alloys.
3. Ratios of fracture mechanical (plastic zone size, crack opening displacement) and microstructural (grain size, particle spacing) dimensions.
4. Effective crack extension forces as modified by strain induced martensitic transformation, crack branching, or pre-existing microcracks, and inclusions.

KEYWORDS

Fracture, fatigue, microstructure, strain localization, partial properties, plastic zone, transformed zone, crack branching, crack initiation, crack propagation.

INTRODUCTION

This paper deals with effects of microstructure on fracture in metals. Fracture is sometimes intended, as in grinding, more often it is undesired and, if unforeseen, may lead to catastrophies.

Fracture is always a consequence of external or internal stresses. The loading conditions can be subdivided in: continuously raising, constant or cyclic with time. In addition the environment of the material is of great importance: temperature, chemical environment and radiation. Four stages have to be

Fig. 1. 4 Stages of crack formation (N number of cycles).

    w work hardening/softening
    i crack initiation
    p subcritical crack propagation
    c critical crack propagation

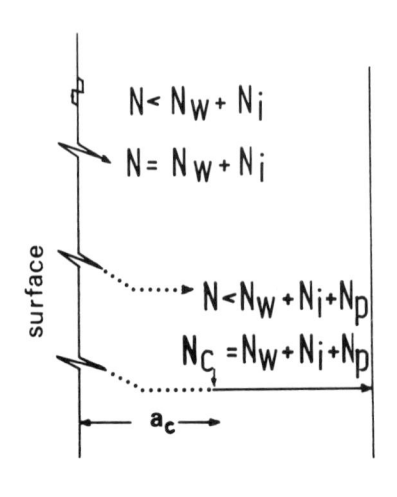

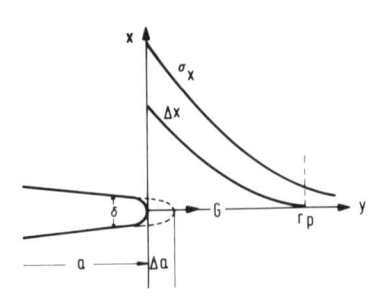

Fig. 2. Definition of the crack extension force G, external stress σ; σ_x, Δx stress field, displacement field.

Six elements of the microstructure

Element No.	Geometr. dimension	Dimension of density ρ_i	examples
1	0	$[m^{-3}]$	vacancy, interstitial
2	1	$[m^{-2}]$	dislocation
3	2	$[m^{-1}]$	grain boundary $\alpha\alpha$, interface $\alpha\beta$
4	3	$[m^{0} = 1]$	particle, pore (p)
5	-	-	crystal anisotropy (texture)
6	-	-	microstructure anisotropy

Fig. 3.

distinguished between formation of a crack and final rapid propagation, by which the life of a specimen or a part will end (Fig. 1).

The number of cycles N of a fatigue experiment will be used to define the principle stages of a crack with a length a:

$$N_w + N_i + N_p = N_c \tag{1a}$$

$$0 + a_i + a_p = a_c = \frac{G_c E}{\sigma^2 \pi} = \frac{K_c^2}{\sigma^2 \pi} \tag{1b}$$

The stages p (subcritical propagation) and c (critical propagation) only can be treated by fracture mechanics (Knott, 1973). Cyclic work hardening or softening (stage w) (Laird, 1974; Mughrabi, 1976) and the formation of crack nuclei (stage i) (Fine, 1980) is left to materials science for interpretation. A crack nucleus must aquire a size a = a_i to propagate at the applied stress σ. In most cases this will occur by a subcritical mechanism (fatigue, stress corrosion, creep). The general feature of all special mechanisms of crack propagation (stage p and c) is a certain value of a crack extension force $G = K^2/E$ which has to be applied to make a crack grow. It can be calculated for a given stress, crack length and specimen geometry (Knott, 1973)[1]. The resistance of the material stems from the energy required to form two surfaces $2\gamma_s$ and from the plastic deformation γ_p pro unit area of new crack (Fig. 2):

$$G = 2\gamma_s + \gamma_p \approx \frac{\int_o^{\Delta x} \sigma_x \Delta x}{\Delta a} dy \tag{2}$$

$\gamma_p \gg \gamma_s$ for most alloys discussed in this paper. The dimension of the specific crack extension energy $G \equiv Jm^{-2}$ is identical with that of a force on a dislocation $f = \tau b$. However not only resistance to plastic deformation σ_y or τ_y, but also the crack opening displacement $\Delta x \equiv \delta$ is responsible for the resistance to crack growth. As a consequence it is more difficult to derive quantitative relations for fracture mechanical properties, as it is for the yield stress of alloys with complex microstructure (Hornbogen, 1979).

Fracture can be treated on three structural levels. Physics is concerned with the breakage of individual atomic bonds, fracture mechanics disregards structure and deals with stress and strain fields in a mechanical continuum. Object of the present paper is a systematic approach indicating the relevance of microstructure for mechanisms of fracture (Hornbogen, 1977a).

MICROSTRUCTURE

The structural level of microstructure is found above that of crystal and glass structures i. e. above that of the phases. Microstructure can be defined in the classical way as arrangement of phases in space. More comprehensive is the definition of structural discontinuities as microstructural elements. Zero- to three-dimensional defects plus crystal-structural and micro-structural anisotropy provide us with six elements. They will serve as hardening mechanisms, if the yield stress is of concern (Fig. 3). Density, distribution, and shape of these elements will allow a full description of a microstructure.

---

[1] Effects of state of stress and specimen dimension are not considered in this paper.

Two-phase microstructures are of special concern for crack formation and propagation. A dispersion, duplex, and net can be defined as principle types of two-phase structures. All of them exist in anisotropic varieties (Fig. 4). For a quantitative description the densities of grain boundaries $\rho_{\alpha\alpha}$, $\rho_{\beta\beta}$ and interfaces $\rho_{\alpha\beta}$ are useful. The grain boundary density is inverse proportional to the diameter $S_{\alpha\alpha}$ of α-grains. The following ratios of these densities characterize the types of two-phase microstructures (Hornbogen, 1981):

$$\frac{\rho_{\alpha\alpha}}{\rho_{\beta\beta}} = 2\frac{\rho_{\alpha\alpha}}{\rho_{\alpha\beta}} \qquad f_\alpha = f_\beta = 0,5 \quad \text{Duplex} \qquad (3a)$$

$$\frac{\rho_{\beta\beta}}{\rho_{\alpha\beta}} = 0 \qquad 0 < f_\alpha < 1 \qquad \text{Dispersion of } \beta \text{ in } \alpha \qquad (3b)$$

$$\frac{\rho_{\alpha\alpha}}{\rho_{\alpha\beta}} = 0 \qquad 0 < f_\beta < 1 \qquad \begin{array}{l}\text{Net of } \beta \text{ formed at} \\ \alpha\alpha\text{-grain boundaries}\end{array} \qquad (3c)$$

The volume fractions of the α and β phase in an ideal duplex structure is $f_\alpha = f_\beta = 0.5$. For the other microstructures f may assume any value: $0 < f_\beta < 1$. Transformations can occur from one type of microstructure to another if the volume fraction (or shape, distribution) of the phases change. Such microstructural transformations lead to changes in the percolation behaviour of the phases. A dispersion percolates in the matrix α, a net along the net-phase β and a duplex structure through both phases. Two-dimensional percolation is required for a crack to propagate exclusively through one phase. In all other cases the crack has to interact with both phases of a microstructure. Thus qualitative differences in crack propagation behaviour can be associated with transformation of the type of microstructure (Fig. 5).

There are four principal aspects which are necessary for considerations of microstructural effects on fracture. Any of them may dominate the mechanism of fracture. In most cases more than one has to be considered for a full understanding and derivation of quantitative models for the relationship between microstructure of a particular alloy and its fracture behaviour.

1. PRINCIPLE: LOCALIZATION OF PLASTIC STRAIN

In fracture mechanics a microscopically homogeneous strain is assumed to be found in the plastic zone. The radius r_p is defined by the contour ahead of a crack tip at which the stress is equal to the yield stress σ_y of the alloy (Fig. 6) (Hornbogen, Minuth and Stanzl, 1980).

$$r_p \sim \frac{GE}{\sigma_y^2} \qquad (4a)$$

From geometrical considerations follows the crack opening displacement $\Delta x \equiv \delta$ again assuming homogeneous strain at $r < r_p$ (Fig. 2).

$$\delta \sim \frac{G}{\sigma_y} \qquad (4b)$$

In real alloys this assumption is often not well fulfilled. In the extreme case plastic strain in a crystal is localized in only one slip plane. Other types of localization are found in metallic glasses or in precipitation hardened alloys which contain particle free zones (PFZ) at grain boundaries (intercrystalline localization).

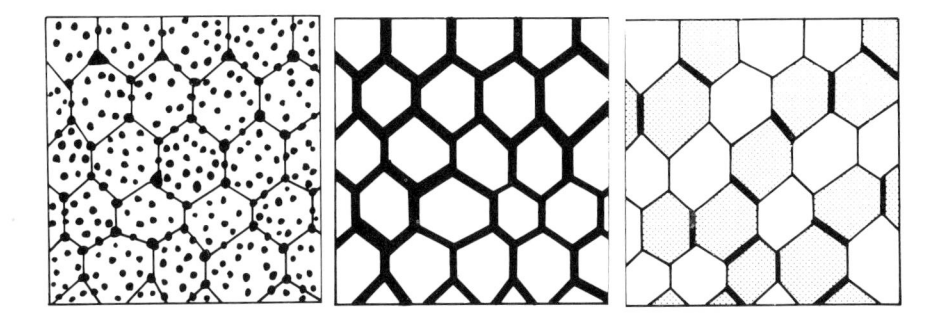

Fig. 4. Types of two phase microstructures:
dispersion, net, duplex, provide different crack paths.

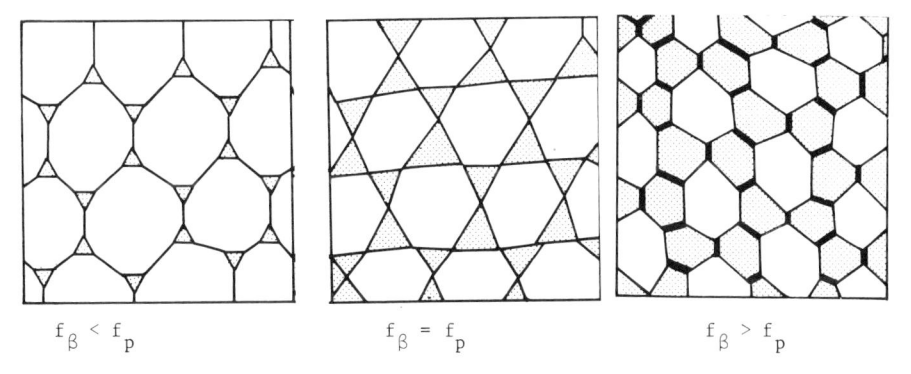

$f_\beta < f_p$ $f_\beta = f_p$ $f_\beta > f_p$

Fig. 5. Microstructural transformation with increasing volume fraction at f_p (point of percolation), enforces a change of crack path.

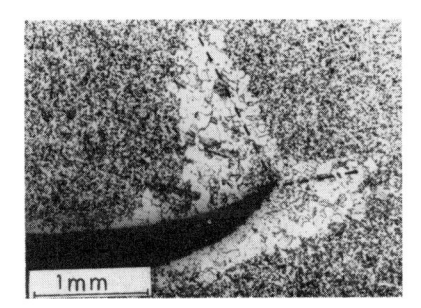

Fig. 6. Contour of the plastic zone preceding a fatigue crack in α-Fe + 1.8 wt% Si. The strain field has been analysed by a recrystallization method.

Transcrystalline localization of strain is favored by (Hornbogen and Zum Gahr, 1975):

1. a small number of dislocation sources
2. a small number of equivalent slip systems
3. the difficulty for dislocations to leave their slip plane by cross-slip.

It can be concluded that high stacking fault energy metals (Al, α-Fe) containing a dispersion of hard (unshearable) particles favor a homogenisation. Therefore tempered steels and some overaged Al-alloys conform to the greatest extend with the assumption of homogeneous plastic deformation in fracture mechanics.

Besides a low stacking fault energy of the matrix, the shearing of obstacles such as coherent particles is an important source for transcrystalline strain localization (Hornbogen and Zum Gahr, 1975). Its tendency can be quantitatively expressed by the local change in critical resolved shear stress per number n of passing dislocations (Fig. 7):

$$-\frac{d\Delta\tau}{dn} = \frac{\gamma_{APB} f_p}{d_p} \qquad (5)$$

Equation 5 applies to a dispersion of coherent particles with diameter d_p, volume fraction f_p, and energy of an antiphase domain boundary γ_{APB}. A negative value of $\frac{d\Delta\tau}{dn}$ is a prerequisite for localization.

Strain localization favors crack initiation in the surface for all loading conditions. This is shown for fatigue in ($\gamma+\gamma'$)-superalloy microstructures in Fig. 8 (Gräf and Hornbogen, 1977).

The transcrystalline formation of a fatigue crack nucleus of the size a_i will occur in different alloys between the following extreme mechanisms:

A. cyclic deformation and workhardening, strain localization by formation of softened persistant bands, and consequent crack initiation, for homogeneously deforming alloys.
B. primary formation of high slip steps at softened slip planes, formation of intrusions in connection with very limited cross slip, in alloys with localized strain (Partridge, 1965).

At first sight it is surprising that crack propagation is retarded by localized strain in crystals (Fig. 9). The explanation is based on a portion of reversible slip which can take place if dislocations stay in their original slip plane. As crack progress Δa per cycle is a part $A \sim 1/2$ of the crack opening displacement (equ. 4b), partial or complete reversibility can reduce Δa as a function of the number of dislocations n_R, which move reversibly at the tip of a crack (Hornbogen and Zum Gahr, 1976).

$$\frac{d\Delta a}{dN} = A\delta = A\frac{\Delta G}{\sigma_y} \sim A\, n\, b \qquad \text{homogeneous strain} \qquad (6a)$$

$$\frac{d\Delta a}{dN} \sim A\,(n-n_R)b \qquad \text{localized strain} \qquad (6b)$$

Thus localization of plastic strain promotes crack initiation, however retards propagation. This is taken into account by applying surface treatments (for example by shot peening) to a microstructure which produces localized strain in the undeformed surface (Hornbogen and Verpoort, 1980; Hornbogen, Thumann and Verpoort, 1981).

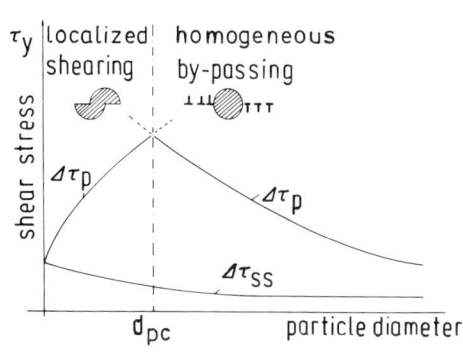

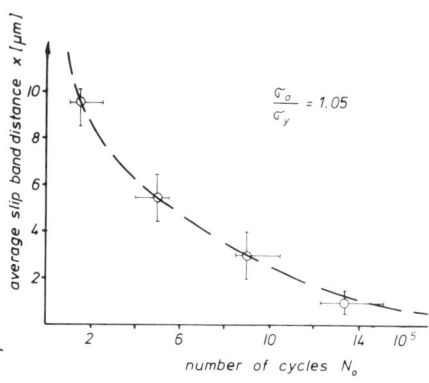

Fig. 7. Shearing or bypassing of particles by dislocations as the cause of localized or homogeneous strain.

Fig. 8. Number of cycles (N_w+N_i) required for crack initiation, as a function of strain localization in a 20 Cr, 2.4 Ti, 1.4 Al Ni-superalloy (wt%) with different microstructures.

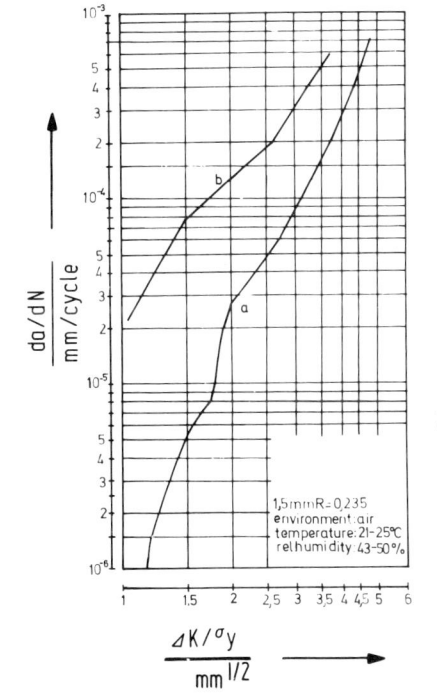

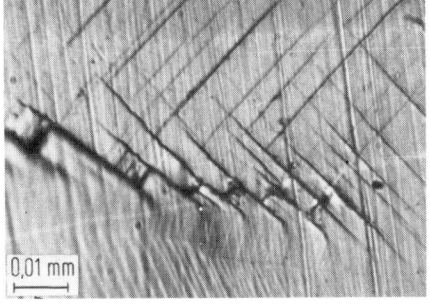

Fig. 9. Fatigue crack propagation in a γ-Fe, 36 Ni, 12 Al (at.%). alloy:

a) Crack tip preceded by highly localized slip on {111}-planes.

b) Crack growth rates in different microstructures with equal yield stress σ_y, but localized a) and homogeneous b) strain.

2. PRINCIPLE: PARTIAL FRACTURE MECHANICAL PROPERTIES

Many alloys are composed of two or more microstructural components with different mechanical properties. For the interpretation of bulk fracture mechanical properties these partial properties of the microstructural constituents have to be used in addition to volume fraction f and type of microstructure (Chapter 2, figs. 4, 5). A way to derive quantitative relations between microstructure and a bulk property P is the arrangement of microstructural components as mechanical elements parallel or in sequence to the direction of the external load. For the resulting equations the following three limiting conditions are found (Fig. 10) (Friedrich and Hornbogen, 1980).

$$P = \bar{p}_\alpha f_\alpha + \bar{p}_\beta f_\beta \tag{7a}$$

$$P = \bar{p}_\alpha f_\alpha \tag{7b}$$

$$P = \bar{p}_\alpha \tag{7c}$$

The bulk property is either a function of the volume fractions f_i and partial properties p_i of all phases, or of f_α and $\bar{p}_\alpha$ only, or of $\bar{p}_\alpha$ independent of its volume fraction. For the application of these relations a microstructure shall be considered in which the tensile strength of α is less than the yield stress σ_y of β (Fig. 11). Dispersion and net structures can be treated approximately as sequential, duplex structures as parallel arrangements. Consequently we find the following examples for equations 7a, b,c: For a $(\alpha+\beta)$-duplex structure (Fig. 12) (Hornbogen and Stratmann, 1976).

$$G_c = \bar{G}_{c\alpha} f_\alpha + \bar{G}_{c\beta} f_\beta \tag{8a}$$

For a coarse dispersion of hard β in α (for example dual phase steel) plastic strain and consequently the fracture strain ε_f is resistricted to α:

$$\varepsilon_f = \bar{\varepsilon}_{f\alpha} f_\alpha \tag{8b}$$

The fracture stress of the same type of microstructure is equal to that of the α component.

$$\sigma_f = \bar{\sigma}_{f\alpha} \tag{8c}$$

An important case is a structure, which is formed by a very soft PFZ which surrounds the precipitation hardened interior of grains. Separation takes place by a pseudo-intercrystalline rupture inside the PFZ, while the hardened interior of the grain does not participate in plastic deformation and consequently in dissipation of energy. A net-structure model is suitable to deal with this mechanisms (Fig. 13; equ. 7b, 8b):

$$G_c = \bar{G}_{cPFZ} f_{PFZ} = G_{cPFZ} \frac{d}{S} \tag{9}$$

d is the thickness of the PFZ, S the grain size. The measured grain size dependence for microstructures with constant d follows from this model (Fig. 13). The subltety of the treatment is increased by taking into account that the crack path is not exclusively intercrystalline, but becomes transcrystalline if a critical angle between external stress and orientation of the grain boundary is exceeded (Gräf and Hornbogen, 1977; Hornbogen and Kreye, 1982).

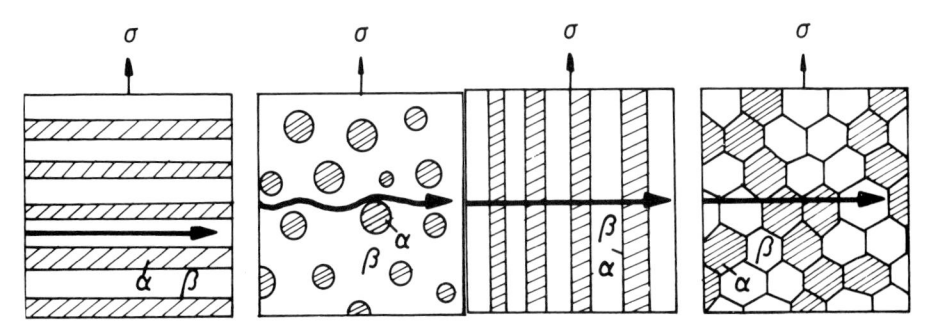

Fig. 10. Sequential and parallel arrangement of phases α and β in two-phase microstructures.

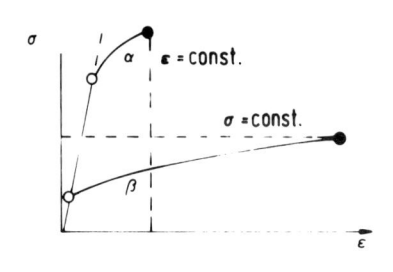

Fig. 11. Schematic representation of partial tensile properties of microstructural components α and β.

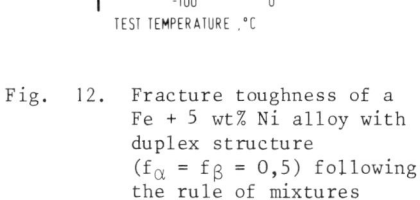

Fig. 12. Fracture toughness of a Fe + 5 wt% Ni alloy with duplex structure ($f_\alpha = f_\beta = 0,5$) following the rule of mixtures equ. 7a, 8a.

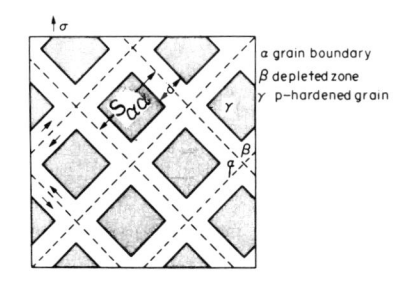

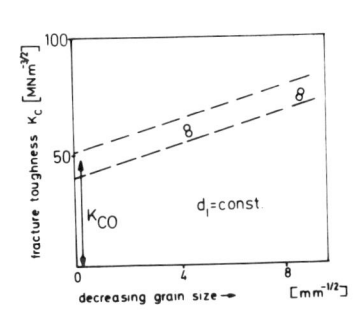

Fig. 13. a) Model for pseudo-intercrystalline crack growth in particle-free zones.

Fig. 13. b) Grain sizes dependence of fracture toughness K_c of an precipitation hardened Al, 4.6 Zn, 1.2 Mg.

3. PRINCIPLE: RATIO OF FRACTURE MECHANICAL AND MICROSTRUCTURAL DIMENSIONS

Fracture mechanical dimensions are the extension of the plastic zone r_p (equ. 4a), the amount of crack opening displacement δ (equ. 4b), or the critical specimen thickness. The ratio with microstructural dimensions, such as grain diameter $S_{\alpha\alpha}$, particles spacing S_p, or lamellar spacing, are of relevance for the fracture mechanisms. If a small fatigue crack grows at a constant load amplitude, the ratio $r_p/S_{\alpha\alpha}$ will assume the values (Fig. 14):

$$\frac{r_p}{S_{\alpha\alpha}} \lesssim 1 \qquad (10)$$

Only for $r_p > S_{\alpha\alpha}$, and a duplex-structure the crack extension energy will follow the rule of mixtures (equ. 7a, fig. 12). For $r_p < S_{\alpha\alpha}$ the crack can pass through the softer α-phase exclusively.

In single phase polycrystals a crack will behave as in a single crystal. For $r_p \gg S$ many grain boundary sources can be activated and plastic strain becomes more homogeneous. This difference is especially significant in microstructures with a high tendency for transcrystalline localization of plastic strain (Fig. 14).

A relation between the critical crack opening displacement δ and the particle spacing S_p has been used to derive a quantitative relation for the fracture toughness of alloys with a dispersion of non-coherent undeformable particles. The mechanism implies decohesion of the α-β-interfaces, formation of pores, and consequent local shear and necking of the ductile matrix (Fig. 15, equ. 11a) (Brown and Embury, 1973)

$$S_p \approx \delta_c \qquad (11a)$$

$$G_c = \frac{K_c^2}{E} = \sigma_y \delta_c = C \frac{\sigma_y d_p}{f^{1/3}} \qquad (11b)$$

This equation describes well the dependence of fracture toughness on volume fraction f of ductile alloys which contain for example slag inclusions.

4. PRINCIPLE: EFFECTIVE CRACK EXTENSION FORCES

Critical or subcritical crack extension takes place if the external loading conditions (K, G) provide sufficient energy for continued separation at the crack tip. Certain microstructural features are able to raise or lower the effectiveness of the loading conditions, and therefore accelerate or retard crack growth. Fracture mechanics usually deals with the growth of one crack. There are however microstructures which allow for crack branching or the formation of satellite cracks which run parallel to the main crack (Fig. 16). If the plastic zones of these crack tips do not overlap (equ. 4a) the energy required for crack growth is raised (1+n) times, where n is the average number of cracks in branches and satellites and G_{c0} valid for the movement of a single crack (Fig. 17) (Erven, 1982).

$$G_c = G_{c0} (1+n) \qquad (12a)$$

$$K_c = K_{c0} \sqrt{1+n} \qquad (12b)$$

Fig. 14.

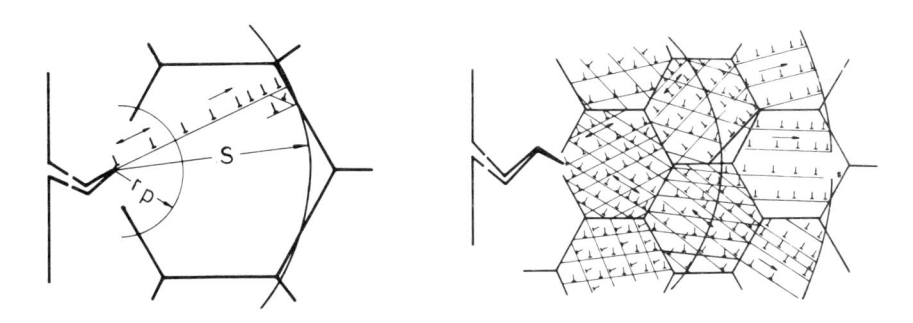

a) Schematic representation of the condition $r_p < S_{\alpha\alpha}$ and $r_p > S_{\alpha\alpha}$ and its consequence of strain localization.

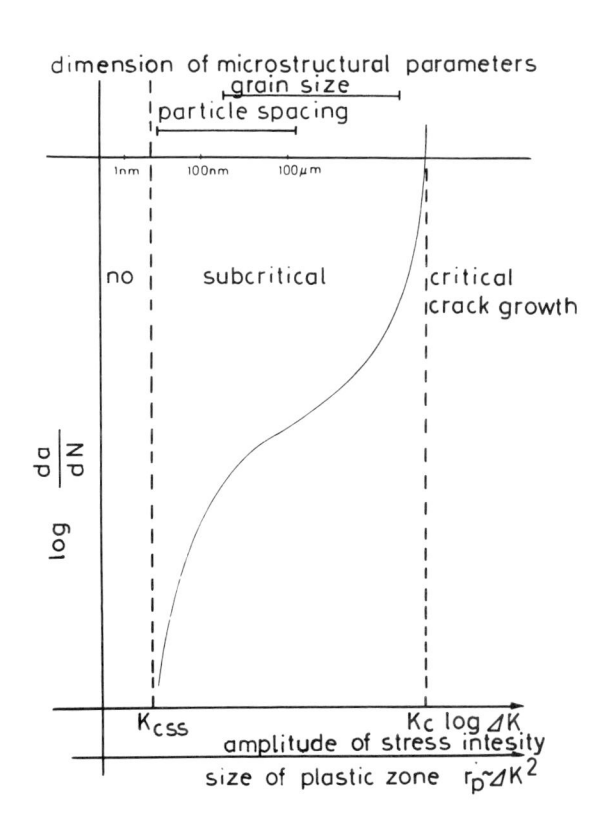

b) Changing ratio of the dimension of the plastic zone r_p and various microstructural dimensions, for a $d\Delta a/dN$-$f(\Delta G)$-curve, schematic.

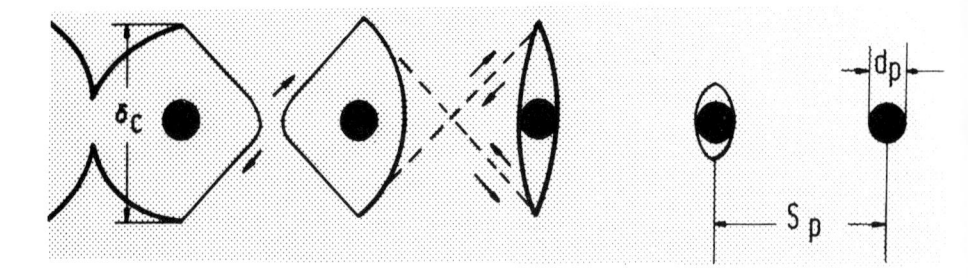

Fig. 15. a) Model for dimple fracture originating by decohesion of particle-matrix interfaces.

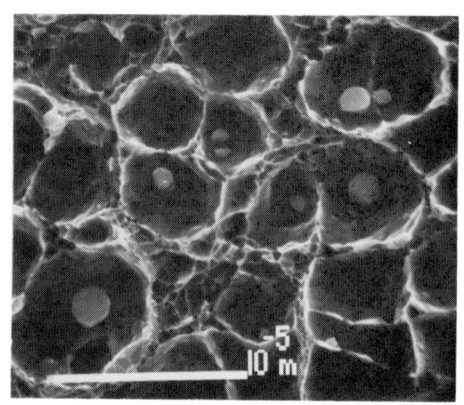

Fig. 15 b) Dimple fracture of a tempered Ni, Cr, Mo-steel 0.25 wt% (K_{IC}-test with 25 mm CT-specimen).

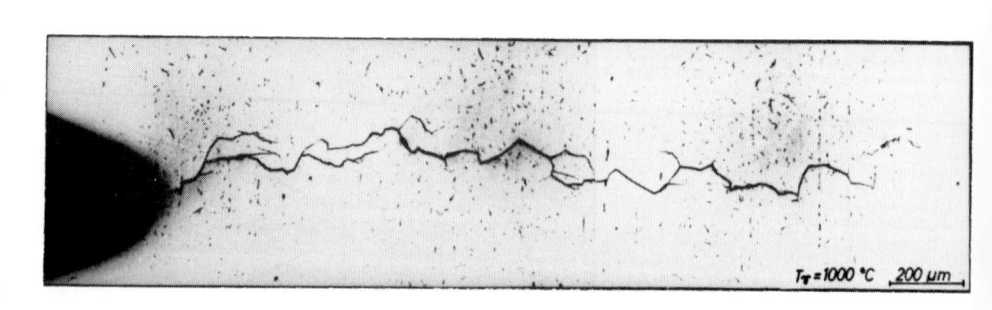

Fig. 16. Crack branching in a α-Fe, 0.96 C, 2.0 Mn, 0.3 Cr (wt%) tool steel as quenched from 1100 °C.

A different reason for modification of the loading conditions can be due to local changes in specific volume at the crack tip. This occurs by strain induced martensitic transformation of a metastable (austenitic) phase. In addition to the plastic zone in which the volume stays approximately constant (equ. 4a) a transformed zone arises in which the volume will be increased considerably (3 - 5 % for example by the $\gamma \to \alpha_M$ of iron alloys). If the transformed zone is surrounded by untransformed structure an internal compressive stress $-\Delta\sigma_M$ arises, which reduces the external tension at the crack tip (Fig.2). The additional internal stress field can be treated with the concept of a continuum edge dislocation situated in the boundary of the internal transformed zone (Fig. 18) (Hornbogen, 1977b, 1978).

If a transformation would be associated with a negative volume change, or if an alloy contains microcracks, pores, or inclusion of nil-strength phases such as graphite, the external stress will be raised by $+\Delta\sigma_M$. A suitable averaging method will lead to quantitative relations between their microstructural features and macroscopic crack extension properties. Such relations can be derived by modifying the nominal loading conditions

$$G = \frac{K^2}{E} = \pi a \sigma^2 \qquad (13a)$$

by the microstructural effects $\pm\Delta\sigma_M$, to obtain the effective values:

$$G_{eff} = \frac{K^2_{eff}}{E} = \pi a \, (\sigma \pm \Delta\sigma_M)^2 \qquad (13b)$$

An increase of decrease or the effective crack extension energy (or stress intensity) will cause an inverse change of the critical values for example the critical crack extension energy G_c (or the fracture toughness K_c):

$$\frac{G}{G_{eff}} = \frac{G_c}{G_{c0}} = \frac{K^2_c}{K^2_{c0}} \qquad (14)$$

G_{c0} or K_{c0} are the properties of the matrix (i.e. untransforming, crack free, inclusion free). Along this line a relation for the fracture toughness of grey cast iron can the derived in which fracture toughness of the metallic matrix $K_{c\alpha}$, volume fraction f_β, length l, and edge radius ρ of graphite and a statistical factor C are contained (Hornbogen and Motz, 1977):

$$(G_c E)^{1/2} = K_c = \frac{K_{c\alpha}}{f_\alpha + f_\beta C(1+2\sqrt{l/\rho})} \qquad (15)$$

A similar equation will describe the properties of a material which contains microcracks or pores.

It may be noted here that existing microcracks are reducing the resistance to crack growth, while formation of satellite cracks, their growth and branching are increasing it. Both effects are sometimes interrelated (Fig. 17). In ceramics and brittle metals the mechanisms reported in this chapter provide the only sources for toughening (Garvie, Hammik and Pascoe, 1975). Instead of a plastic zone (equ. 4) the concept of a reaction zone should be used for the analysis of crack propagation in such materials in which energy is not primarily dissipated by plastic deformation.

Only a few of the enormous number of phenomena, in which microstructure plays a role in fracture have been discussed in this paper. It provides an attempt to systematize these effects, and to obtain an improved qualitative and quan-

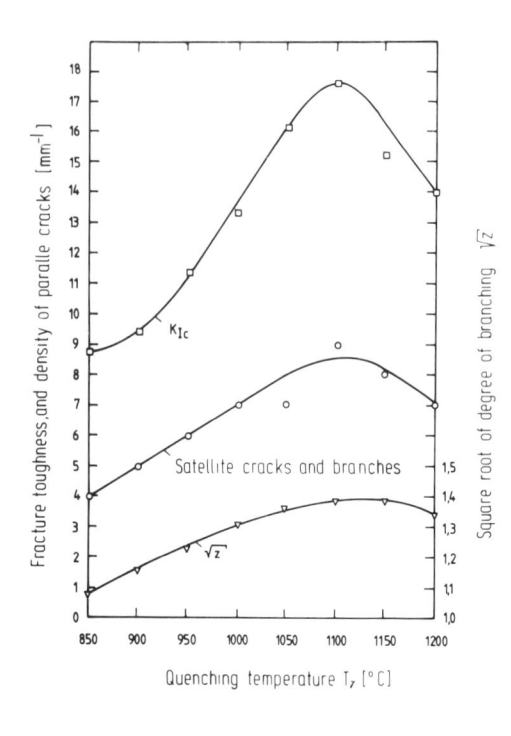

Fig. 17. An increase in fracture toughness of the tool steel as quenched from different temperatures is caused by crack branching, composition as in Fig. 16.

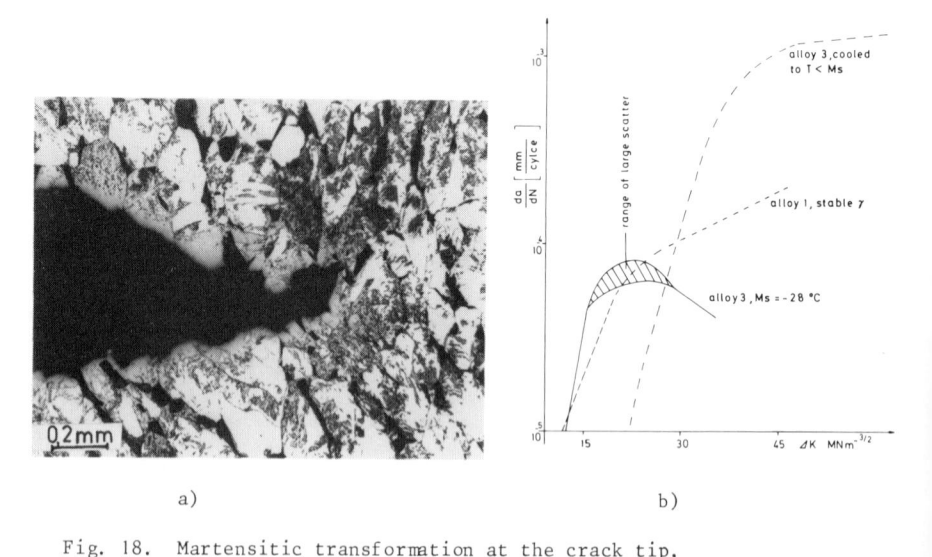

Fig. 18. Martensitic transformation at the crack tip.
a) Transformed zone in an fatigue γ-Fe, 20 Ni, 7 Al (wt%) alloy.
b) Fatigue crack growth in stable austenitic, transformed and transforming γ-Fe-Ni-Al alloys.

titative understanding. An application of these principles should allow a better insight into integrated phenomena such as fatigue life, it should also help to develop new or improved metallic materials.

ACKNOWLEDGEMENT

Most of the work reported here was conducted with support by The German Science Foundation (DFG-Schwerpunkt, Bruch Ho 325-9). Helpful for the initiation of my work on fracture were many stimulating discussions during a sabbatical leave, spent with the Metals Science Group of Battelle, Columbus, Ohio 1974/75. Important individual contributions to the research project on microstructure and fracture came from the following coworkers at the Ruhr-University Bochum: J. Becker, U. Bruch, F. Erven, K. Friedrich, M. Gräf, H. Kreye, P. Stratmann, M. Thumann, C. Verpoort, K. H. Zum Gahr.[2]
The attempt to a systematic approach to microstructure and fracture was first developed for lectures given 1978/79 at the Groupe de physique du Solide, École des Mines, Nancy, France.

REFERENCES

Brown, L.M., J.D. Embury (1973). Proc. ICSMA 3, Cambridge, Vol. 1, p. 164.
Erven, F.(1981), PHD-Thesis, Ruhr-University Bochum, to be publ. HTM 1983.
Fine, M.F. (1980). Met. Trans 114, p. 365.
Friedrich, K., E. Hornbogen (1980). J. Mat. Sci. 15, 2175.
Garvie, R.C., R. H. Hammik, R.T. Pascoe (1975). Nature 258, p. 708.
Gräf, M., E. Hornbogen (1977). Acta Met. 25, p. 877, 883.
Gräf, M., E. Hornbogen (1978). Scripta Met. 12, p. 147.
Hornbogen, E., K. H. Zum Gahr (1975). Metallography 8, 181.
Hornbogen, E., P. Stratmann (1976). Proc. ICSMA 4, Nancy, Vol. 2, p. 837.
Hornbogen, E. (1977). Proc. ICF4, Waterloo, Vol. 2, p. 837.
Hornbogen, E. (1977a). Z. Metallkde. 72, p. 445.
Hornbogen, E. (1977b). Proc. ICF4, Waterloo, Vol. 2, p. 149.
Hornbogen, E., J. Motz (1977). Int. Cast Metals J. 2., No. 4, p. 31
Hornbogen, E. (1978). Acta Met. 26, p. 147.
Hornbogen, E. (1979). Proc. ICSMA 5, Aachen, Vol. 2, p. 1337.
Hornbogen, E., C. Verpoort (1980). Proc. 4th Int. Conf. Superalloys, ASM, Metals Park, p. 585.
Hornbogen, E., E. Minuth, S. Stanzl (1980). Mat. Sci. Eng. 43, p. 145.
Hornbogen, E. (1981). Z. Metallkde. 72, p. 739.
Hornbogen, E., M. Thumann, C. Verpoort (1981). 1 · Int. Conf. Shot Peening, Paris, p. 381.
Hornbogen, E., H. Kreye (1982). J. Mat. Sci. 17, p. 979.
Knott, J.F. (1973). Fundamentals of Fracture Mechanics. Butterworth, London.
Laird, C., C. Calabrese (1974). Mat. Sci. Eng. 13, p. 141 and 159.
Mughrabi, H., Ch. Wütrich (1976). Phil. Mag. 33, p. 903.
Partridge, P. G. (1965). Acta Met. 13, p. 517

[2] J. Albrecht, B.B.C. Baden, Switzerland, contributed the micrograph Fig. 18b.

Routes to Higher Strength and Ductility of Steels

T. Gréday

Centre de Recherches Métallurgiques, Liège, Belgium

ABSTRACT

In this paper, an overview is given of the properties and microstructures that can be achieved by controlled rolling and control cooling of steels and products such as HSLA, Dual-Phase, Re-bars and wires.

KEYWORDS

Controlled rolling ; controlled cooling ; allotropic transformation ; microstructure ; structural steels ; HSLA steels ; dual-phase steels ; sheets ; strips ; bars ; wires.

INTRODUCTION

Industrial hardening of a structural steel exhibiting a ferritic microstructure was realized for the first time twenty years ago. It consisted in the precipitation hardening by fine carbide or nitride.particles formed by elements such as niobium or vanadium. Those steels, which since the sixties have known an important development,are called high strength low alloyed steels (HSLA).

Encouraged by this positive result, research metallurgists have then studied more carefully how to use practically some other principles of Physical Metallurgy which could also improve the mechanical properties.

A schematic representation of the possible mechanisms and their main governing parameters is given in Fig. 1. One can immediately conclude that the achievement of maximum profit-making and reproducibility requires a perfect knowledge of the effects produced by hot working and subsequent cooling on the morphology of the existing phases and on the allotropic transformation of austenite.

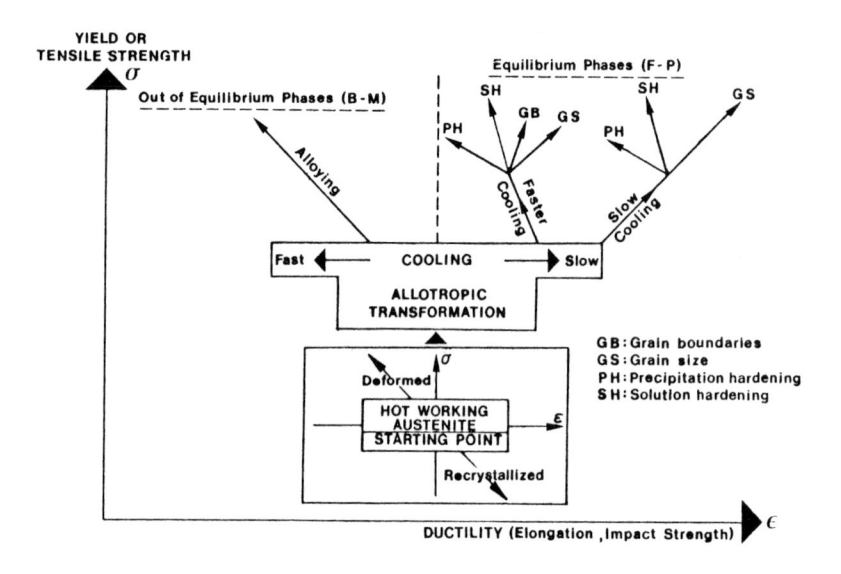

Fig. 1. Schematic representation of the main hardening processes.

There are three parameters to control : temperature - deformation and cooling. Nowadays, the metallurgist must absolutely be able to make the best possible use of as simple alloys as possible - which are very often the cheapest ones - in order to avoid using some elements, the disponibility of which is limited on earth.

Our purpose here is to show by practical examples the different ways chosen by the steelmaker of the seventies to reach the above-mentioned targets.

Therefore, we shall review the progress realized in the field of thermomechanical treatments (TMT) thanks to the knowledge acquired on the recrystallization of deformed austenite. Then, we shall discuss the interest of controlling the cooling rate after deformation, which will lead us quite naturally to analyse more complex microstructures such as those obtained in dual-phase steels.

HOT WORKING OF AUSTENITE

Though the term thermomechanical treatment (TMT) itself is clear enough, it might be useful yet to recall its definition by Duckworth (1966):
"The thermomechanical treatment consists in achieving plastic deformation prior to or during an allotropic change so as to obtain improved mechanical properties".

On this basis, the various TMT can be classified into three large groups (Lamberigts, 1977), according to whether the deformation happens before or during the allotropic transformation, or whether it is applied to the products of transformation. These three large classes can be sub-divided further in order to take the structural state of the metallic phase which

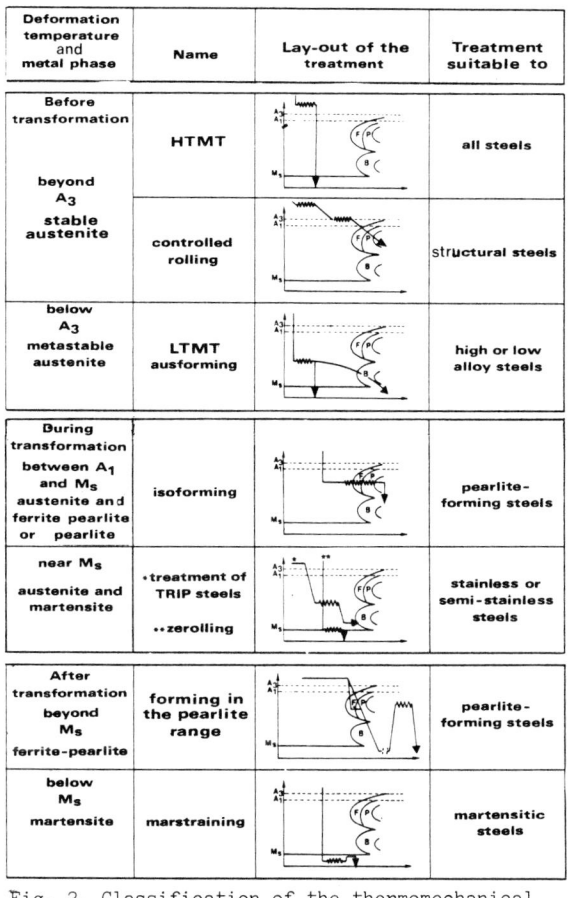

Deformation temperature and metal phase	Name	Lay-out of the treatment	Treatment suitable to
Before transformation beyond A_3 stable austenite	HTMT		all steels
	controlled rolling		structural steels
below A_3 metastable austenite	LTMT ausforming		high or low alloy steels
During transformation between A_1 and M_s austenite and ferrite pearlite or pearlite	isoforming		pearlite-forming steels
near M_s austenite and martensite	*treatment of TRIP steels **zerolling		stainless or semi-stainless steels
After transformation beyond M_s ferrite-pearlite	forming in the pearlite range		pearlite-forming steels
below M_s martensite	marstraining		martensitic steels

Fig. 2. Classification of the thermomechanical treatments of steels.

undergoes the plastic deformation (Fig. 2) into account. Now, two of these treatments are much more widely used : they are the HTMT and controlled rolling.

When hot working is applied to austenite at high temperature (> Ar_3) the phase structure can undergo two types of modifications :
1. Modification of the morphology and density of crystalline defects as well as of the kinetics of carbides and nitrides precipitation.
2. Recovery, polygonization or recrystallization, resulting from the first effect because of the energy stored during hot working.

Recrystallization of hot worked austenite has been the subject of many studies. Two syntheses of those studies have already been presented from this platform by Rossard(1973) and Jonas(1976). Increasing the strain of austenite by applying multiple hot working steps as it does in hot rolling, also increases the density of crystalline defects. Such a process must bring about a progressive and continuous hardening of the phase. It is well known that there is a critical strain ϵ_c, above which sufficient energy density has built up to promote the nucleation of new grains together with a softening of austenite. This process can develop during or after hot working, and it is called dynamic or static recrystallization.

While using quite different products and methods to measure the deformation the different schools (Jonas, 1976 ; Lamberigts, 1974 ; Le Bon, 1973 ; Weiss, 1982) which are studying these processes converge reasonably well in their estimates of the general effects of the following parameters : temperature and ratio of deformation - distribution of deformation between successive steps - microprecipitation of carbides and nitrides.

For example, one usually agrees that in niobium microalloyed steels dynamic recrystallization could occur to higher temperature, while static recrystallization is inhibited. The precise manner of action of the niobium has been extensively discussed: niobium in solution or precipitated (Jonas, 1979 ; Lamberigts, 1974 ; Le Bon 1973). Finally, it seems that niobium in solution by slowing down the rate of static recovery is retarding static recrystallization (Luton, 1980). The effect on dynamic recrystallization is related directly to the balance between niobium in solution or precipitated (Lamberigts, 1977).

The complexity of those phenomena clearly appears in the figures proposed in a report on TMT research carried out in the CEC (Lamberigts, 1977).

The purpose of steelmaking is the production, at the lowest cost, of steels characterized by an optimal combination of their mechanical properties. There is a direct relationship between the latter and the final microstructure which, owing to allotropic transformation, is at least indirectly the result of the complete metallurgical story of the product. Unfortunately, it is difficult to meet the optimal values of each mechanical property at the same time because of the complex interactions existing between the mechanisms they are resulting from. For example, one knows that in the case of a Nb-microalloyed steel, it is not possible to achieve both the finest ferritic microstructure and the largest precipitation hardening, since the former can be obtained by controlled rolling at low temperature, while the latter is only reached when rolling is completed at high temperature.

The main aspects of recrystallization of strained austenite are shown in Fig. 3 to 5 which give a three-dimensional representation (T, ϵ, t) of the changes expected to appear in the austenite during assumedly instantaneous deformations and subsequent holding periods.

The (T, ϵ) area is divided into two zones for which dynamic recrystallization either can or cannot be inhibited by deformation, depending on whether $\epsilon \gtrless \epsilon_c$. The time scale measures the holding period necessary to give a stable microstructural state to austenite i.e. after complete recrystallization. It is obvious that such a representation corresponds to a simplified model in which an important practical factor has been omitted : the deformation rate. Three distinct cases can be identified : steels not giving rise to micro-precipitation (Fig. 3), microalloyed steels where microprecipitates still exist at the reheating temperature prior to deformation (Fig. 4), and microalloyed steels in which reheating dissolves the microprecipitates (Fig. 5).

The first case is that of carbon steel (Fig. 3). When dynamic recrystallization does not occur, the lower the deformation temperature and the greater the reduction, the more important the work hardening. When dynamic recrystallization is sanctioned by an appropriate combination of the T_D/ϵ, maximum structural refinement is achieved by a complete dynamic recrystallization, while static recrystallization leads to a coarser final microstructure.

In the second case (Fig. 4) the precipitated particles prevent the coalescence of the austenite grains and, with the atoms of dispersoid elements in solid solution, retard the kinetics of dynamic and static recrystallizations. The recrystallized grains are always at least as fine as those of C-Mn steels treated under the same conditions, because the microprecipitates tend to increase the energy of activation of dynamic recrystallization and to prevent the coalescence of statically recrystallized grains. The degree of refinement of ferrite depends as much on the work hardening of the austenite grains when the γ-α transformation starts, as on their average size.

Schematic representation of recrystallization of deformed austenite

Fig. 3. In C-Mn steels.

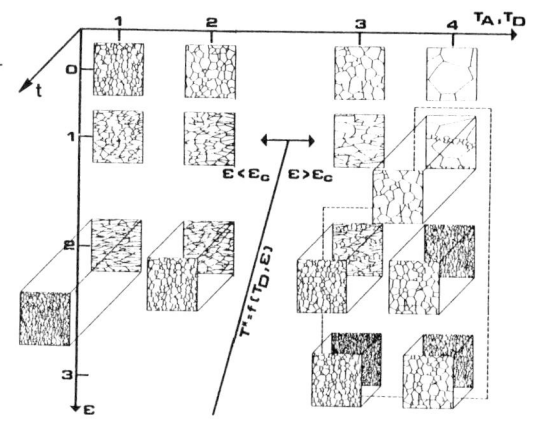

Fig. 4. In Nb-steels reheated below the temperature of dissolution of the microprecipitates.

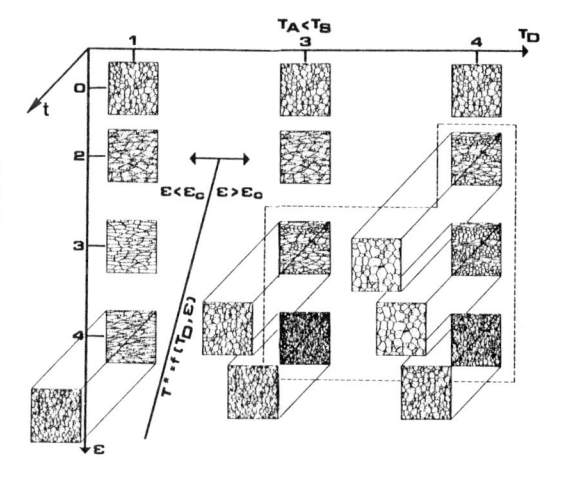

Fig. 5. In Nb-steels reheated above the temperature of dissolution of the microprecipitates.

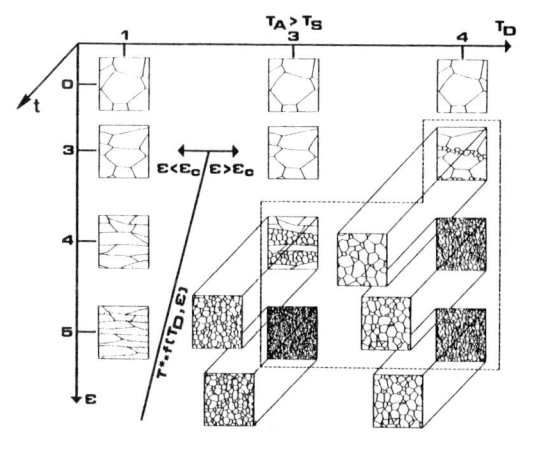

Finally, in the third case (Fig. 5), when recrystallization does not occur, the austenite remains work hardened because of the combined effect of strain-induced microprecipitation and of the atoms of microalloying elements in solution. At higher temperature, there can be a dynamic recrystallization ; however, it is delayed because the energy of activation of the phenomenon can be considerably increased by the action of microalloying elements in solid solution.

In order to understand the mechanical properties of a steel product treated in the conditions given in this general scheme, it is necessary to determine which relationship can exist between the austenitic structure prior to transformation, the conditions of transformation process and the mechanical properties, i.e. yield strength and resistance to brittle fracture.

In static recrystallization, three different types of softening can be observed (Fig. 6). Mode I is static recovery, mode II real static recrystallization, and mode III metadynamic recrystallization. The higher the temperature, the faster the static recrystallization, but its kinetics also depends on the amount of reduction and on the austenitizing conditions. There is an empirical relationship between temperature and deformation on the one hand, and the recrystallization rate G on the other hand : (Kozasu, 1971)

$$G = \alpha \exp. Q/_{RT} \cdot \exp\left[13\ (\epsilon - 0,33)\right], \text{ when } \epsilon < \epsilon_c \quad (1)$$

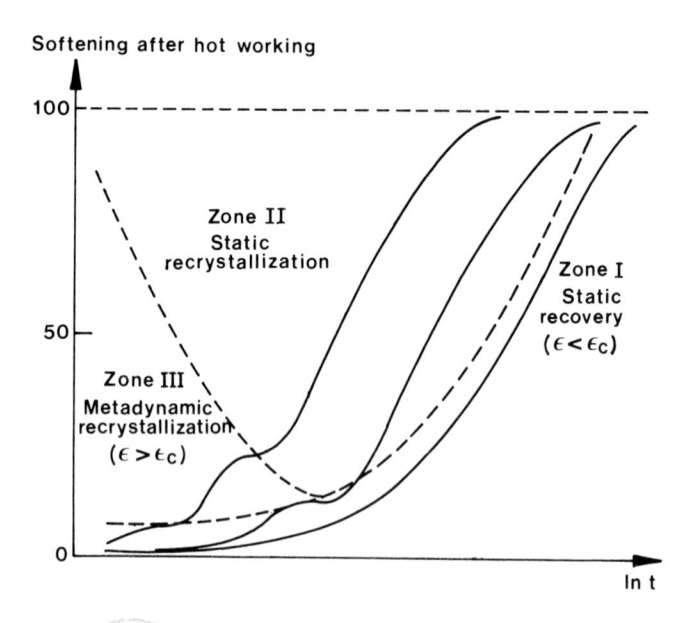

Fig. 6. The three softening mechanisms operative after hot deformation of austenite.

Moreover, the evolution in time of the recrystallized volume fraction can be expressed theoretically by the model of nucleation at random :

$$\ln \left(\ln \frac{1}{1-x}\right) = \ln D_o - \frac{nQ}{RT} + n\ln t \qquad (2)$$

Q is the apparent activation energy, R the perfect gas constant, D_o the diffusion coefficient and t, the time elapsed from the beginning of the process; n often has a value near the unity.

The static recrystallization rate depends on the chemical composition, especially on the presence of elements able to form precipitates which would remain stable at the temperatures considered and restrain the movement of grain boundaries. In each case, the values of Q and n, and the incubation time must be measured. For instance, we have obtained Q = 15kcal/mole for a C(.15%) - Mn (1.2%) steel and Q = 34 kcal/mole for the corresponding Nb (.02%) steel.

Recently, another aspect has proved to be important in such studies. It is the type of mechanical testing used to measure the characteristics of dynamic recrystallization, especially the critical strain ϵ_c. Results published recently by Sakai and his co-workers (1982) clearly show that the values of ϵ_c obtained by twist tests are always greater than those measured by use of homogeneous deformation tests such as tension or compression (Fig. 7). Such a deviation is of practical importance when one tries to build a mathematical model of the plastic strain behaviour. The value of ϵ_c must be introduced in such a model to improve the process of controlled rolling, to realize the automation of the process, or to calculate a new rolling mill.

Such directions have been followed recently by various research teams such as those of Saito (1980) and also Ouchi (1980). Let us mention, for instance, the results achieved by Ouchi(1980) while studying Nb and Nb-Mo containing steels to be used for pipe in the X-60 and X-70 grades, the thickness of the sheets being in the range of 12.5-19mm. About the effects of the different parameters of the rolling sequence on both ultimate tensile and

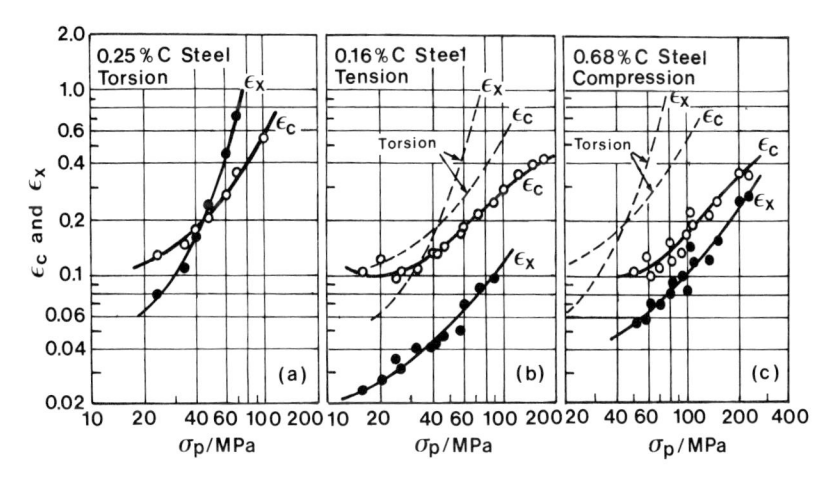

Fig. 7. Dependence of ϵ_c, initiation strain for recrystallization, and growth strain, on the critical (or peak) stress σ_p.
a. from torsion flow curves - b. from tension data - c. from compression data. (Sakai, 1982).

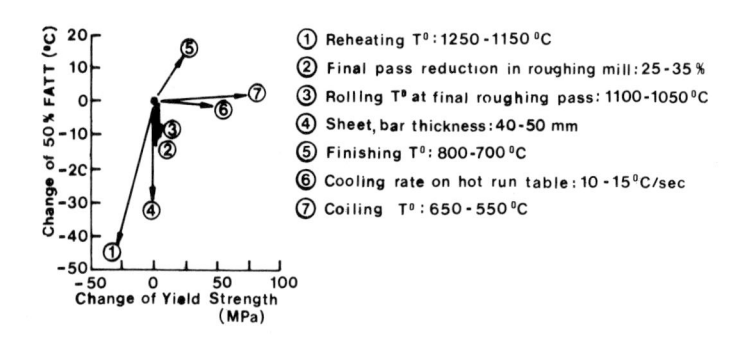

Fig.8. Effect of processing parameters in hot-strip mill on yield strength and FATT (Ouchi 1980).

impact strengths, Ouchi shows (Fig. 8) that the most active factors are : for the impact strength, the reheating temperature of the slab and the increase of its thickness, and for the U.T.S. the reduction of the coiling temperature and the increase of the cooling rate after rolling.

COOLING RATE AFTER HOT WORKING

Another factor which has also turned out to be interesting is the choice of cooling rate and its control after hot working and before and during allotropic transformation.

Our first example will be related to products where an increase of this rate does not modify the final equilibrium microstructure which remains of the ferrite-pearlite type. Mathy and co-workers (1982) show that between 800 and 500°C, an increase of the cooling rate from .15 to 2.5°C/sec improves by 25-30 MPa the yield strength of 80mm thick plates made of a C-Mn structural steel ; simultaneously, the 35 J/cm^2 transition temperature has also been improved by 15°C. Explanation of those changes will not be found in a satisfying way when considering only the grain refinement ; changes in the constitution of grain boundaries must also be taken into account.

A second example will be the new process for the production of bars developed by Economopoulos and his colleagues (1975). The bar is drastically cooled through a water-cooling device in such a way that its superficial layers transform into martensite while the centre of the bar remains austenite. After this quenching state and during natural air-cooling, the superficial layers undergo a tempering due to the heat flow emitted by the core of the bar. Those bars are characterized by a mixed microstructure (Fig. 9.a). The yield strength of the product can be largely modified (Fig. 9.b) by changing the treatment variables such as the standing temperature and the duration of the quenching stage. The main result of such a process is the achievement of high yield strength and ductility levels with a weldable carbon steel grade. Some of those levels of properties are impossible to achieve even when optimum use is made of the precipitation hardening ability of the niobium or vanadium containing steels.

A third example can be found in processes where high carbon wire rods undergo an appropriate and controlled mild cooling at the end of rolling. It is now known that the convenient microstructure giving the best drawability properties to the wire is composed of a fine pearlitic phase avoiding any proeutectoïd ferrite and above all, bainite pools. Such a microstructure is readily achieved by a lead-bath patenting. There are technical and economical factors which justify the search for another treatment easier to control and cheaper to realize. Various solutions have been proposed (Sannomiya, 1981; Takeo, 1975; Wilson, 1974).

They are all based on the use of a controlled cooling device the fluids of which range from air to water salt baths and fluidized bed. The simplest way, in this case is to use a liquid at boiling point, and water is the most appropriate liquid given its cost and its high heat of vaporization. Applying such a cooling to high carbon wires shows that the cooling rate lies between those of lead patenting and air cooling (5.5mm in diameter rods) and the same conclusion has been obtained when measuring the tensile strength (Fig. 10).

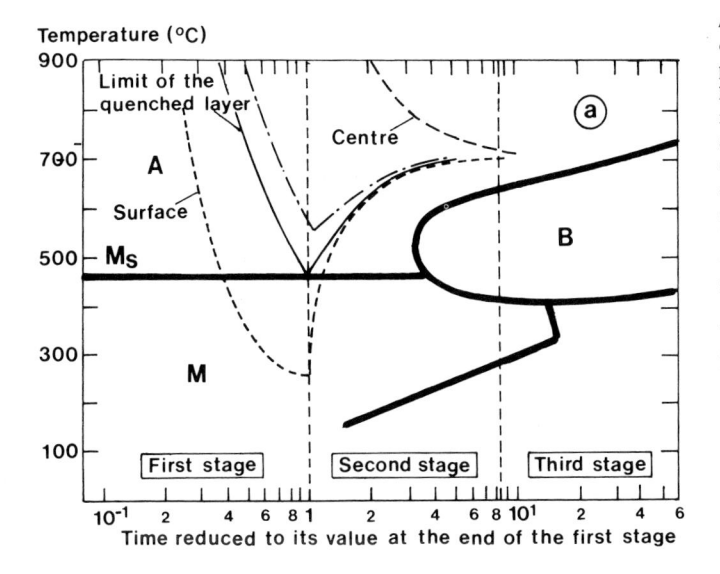

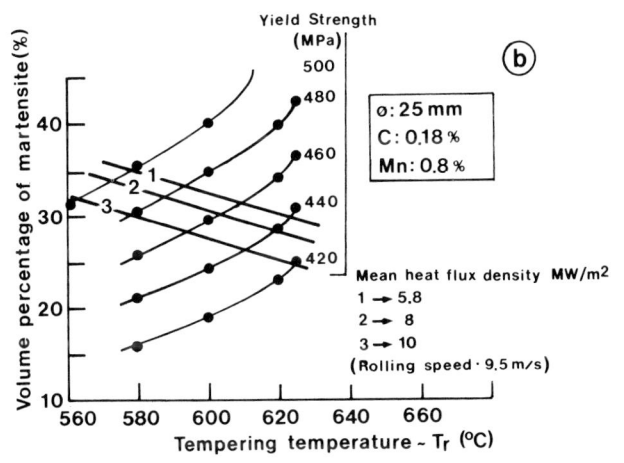

Fig. 9. a - Changes of temperature with time at various points of the cross section of a 20mm rod related to the CCT diagram (finishing temperature 1070°C).
b - Effects of the heat flux density (finishing temperature 1070°C)(Economopoulos, 1975).

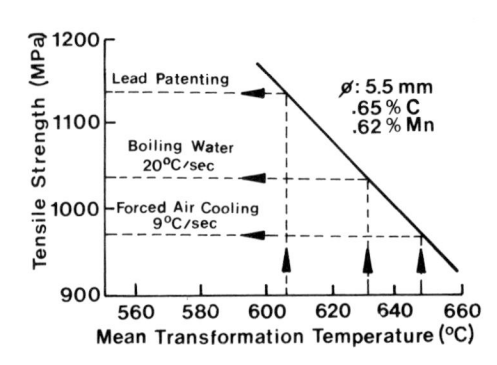

Fig. 10. Relation between the mean transformation temperature and the tensile strength of a wire, (Economopoulos, 1981).

Moreover, industrial use of such cooling has shown that the consistency of properties in the same reel is exceptional (Economopoulos, 1981).

DUAL-PHASE STEELS

So called by Hayami and Furukawa (1975), they form a new class of HSLA steels characterized by a mixed microstructure formed by a soft ferritic matrix in which a small amount (generally less than 10%) of pools is finely dispersed, consisting essentially of martensite, but sometimes also retained austenite and even bainite. The tensile stress-strain curve of a D.P. steel is continuous, the $Y.S./U.T.S.$ ratio being around .5 and the hardening coefficient reaches a high level, e.g. $n \geq .2$.

The automobile industry was the first to be sensibilized by the promises of those steels. The particular combination of their mechanical properties allows the forming of products (wheels, bumpers ...) by application of stresses approximately equal to the ones necessary for the same shaping of a classical steel. The advantage of D.P. steel is its strong hardening after straining, which leads to the use of thinner sheets to meet the same requirements in mechanical properties as well as in security. The final gain then is a lowering in weight of a given piece.

It can be easily understood that the search to produce industrial D.P. grades has been centered on automobile uses : for thin sheets, because of the possibilities of continuous annealing, and for strips in rolled or annealed state.

D.P. steels can be classified into three types :

- the I.A.P.P. steels, intercritically annealed dual-phase steels. This is the case of thin sheets continuously annealed at a temperature above A_1 ;
- the A.R.D.P. steels, austenite rolled dual-phase steels. This is the case of strips rolled in the austenitic temperature range, and for which the dual-phase character is achieved by using an appropriate cooling cycle after rolling;
- the I.R.D.P. steels, intercritically rolled dual-phase steels. Those strip or plates are rolled with one or more finishing passes at a temperature between those of A_3 and A_1 transformation points.

Several symposia have been devoted to research in the field of D.P. steels (Davenport, 1979 ; Kot, 1979, 1981). The knowledge acquired during those studies can be summarized as follows.

		Thickness mm (in)	Anneal. time
Cold-rolled	A	.8-1.05 (.031-.041)	2'30"
	B	1.4 (.055)	2'30"
Hot-rolled	C	2.4 (.094)	4'30"
	D	3.2 (.126)	6'30"
	E	4.0 (.157)	6'30"

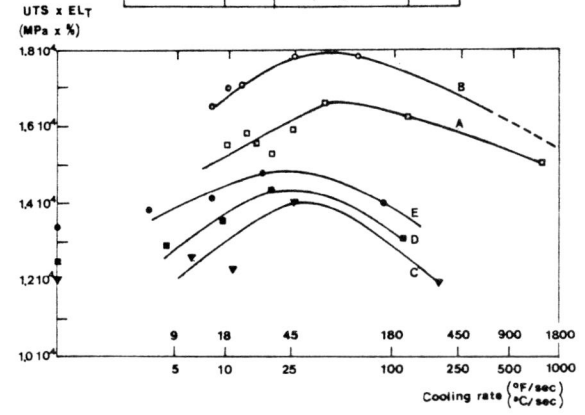

Fig. 11. Influence of the cooling rate on the product tensile strength (UTS)×total elongation (EL$_T$). Intercritical annealing performed at 825°C (1517°F).

The I.A.D.P. are produced using continuous annealing. The kinetics of austenite formation during a reheating for a short time (less than one minute) at a temperature between A$_1$ and A$_3$ has been studied among others by Speich (1979), Garcia (1979) and Thomas (1979).

Several factors influence this formation : the distribution of cementite particles and the deformed or recrystallized state of the ferrite before annealing, the parameters of annealing such as heating rate - the latter being essential to achieve the appropriate microstructure as shown by Messien and co-workers (1981). Increasing the cooling rate raises both the UTS and the parameter (UTS x El) as long as the cooling rate does not exceed 100°C/sec (Fig. 11). Using faster cooling rates, while still increasing UTS, also brings about an important loss of ductility. This behaviour is correlated with the replacement of equiaxed ferrite by bainite. Indeed, regression relationship between cooling rate and chemical composition can be calculated. For instance, for 5% martensitic pools dispersed in a fine ferrite (12-13 ASTM) matrix (Messien 1981) :

$$\log C_R^{5M}(\degree C/sec) = 4.93 - 1.7\,[\%\,\mathrm{Mn}] - 1.34\,[\%\,\mathrm{Si}] - 5.68\,[\%\,\mathrm{C}] \quad (3)$$

$$\sigma = .11 \qquad r = .972$$

Thus, a cooling rate between 10 and 70°C/sec (measured between 800 and 500°C) promotes an optimum dispersion of fine martensitic pools in the ratio of one pool for 2 ferrite grains (ASTM 12). The content in martensite is then between 5 and 10%, and the mean size of the pools between 1 and 2 square microns. Moreover, secondary ferrite, resulting from the transformation of a part of the austenite during cooling, contributes to the grain refinement. Special metallographic etch (Fig. 12) reveals the different phases existing in such a microstructure.

The A.R.D.P., especially coiled strips, are characterized by a dispersion of martensite-bainite pools in a ferritic matrix. An increasing amount of bainite in these pools alters the Y.S./U.T.S. ratio from .5 to .75. The mechanism of formation of this D.P. microstructure is now well understood.

Fig.12. Primary and secondary ferrite and martensite pools. Dual-phase steel intercritically annealed, cooled at 50°C/sec (90°F/sec).

The use of appropriate cooling cycles favours the appearance of a window between the ferritic and bainitic transformations of a C.C.T. diagram. As shown by Coldren (1979), this window indicates which coiling temperature range must be chosen to achieve the desired D.P. microstructure. Experimental determination of the C.C.T. curves by thermal analysis and dilatometry clearly shows more transformation points than usual. These observations enable to draw a double diagram (Fig. 13), which confirms that, after appropriate treatment, the steel is divided into two parts, a carbon-rich and a low-carbon one, behaving separately (Gréday,1979).

Some elements facilitate the achievement of the DP microstructure. It is the case of molybdenum and vanadium when the coiling temperature is low. Manganese($\leq 1.5\%$), chromium ($\leq$ 1 %) and silicon ($\leq$ 1 %) have interesting effects too. The UTS which can be achieved ranges from 400 to 1000 MPa.

Let us conclude with a few words about I.R.D.P. It has been shown by Mathy (1981) that a good set of mechanical properties, tensile and impact strengths, depends on the development in the ferrite, of a sub-structure that impedes the propagation of cleavage by stopping or deviating the microcracks (Fig. 14). This process explains the high level of impact strength. With a substructurated ferrite containing from 20 to 30% of martensite-bainite pools, it is possible to reach UTS from 650 to 900 MPa, a total elongation from 20 to 30% and a transition temperature of -110°C at energy level of 35 J/cm^2.

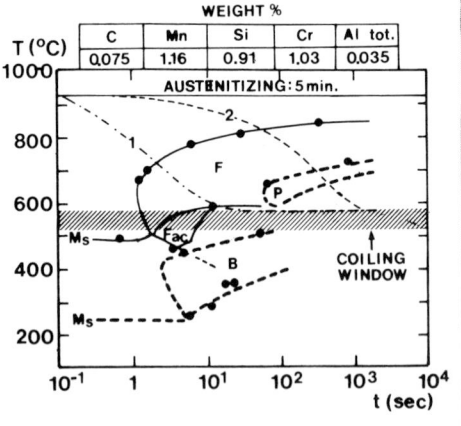

Fig. 13. C.C.T. diagram of a dual-phase steel.

Fig. 14. a - Grain and subgrain boundaries are possible sites for dislocation sources.
b - Deformation bands associated with rupture appeared in a martensite pool.

BAINITIC STRUCTURES

The purposes of having recourse to "out-of-equilibrium" structures can be vastly different. Bainite in very low carbon and low alloyed steels has proved its interest in increasing the yield strength or the toughness. Boron is known as being one of the elements which retains most the ferritic transformation by lowering the start temperature of the bainitic transformation without increasing the martensitic hardenability of the steels (Pickering, 1975), a fact which provides the best guaranties as regards the weldability of the product.

Alloying with boron alone does not seem to be conducive to the desired result. In the case of medium-gauge (12mm) low-carbon steel plates, a homogeneous bainitic structure can be obtained, without making use of drastic cooling, merely by alloying with boron and niobium (Lamberigts, 1976).

This combined alloying leads to yield strength in the range 400-450 MPa in the case of 0.07% C steels but the fracture toughness depends very much on the structural fineness, itself dependent on the conditions of hot-rolling (Fig. 15).

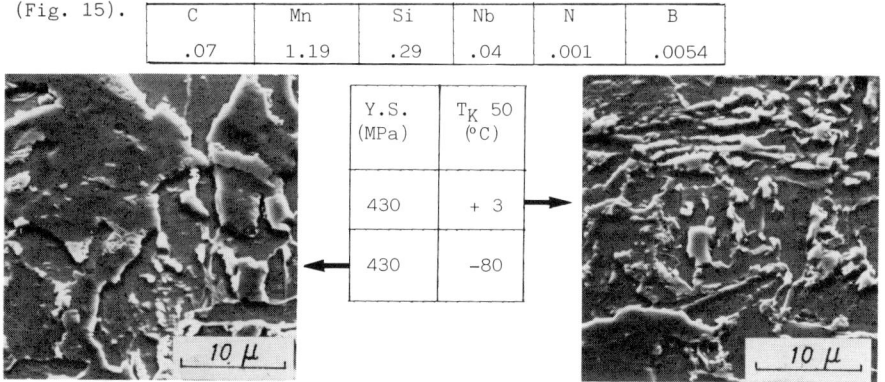

C	Mn	Si	Nb	N	B
.07	1.19	.29	.04	.001	.0054

Y.S. (MPa)	T_K 50 (°C)
430	+ 3
430	-80

Fig. 15. Influence of fineness of the bainitic microstructure on the impact transition temperature.

CONCLUDING REMARKS

Both producers and users of metals and alloys know about the shortages of raw materials. The end of this century at least, will be governed by a much greater efficiency than before in using metals and alloys. This trend has more and more appeared in the research performed in Physical Metallurgy, over the ten last years.

Even if completely new alloys will emerge within the next twenty years, steel remains the most competitive material to face increasing prices of production by property improvement, volume reduction and better reproducibility. The various examples presented here show clearly the routes which have been chosen by the steel industry to fulfil the above mentioned requirements for carbon and low alloy structural steels.

ACKNOWLEDGEMENTS

The author is indebted to the Institut pour la Recherche dans l'Industrie et l'Agriculture (I.R.S.I.A.) Brussels, and to the Commission of the European Communites (CEC) for their financial support to the research performed at CRM in most of the topics examinated in this presentation.

REFERENCES

Coldren, A.P., Tither, G., Cornford, H. and Hiam, J.R. (1979). Proceedings on Formable HSLA and Dual-Phase Steels. TMS-AIME. New York, 207-228.
Davenport, A.T. (1979). Editor. Proceedings on Formable HSLA and Dual-Phase Steels. TMS-AIME. New York
Duckworth, W.E. (1966), J. of Metals, 915-922
Economopoulos, M., Respen, Y., Lessel, G. and Steffes, G. (1975) C.R.M. Reports. 45-1-17.
Economopoulos, M. and Lambert, N. (1981). Wire Journal. 90-95.
Garcia, C.I. and De Ardo, A.J. Editors. Proceedings on Structure and Properties of Dual-Phase Steels. TMS-AIME. New-York, 40-61
Gréday, T., Mathy, H. and Messien, P. (1979) Editors. Proceedings on Structure and Properties of Dual-Phase Steels. TMS-AIME. New-York, 260-280.
Hayami, S. and Furukawa, T. (1975) Microalloying 75. M. Korchynsky. Editor. Union Carbide. New York.
Jonas, J.J. (1976). Proceedings 4th Int. Conf. on Strength of Metals and Alloys. Nancy. 976-1002.
Jonas, J.J. and Weiss, I. (1979). Metals Sci. 13, 238-245.
Kot, R.A. and Morris, J.W. (1979). Editors. Proceedings on Structure and Properties of Dual-Phase Steels. TMS-AIME. New-York.
Kot, R.A. and Bramfitt, B.L. (1981). Editors. Proceedings on Fundamentals of Dual-Phase Steels. TMS-AIME. New-York.
Kozasu, I. and Shimizu, T. (1971) Trans. I.S.I.J., 359-366.
Lamberigts, M. and Gréday T. (1974). C.R.M. Reports, 24-38.
Lamberigts, M. and Gréday T. (1976). C.R.M. Reports, 49, 24-35.
Lamberigts, M. (1977). Steel Research Reports Eur. 5828, Com. Europ. Communities Brussels.
Le Bon, A., Rofès-Vernis, J. and Rossard, C. (1973). Mém.Sc.Rev. Met., LXX. 7-8, 577-588.
Luton, M.J., Dorvel, R. and Petkovic, R.A. (1980) Metall. Trans., 11 A, 411-420.
Mathy, H., Gouzou, J. and Gréday, T. (1981) Proceedings on Fundamentals of Dual-Phase Steels TMS-AIME. New-York, 403-426.
Mathy, H., (1982) Internal Report, C.R.M. in press.
Messien, P., Herman, J-C., Gréday, T. (1981) Proceedings on Fundamentals of Dual-Phase Steels TMS-AIME. New-York, 161-180.
Ouchi, C. Okita, T. Ichihara, T. and Ueno, Y. (1980). Trans. ISIJ 20,833-841.
Pickering F.B. (1975). Microalloying 75. M. Korchynski, Editor. Union Carbide. New York.
Rossard, C. (1973). Proceedings 3rd Int. Conf. on Strength of Metals and Alloys. Cambridge. 175-203.
Saito, Y., Koshizuka, N., Skiga, C. Sekine, T., Yoshizato, T. and Enami, T. (1980). Proceedings of Int. Conf. Steel Rolling I.S.I.J. Tokyo.
Sakai, T., Akben, M.G. and Jonas, J.J. (1982). The Thermomecanical Processing of Microalloyed Austenite. A.J. De Ardo and P.J. Wray, editors. TMS-AIME(Warrendale, P.A.).
Sanomiya, A., Takakashi, E. and Shimazu, S. (1981) Kobe Steel Tech. Bul. 1023.
Speich, G.R. and Miller, R.L. (1979). Proceedings on Structure and Properties of Dual-Phase Steels. TMS-AIME. New-York, 146-182
Takeo, K., Maeda, K., Kamise, T., Iwata, H., Sutoni, Y and Nakuta, H.(1975) Trans. I.S.I.J. 15, 422-428.
Thomas, G. and Koo, J-Y. (1979) Proceedings on Structure and Properties of Dual-Phase Steels. TMS-AIME. New-York, 183-201.
Weiss, I., Fitzsimmons, G., and De Ardo, A.J. (1982). The Thermomechanical Processing of Microalloyed Austenite. A.J. De Ardo and P.J. Wray, editors. TMS-AIME (Warrendale, PA).
Wilson, N.A. (1974) U.S. Patent 516.767.

Ductile Fracture

J. D. Embury

Department of Metallurgy and Materials Science, McMaster University, Hamilton, Ontario, Canada

ABSTRACT

This review deals with the succession of processes of void nucleation growth and linkage which lead to ductile fracture. The experimental evidence is related to simple theoretical models which permit the process of damage accumulation which constitutes ductile fracture to be considered for a variety of stress states. In order to indicate the competition between ductile fracture and other failure modes a series of failure maps for axisymmetric deformation are presented. In the final section of the paper the process of damage accumulation in some forming processes is considered. The review emphasizes the combined influence of microstructural parameters and stress state needed to give a comprehensive description of ductile fracture.

KEYWORDS

Ductility; void nucleation; void growth; stress state; failure maps.

INTRODUCTION

The process of ductile fracture is of importance both from the fundamental viewpoint of how microstructural damage, in the form of voids, is initiated and grows during plastic flow and from the more practical viewpoint that ductile fracture may terminate the useful strain which can be applied in a variety of metal forming operations. Clearly, inherent in the ductile fracture process are phenomena which can be considered on a continuum basis such that the scale of events is of no consequence, but the process, e.g. void growth, can be modelled in terms of the operative rate of strain and the stress state (McClintock, 1968; Thomason, 1968). However, as in most fracture processes there are other events which are intrinsically heterogeneous and depend on local microstructural conditions. These processes are not readily modelled by a continuum method but must be subject to extensive experimental study in order to permit the controlling variables to be delineated.

All ductile fracture processes whether high temperatures or low temperature processes can be considered as the end results of a sequence of three processes involving voids, viz a) nucleation, b) growth, and c) linkage. It is imperative to note that all three of these stages is in competition with both general plastic and localized plastic flow events. Thus, depending on the flow stress level, the stress state and the microscopic criteria which govern the alternative flow and fracture modes, ductile fracture may be effectively controlled by any of the three stages in the sequence. Ductile fracture can thus be considered as involving a process of damage accumulation due to void growth which accompanies the evolution of the total strain path. The approach taken in the present article will be to first describe the processes leading to damage accumulation by emphasizing both the microstructural and stress state factors which influence void nucleation and growth. In order to generalize the process, failure maps will be developed which enable the competition of ductile fracture and other processes to be viewed in a more generalized framework. In the final section consideration will be given to a broad range of strain histories in which ductile failure may represent the limiting processes.

NUCLEATION OF VOIDS

In general void nucleation occurs at second phase particles and a variety of criteria (Argon, 1972; Goods and Brown, 1979) have been considered in order to describe the nucleation condition. Although void nucleation can be considered in terms of the exchange of elastic energy in the particle or its surrounding plastic field and the creation of new surfaces at the void, the simple energy condition is not usually sufficient because a mechanism for rupturing the particle matrix interface is also a necessary condition for void nucleation. In order to provide a macroscopic criteria for void nucleation a number of authors have reported the imposed strains at which nucleation occurs. Thus Gould and Humphreys (1973) report strain values of 0.1 to 0.2 for Al_2O_3 in Cu and 0.2 to 0.43 for BeO in Cu whereas a variety of workers (Inoue and Kinoshita, 1976; Brown and Embury, 1973) using the $Fe-Fe_3C$ system report values of the order of 0.4 to 0.6.

This data can be rationalized using the model of plastic relaxation at second phase particles proposed by Brown and Stobbs (1971) which predicts a weak size dependence for void nucleation. Although it is convenient to consider an average macroscopic strain at which nucleation occurs, the nucleation event depends on local conditions. These may include the local spacing of particles or whether they are located at grain boundaries (Iricibar, 1980). The use of an average macroscopic strain is not strictly justifiable because as discussed later nucleation is really a process which spreads through the particle distribution over a range of plastic strain. For simplicity we can consider two possible extremes for the nucleation, a) the situation where the nucleation strain reflects the average volume fraction of particles and decreases slowly with average volume fraction as shown in Fig. 1; b) a condition where nucleation is dominated by the extreme value of the particle or inclusion distribution. An example of this is in through thickness ductility of rolled products where local aggregates of inclusions may severely reduce the ductility as illustrated by the fracture surface in Fig. 2.

In order to model the nucleation condition let us consider that a critical normal stress σ_c must be exceeded at the particle matrix interface. If we include the action of the hydrostatic stress σ_m (where $\sigma_m = 1/3\ \sigma_{ii} - p$), we can express the local stress for nucleation σ_{loc} as

$$\sigma_{loc} + \sigma_m = \sigma_c \tag{1}$$

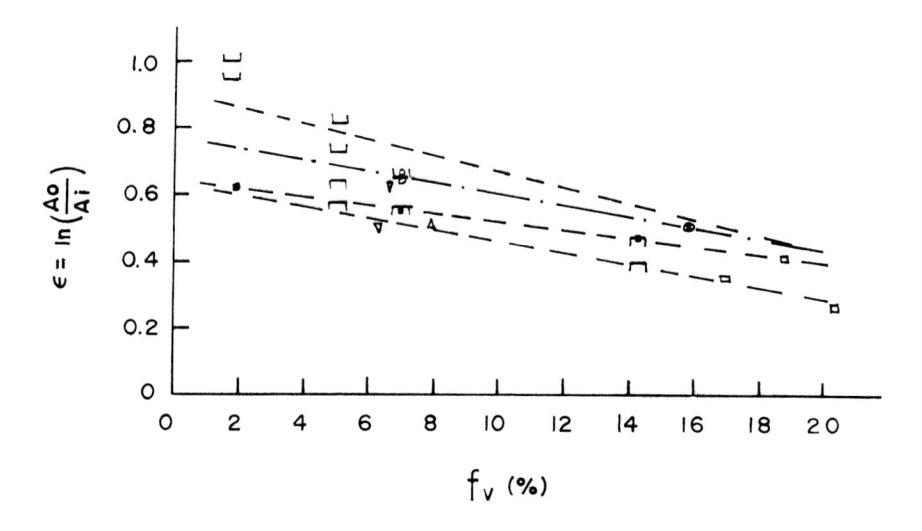

Fig. 1. A diagram indicating the experimental evidence for the void nucleation strain as a function of volume fraction in the system Fe-Fe$_3$C.

The value of σ_{loc} is related to the local dislocation density ρ_{loc} which in a simplified model can be assumed to increase linearly with strain (Brown and Stobbs, 1971). Thus if $\sigma_{loc} \propto \sqrt{\rho_{loc}}$ the nucleation strain ε_N can be expressed as

$$\varepsilon_N^{\frac{1}{2}} = \frac{1}{H_2}(\sigma_c - \sigma_m) \qquad (2)$$

where H_2 is a constant. The data in Fig. 3 indicates that the experimental data is in accord with equation (2) and that the value of σ_c (estimated at $\varepsilon_N = 0$) is of the order of 1200 MPa (E/150). The practical importance of this approach is also indicated in Fig. 3. If the interface between the matrix and the second phase particles which govern the ductility is weak the line representing nucleation will slide toward the origin. However, for an idealized dispersion if the interface is strong and the dispersion of the second phase is refined to promote local recovery rather than the build up of σ_{loc} which promotes decohesion, the nucleation condition will be delayed to higher strains. This may be an attainable condition in techniques such as rapid solidification processing or other techniques where finer dispersions of inclusion can be produced.

In materials which contain a high volume fraction of inclusions or second phase particles, the nucleation strain may represent a major portion of the

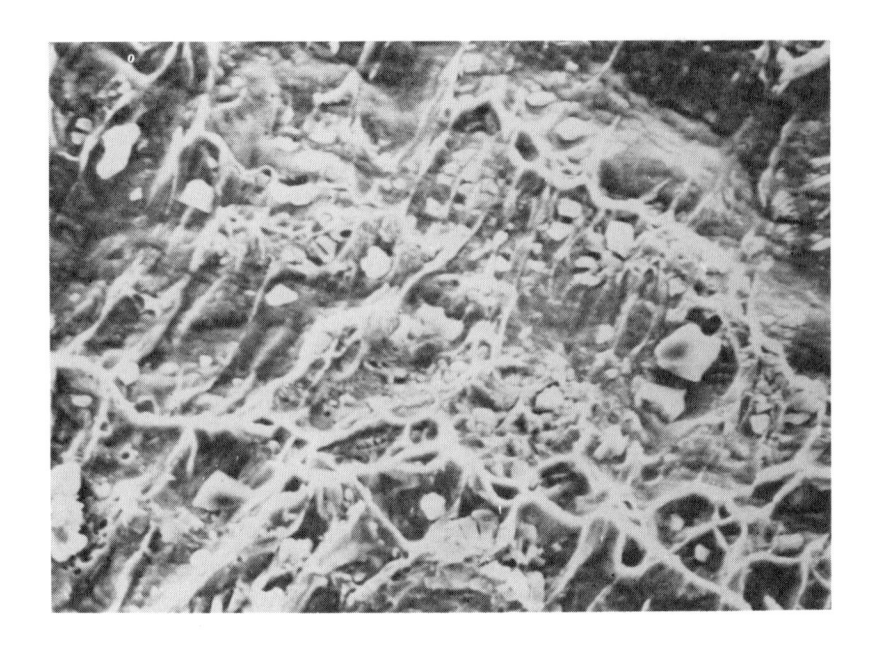

Fig. 2. A scanning electron micrograph showing the fracture surface of a low ductility sample containing aggregates of calcium aluminates.

total ductility. Under these conditions the nucleation itself can be viewed not as a unique event but as a cumulative damage event in which the process of decohesion spreads through the existing particle dispersion. The recent work of Brownrigg and Spitzig (1981) indicates that with superimposed pressure the number of particles which exhibit voiding increases more slowly as shown in Fig. 4. This dependence on imposed pressure suggests that the nucleation can be considered as a continuous function of strain with a nucleation front moving through the particle distribution at a rate dependent on the applied hydrostatic pressure as shown schematically in Fig. 5.

From a practical viewpoint we can consider that for some materials with high volume fractions of second phase particles the ductility can be controlled via the nucleation event either by adjusting the strain path so that the hydrostatic component is compressive or controlling the particle size distribution so that the passage of the nucleation front is distributed over a wide range of strain.

It is also pertinent to consider the inverse of this argument, namely that under conditions of high hydrostatic tension the progress of the nucleation may be accelerated and ductility will be drastically reduced.

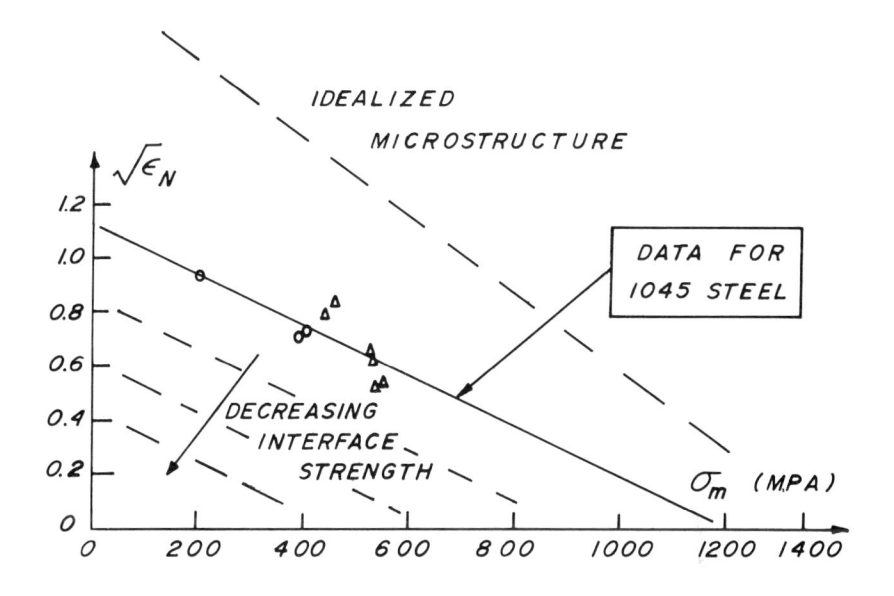

Fig. 3. A diagram indicating the relationship between nucleation ε_N, strain and hydrostatic pressure σ_m. The data points are taken from LeRoy et al. (1981) and the dotted lines represent hypothetical conditions.

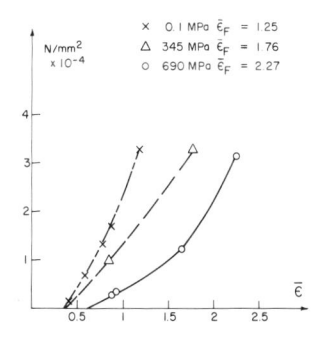

Fig. 4. A diagram showing the development of void density (number of voids per mm^2) as a function of effective strain for various superimposed pressures. Data for 1045 steel taken from Brownrigg (1981).

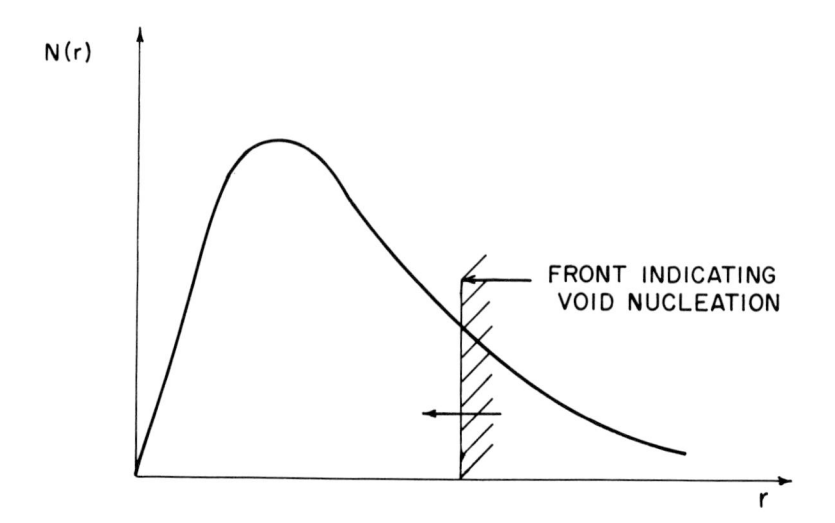

Fig. 5. A schematic diagram indicating the progress of a nucleation front through a hypothetical particle distribution. The nucleation front moves to the left with increasing strain. At high imposed hydrostatic pressure the front moves more slowly in accord with the data of Fig. 4.

VOID GROWTH

The process of void growth has been considered by a variety of authors using continuum models. In essence the problem is a geometric one dealing with the rate of change of the radii of an isolated void in an imposed strain rate field and generalizing this to consider the allowable strain in terms of the inter-particle spacing. Hence, the dominant microstructure feature which appears in the description of growth is the volume fraction of the second phase.

In the elegant solution developed by Rice and Tracey (1969) the void radii in the principal directions R_1 and R_3 can be related to their rates of increase by the equations

$$\dot{R}_1 = R_1 \left(\frac{-\gamma_a}{2} \dot{\varepsilon}_3^\infty + 0.56 \, \dot{\varepsilon}_3^\infty \sinh \left(\frac{3\sigma_m}{2y} \right) \right)$$

$$\dot{R}_3 = R_3 \left(\gamma_a \dot{\varepsilon}_3^\infty + 0.56 \, \dot{\varepsilon}_3^\infty \sinh \left(\frac{3\sigma_m}{2y} \right) \right)$$

(4)

where y is the tensile flow stress, σ_m the mean stress, $\dot{\varepsilon}_3^\infty$ the strain rate remote from the void and γ_a is a factor which describes the amplification of the growth rate of the void relative to the strain in the matrix. For the tensile case it can be shown that

$$\gamma_a = \frac{2 \exp \frac{3}{2} \bar{\varepsilon}}{2 \exp \frac{3}{2} \bar{\varepsilon} - 1} \tag{5}$$

This formulation can be developed to provide a model for void growth or damage accumulation in the tensile test where the ratio $3/2\ \sigma_m/y$ systematically increases from its value of ½ in the region of uniform tensile strain to much higher values as the neck develops. The agreement between theory and experiment is shown for various spherodized steels in Fig. 6.

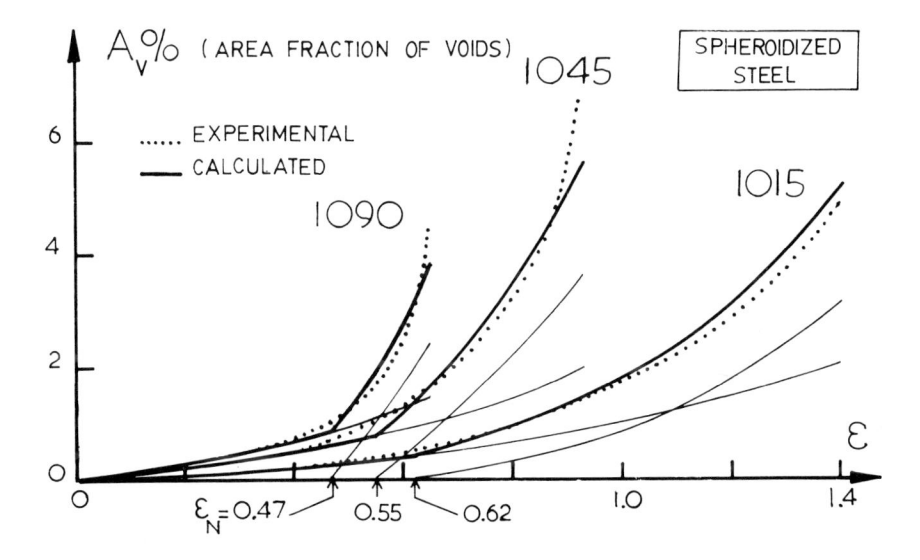

Fig. 6. A comparison of experimental data and the theoretical model (taken from LeRoy and coworkers (1981). The model assumes two void populations, one of which is present at zero strain and one which arises at the values of ε_N indicated on the ε axis.

Clearly the void growth process is dependent on stress state and a table of the relationships between maximum tensile stress, shear stress and mean stress for various simple stress states is given below.

Stress State	Maximum tensile stress	Shear stress	Mean stress
Uniaxial tension	1	$1/\sqrt{3}$	$-1/3$
Biaxial tension	1	$1/\sqrt{3}$	$-2/3$
Uniaxial compression	0	$1/\sqrt{3}$	$+1/3$
Plane strain compression	0	$1/2$	$+2/3$

The classical method of increasing tensile ductility is to apply a large superimposed pressure (Bridgeman, 1952; Brandes, 1970). The results of such experiments can be rationalized in a quantitative manner using the void growth model. The treatment of void growth due to LeRoy et al. (1981) which is based on the formulation of Rice and Tracey (1969) expresses the growth of the void radius in the tensile direction as

$$\dot{R}_3 = R_0 \left(\exp D\bar{\varepsilon}(2 \exp \tfrac{3}{2} \bar{\varepsilon} - 1)^{2/3} \right) \qquad (6)$$

where R_0 is the initial void radius and D is $0.56 \sinh(\tfrac{3}{2} \tfrac{\sigma_m}{\bar{\sigma}})$. If it is assumed that a simple geometric condition limits void growth such that at fracture R is of the order of the mean interparticle spacing and the sinh term is linearized, the ratio of the strain to failure under pressure $\varepsilon_f(P)$ to that recorded in uniaxial tension $\varepsilon_f(0)$ can be expressed as

$$\frac{\varepsilon_f(P)}{\varepsilon_f(0)} = 1 + \frac{1.68\ P}{2\ \bar{\sigma}} \qquad (7)$$

Hence the ductility is predicted to increase linearly with pressure which is in accord with the data productd by a variety of authors and summarized in Fig. 7.

Equation 7 assumes that the dominant influence of pressure is on void growth. However, at large strains both damage accumulation and a decrease in strain hardening occur and other failure modes such as catastrophic shear may become dominant at high pressures (Yajima, 1970; French, 1973). Superimposed hydrostatic pressure can influence a variety of failure modes in addition to those dominated by void growth. In aluminum alloys, in addition to ductile fracture, samples may exhibit shear failure prior to necking or intergranular fracture in uniaxial tensile tests. However, both these modes also involve damage accumulation between the nucleation event and the final failure condition. Superimposed hydrostatic pressure may either suppress the nucleation event or redistribute the damage accumulation. The processes are illustrated in Figs. 8 and 9.

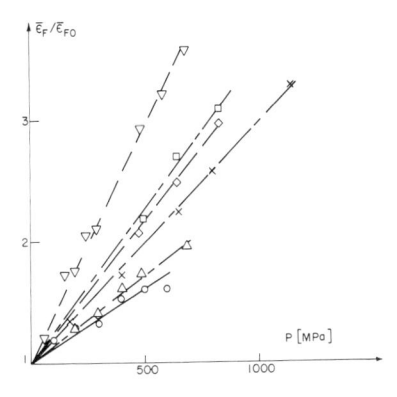

Fig. 7. A diagram indicating the work of previous authors on the ratio of ductility under pressure compared with that in the uniaxial test as a function of imposed pressure.
+ Davidson and Ansell (1968) Fe-0.4% C
0 French and Weinrich (1974) Fe-0.5% C
  Yajima et al. (1970)

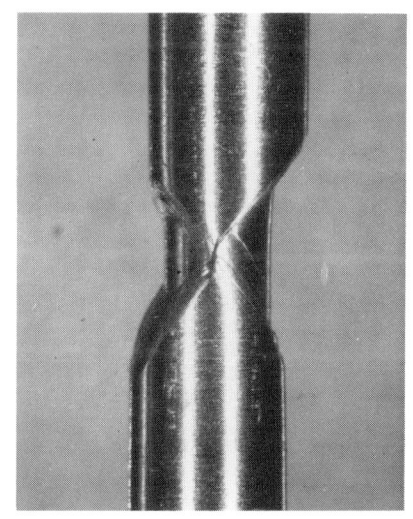

Fig. 8. Micrographs illustrating the behaviour of a 7075 aluminum alloy tested in the quenched condition. The sample on the left failure is uniaxial tension by shear prior to macroscopic necking. The sample on the right was tested in tensile with a superimposed pressure of 690 MPa resulting in redistribution of the shear events.

Fig. 9. Micrographs illustrating the fracture behaviour of an Al-3.6% Cu alloy quenched and aged at 200°C. The micrograph on the left indicates that in uniaxial tension failure involves the nucleation of grain boundary cracks and their linkage to give final failure. In a tensile test conducted with superimposed pressure of 690 MPa the nucleation of grain boundary fracture is supressed and ductile rupture becomes the dominant fracture mode.

FAILURE MAPS

The preceding sections indicate that both the void nucleation and growth events inherent in ductile fracture are dependent on stress state and that both events should be considered in relation to competitive processes of plastic flow and failure which may occur. In order to represent this type of information it is often useful to devise some form of diagram or map of the possible failure processes. In sheet metal forming such maps are commonly used to represent both limit strains and failure processes (Keeler, 1971; Embury, 1981) using a co-ordinate system of the measurements in plane strains. Other types of failure maps have been proposed (Ashby, 1979) which utilized axes of homologous temperature and stress normalized with respect to modulus to divide failure into regimes indicating the dominant failure mode. However, for ductile fracture and the events which are competitive with this process it is useful to develop failure maps which are indicative of loading history and the role of the mean stress.

One method of achieving such a representation is to consider axi-symmetric deformation ($\sigma_1 = \sigma_2 \neq \sigma_3$). The concepts of void nucleation and growth have been considered analytically for this stress system by LeRoy and co-workers (1981) and the results are shown in Fig. 10.

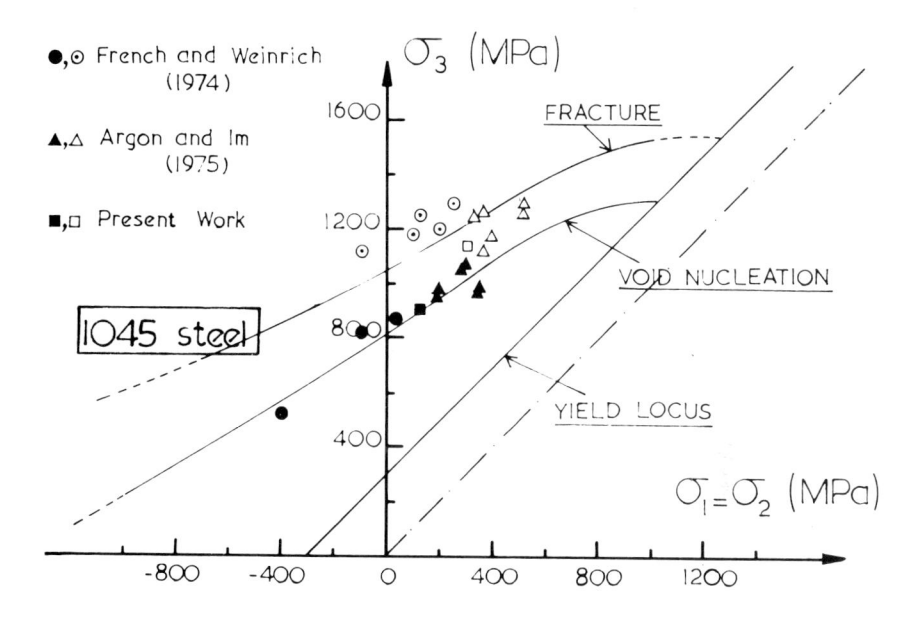

Fig. 10. A diagram showing the loci for void nucleation and fracture strain generated for axi-symmetric deformation.

In this diagram the strains for nucleation and failure are represented in stress space by the separation from the traces of the yield surface which are to a first approximation parallel to the hydrostatic stress line $\sigma_1 = \sigma_2 = \sigma_3$. The form of the loci in Fig. 10 indicate some important physical features of the ductile fracture process. At large hydrostatic tensions the void nucleation strain reduces to zero and void growth becomes catastrophic. However, at large superimposed pressure the nucleation and failure lines diverge from the trace of the yield surface indicating enhanced ductility.

This type of failure map has been extended by the author and Professor M.F. Ashby(1982) to represent other failure modes such as brittle failure and localized shear failure. For simplicity of representation the brittle failure can be represented by the criterion due to Murrell (1963),

$$(\sigma_1-\sigma_2)^2 + (\sigma_2-\sigma_3)^2 + (\sigma_3-\sigma_1)^2 = 24\sigma_f(\sigma_1+\sigma_2+\sigma_3)$$

where σ_f is the fracture stress in simple tension. The shear failure can be represented by a line parallel to the trace of the yield surface. It should be recognized that the localized shear condition is more complex and has recently been treated in a definitive manner by other authors (Pierce, 1982; Anand, 1982) but a diagramatic representation of these models is difficult. In addition, the line representing brittle fracture is an over-simplification because as opined in an earlier section the nucleation events and the final failure event are separated by a period of distributed microcracking which is pressure dependent.

Despite these simplifications the form of the map can be shown as in Fig. 11.

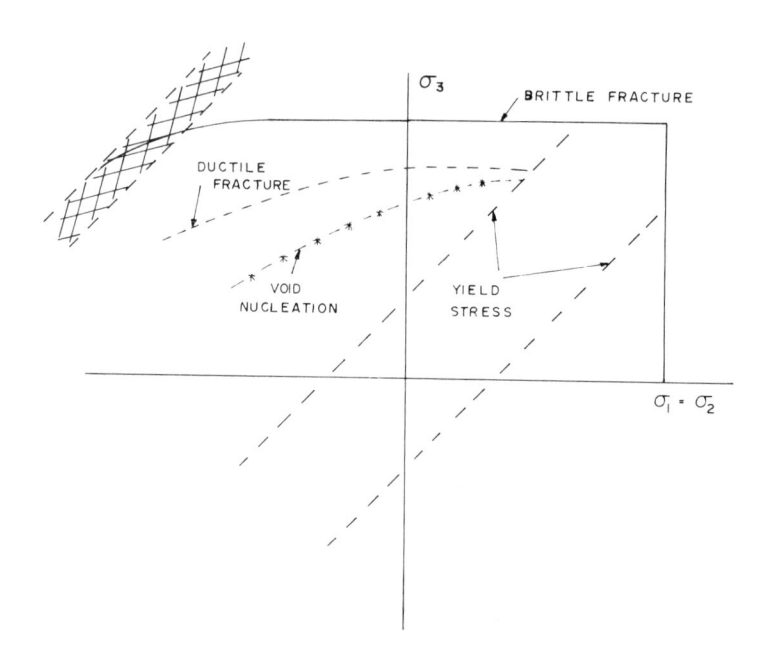

Fig. 11. A diagramatic representation of a failure map containing various failure modes relative to the trace of the yield surface.

The essential feature to consider in considering the competition between failure modes is that the processes inherent in ductile fracture, viz. void nucleation and growth, are achieved by plastic strain. Thus as the yield stress increased the loci for these processes are moved outward in axi-symmetric stress space. Thus as the yield stress increases, ductile failure may be suppressed in favour of other modes. Similarly if we consider the role of increasing applied pressure the ductile failure may be replaced by shear failure at high imposed pressures or modes such as brittle fracture (of which intergranular fracture is a representative) may be suppressed. Diagrams of the form shown in Fig. 11 may be used to represent the influence of temperature, changes in yield stress level due to precipitation hardening or changes in grain size on the fracture modes likely to be achieved for a variety of simple loading paths.

CUMULATIVE DAMAGE IN FORMING OPERATIONS

The treatment of failure maps outlined above indicates the manner in which competitive fracture modes can be considered. However, in many forming processes not only the loading path but the total deformation history must be considered. A variety of previous authors including Oyane (1972) and Osakada (1978) have considered models of this type. Compression tests represent an example where analysis of the strains on the free surface enables the history of the stress components to be derived using the Levy-Mises flow rule. By varying the geometry of the test sample or the frictional conditions a variety of loading paths from homogeneous compression toward plane strain can be achieved. A typical compressional failure and analysis of the deformation history from the surface strains are shown in Fig. 12. Other authors (Kivivuori, 1978) have shown that both tensile and shear fracture mode may be involved in terminating the upsetting test. Recent work (Lahti, 1982) indicates that the critical microstructural parameter involved is the local volume fraction of inclusion at the surface and that the influence of volume fraction of inclusions is greater for loading paths closer to plane strain.

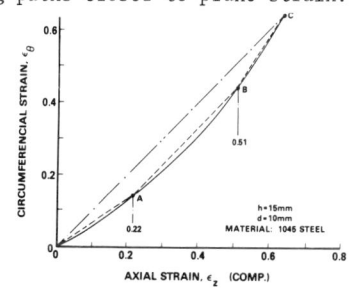

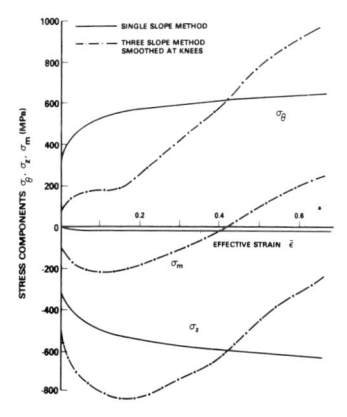

Fig. 12. A composite diagram of an upsetting test showing in the upper diagram the piece wise approximations to the measured strain path and in the lower diagram the variations of the stress components with $\bar{\varepsilon}$ (taken from Chandrasekaran et al. (1982)).

CONCLUDING REMARKS

The description of ductile fracture advanced in the preceding section indicates that the basic description of ductile fracture as a cumulative damage process is sufficient to give reasonable agreement between theory and experiment for a variety of stress states. However, there remain a number of important features which merit further work. In general, descriptions of the nucleation and growth processes are based on average microstructural parameters such as size or volume fraction of second phase particles. It is possible that both the nucleation and the instability which leads to final coalescence are governed by extreme values of the distribution such as the highest local volume fraction. Similarly, the models for void growth are continuum models which do not account for the role of hardening or rate sensitivity. There are clearly examples, such as superplastic materials (Taplin, 1979), where damage accumulation is not limited in the simple geometric manner considered in this review, but is distributed in a manner which reflects the more complex constitutive laws needed to describe the flow behaviour of the matrix between the voids. Finally in attempting to relate ductile fracture to the limit of forming operations a more sophisticated form of damage accumulation map is needed so that the simulation effects of successive increments of imposed strain can be considered in terms of the macroscopic strain distribution, the resultant flow stress and the distribution of damage accumulation in the matrix. These areas represent potential areas of fruitful co-operative research for physical metallurgy and continuum mechanics.

ACKNOWLEDGEMENT

The author wishes to acknowledge the role of the valuable discussions with M.F. Ashby, G. LeRoy, D. Teirlinck, R. Sowerby and O. Richmond during the preparation of this article. The section entitled Failure Maps is the subject of a more comprehensive joint study with Professor M.F. Ashby. Research support received from NSERC and CANMET (Canada) is gratefully acknowledged.

REFERENCES

Anand, L., and W.A. Spitzig (1982). Acta Met., 30, 553.
Argon, A.S., Im, J. and Safogulur (1975). Met. Trans., A6, 825.
Ashby, M.F., Ghandi, C., and D.M.R. Taplin (1979). Acta Met. 27, 699.
Ashby, M.F. and J.D. Embury (1982). Report on Yield and Fracture Surfaces for Brittle Solids, Cambridge University.
Brandes, M. (1970). "Mechanical Behaviour of Materials under Pressure", edited by H.H.D. Pugh, published by Elsevier.
Bridgeman, P.W., (1952). "Studies in Large Plastic Flow and Fracture", McGraw-Hill.
Brown, L.M. and W.M. Stobbs (1971). Phil. Mag. 23, 1201.
Brown, L.M. and J.D. Embury (1973). Proc. 3rd Int. Conf. on Strength of Metals and Alloys, Cambridge, U.K., p. 164.
Brownrigg, A. and W. Spitzig (1981). Private communication.
Chandrasekaran, N., Sowerby, R., Duncan, J.L. and J.D. Embury (1982). Scripta Met. (in press).
Davidson, T.E. and G.S. Ansell (1968). Trans. Am. Soc. Metals, 61, 242.
Embury, J.D. and J.L. Duncan (1981). Ann. Rev. Mat. Sci., 11, 505.
French, I.E. and P. Weinrich (1973). Acta Met., 21, 1533.
French, I.E. and P. Weinrich (1974). Scripta Met., 8, 87.
Goods, S. and L.M. Brown (1979). Acta Met., 27, 1.

Gould, D. and F.J. Humphrey (1973). Microstructure and Design of Alloys, paper 61, published by Inst. of Metals.
Inoue, T. and S. Kinoshita (1976). Tetsu-to-Hogane I.S.I. Japan, 62, 875.
Iricibar, R., LeRoy, G. and J.D. Embury (1980). Metal Sc. J., 14, 337.
Keeler, S.P. (1971). Sheet Metal. Ind., 48, 511.
Kivivuori, S. and M. Sulonen (1978). Annual C.I.R.P., 27, 141.
Lahti, I., and M. Sulonen (1982). Scand. J. Met., 11, 1.
LeRoy, G., Embury, J.D., Edward, G. and M.F. Ashby (1981). Acta Met., 29, 1509.
McClintock, F.A. (1968). "Ductility", published by ASM, p. 255.
Murrell, S.A.F. (1963). "Rock Mechanics", Edited by Fairhurst, C., p. 563, published by Pergamon Press.
Osakada, K., Watadani, A. and H. Sekiguchi (1978). Bull. JSME, 21, 1236.
Oyane, M. (1972). Bull. JSME, 15, 1507.
Pierce, D., Asaro, R.J. and A. Needleman (1982). Acta Met., 30, 1087.
Rice, J.R. and D.M. Tracey (1969). J. Mech. Phys. Solids, 17, 201.
Taplin, D.M.R., Dunlop, G.L. and T.G.Langdon (1979). Ann.Rev. Mat. Sci., 9, 151.
Thomason, P.F. (1968). J. Inst. Metals, 98, 360.
Yajima, M., M. Ishii and M. Kobayash (1970). Int. J. of Fracture Mech., 6, 139.

Deformation at High Temperatures

Terence G. Langdon

*Departments of Materials Science and Mechanical Engineering,
University of Southern California, Los Angeles, California 90089-1453, USA*

ABSTRACT

The steady-state creep of pure metals and solid solution alloys is reviewed, and it is shown that there are both similarities and important differences. The behavior is discussed with reference to the dependence on stress, the substructural characteristics, and the shape of the transient in stress change experiments.

KEYWORDS

Creep, solid solution alloys, climb, viscous glide, subgrains.

INTRODUCTION

The deformation of metals at high temperatures, typically at or above about 0.5 T_m where T_m is the absolute melting point, has received considerable attention over a period of many years. It is now known that, on application of an external load, polycrystalline metals often exhibit an instantaneous strain, a primary stage of creep in which the strain rate changes with strain, a lengthy region of steady-state flow where the strain rate remains essentially constant, and then a tertiary stage of creep leading to final fracture.

Creep at high temperatures is a diffusion-controlled process, and the steady-state strain rate, $\dot{\varepsilon}$, is generally expressed by a relationship of the form

$$\dot{\varepsilon} = \frac{AGb}{kT} \left(\frac{b}{d}\right)^p \left(\frac{\sigma}{G}\right)^n D_o \exp(-Q/RT) \tag{1}$$

where G is the shear modulus, b is the Burgers vector, k is Boltzmann's constant, T is the absolute temperature, d is the grain size, σ is the applied stress, D_o is a frequency factor, Q is the activation energy for the deformation process, R is the gas constant, p and n are constants, and A is a dimensionless constant which incorporates all of the microstructural features other than the grain size. Inspection of equation (1) shows that the significant parameters in high temperature creep are p, n and Q; and the value of p is particularly important because it permits a clear distinction between *boundary mechanisms* which depend on the presence of grain boundaries so that $p \neq 0$ and *lattice mechanisms* which take place entirely within the grains so that p = 0 (Langdon, 1975).

This paper is concerned with deformation at high temperatures. To keep the paper within a reasonable length, attention will be restricted to the creep behavior of pure metals and solid solution alloys under steady-state conditions.

DEFORMATION OF PURE METALS

Several mechanisms usually operate during the creep of metallic materials, and the measured creep rate depends on the interaction between these different processes. In general, the various mechanisms tend to operate independently in pure metals, so that the fastest process is rate-controlling and the stress exponent, n, increases through a series of discrete values as the stress level is increased.

Table 1 lists the major mechanisms occurring during the steady-state creep of pure metals and solid solution alloys, sub-divided into three distinct stress regions: the activation energies are given by Q_ℓ for lattice self-diffusion, Q_{gb} for grain boundary diffusion, Q_p for pipe diffusion and Q_i for the interdiffusion of solute atoms. At low stresses, the behavior is Newtonian viscous with n = 1, and this may arise from diffusion creep either through the lattice (Herring, 1950; Nabarro, 1948) or along the grain boundaries (Coble, 1963) or from Harper-Dorn creep (Harper and Dorn, 1957). As indicated in Table 1, the diffusion creep processes depend on the presence of grain boundaries so that $p \neq 0$ but Harper-Dorn is intragranular and p = 0. At intermediate stresses, pure metals exhibit class M (Metal type) behavior due to a recovery process such as dislocation climb; in practice, climb is achieved by vacancy diffusion either through the lattice at high temperatures (H.T.), giving $Q = Q_\ell$ and $n \simeq 5$, or along the cores of the dislocations at low temperatures (L.T.), giving $Q = Q_p$ ($\simeq 0.6\ Q_\ell$) and an increase in the stress exponent to $(n + 2) \simeq 7$

TABLE 1 Creep Mechanisms in Pure Metals and Solid Solution Alloys

Stress region	Stress dependence		p	Q	Interpretation
Low:					
	$\dot{\varepsilon} \propto \sigma^1$		2	Q_ℓ	Nabarro-Herring creep
	$\dot{\varepsilon} \propto \sigma^1$		3	Q_{gb}	Coble creep
	$\dot{\varepsilon} \propto \sigma^1$		0	Q_ℓ	Harper-Dorn creep
Intermediate:					
†Class M:	$\dot{\varepsilon} \propto \sigma^n$	$(n \simeq 5)$	0	Q_ℓ	Climb (H.T.), recovery
	$\dot{\varepsilon} \propto \sigma^{n+2}$	$(n \simeq 5)$	0	Q_p	Climb (L.T.), recovery
†Class A:	$\dot{\varepsilon} \propto \sigma^n$	$(n \simeq 3)$	0	Q_i	Viscous glide (H.T.)
	$\dot{\varepsilon} \propto \sigma^{n+2}$	$(n \simeq 3)$	0	Q_p	Viscous glide (L.T.)
High:					
	$\dot{\varepsilon} \propto \exp(B\sigma)$		0	Q_p or Q_ℓ	Power-law breakdown

†Pure metals exhibit class M (Metal type) behavior at intermediate stress levels; solid solution alloys may exhibit class M or class A (Alloy type) behavior.

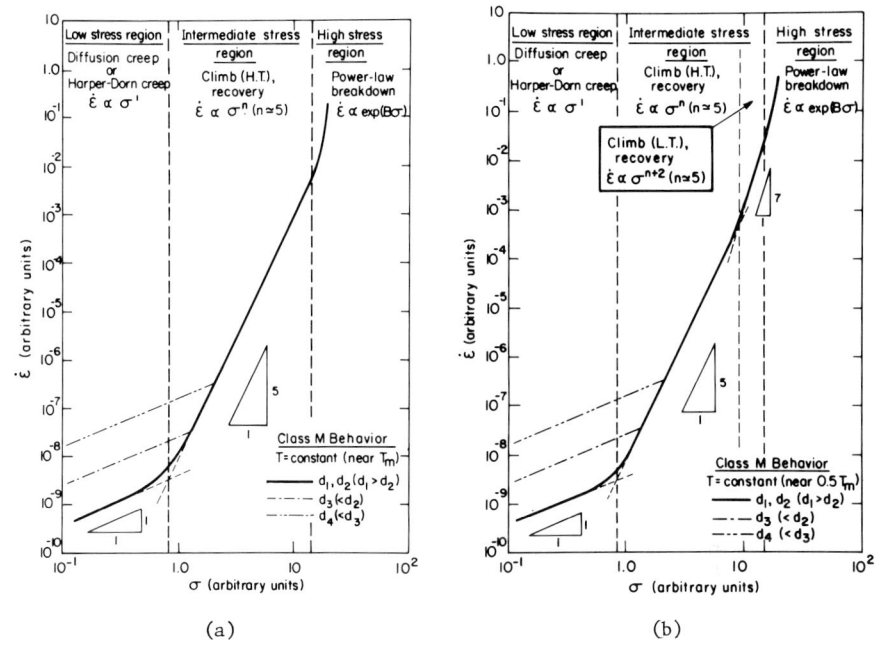

Fig. 1 Schematic illustration of class M behavior in pure metals, showing logarithmic $\dot{\varepsilon}$ versus logarithmic σ at (a) temperatures near T_m and (b) temperatures near $0.5\, T_m$.

because the intragranular dislocation density is proportional to σ^2 (Robinson and Sherby, 1969). Finally, there is a breakdown in power-law behavior at high stresses, and the strain rate depends on stress through an exponential relationship of the form $\dot{\varepsilon} \propto \exp(B\sigma)$ where B is a constant. It should be noted that the dominant creep processes at the intermediate and high stress levels are intragranular in nature with p = 0.

Since the deformation mechanisms in pure metals operate independently, it is possible to schematically illustrate the variation of strain rate with stress in the manner shown in Fig. 1. At temperatures near to T_m, the solid line in Fig. 1(a) shows transitions from n = 1 at low stresses to n = 5 due to climb (H.T.) at intermediate stresses and to power-law breakdown at high stresses. The solid line applies to large grain sizes, such as d_1 and d_2 ($d_1 > d_2$), where Harper-Dorn creep is dominant in the low stress region and all of the processes are independent of grain size (i.e., p = 0 in Table 1). As the grain size is reduced to d_3 (<d_2) or d_4 (<d_3), diffusion creep becomes dominant at low stresses with p = 2 (Nabarro-Herring creep) or 3 (Coble creep). This leads to faster strain rates in the low stress region, as indicated by the broken lines in Fig. 1(a), and a consequent displacement to higher stresses of the point of transition from the low stress to intermediate stress regions. As shown in Fig. 1(b), a similar behavior is observed also at lower temperatures, in the vicinity of $0.5\, T_m$, except that, due to the lower testing temperature, there is an additional transition from climb (H.T.) with a slope of ~5 to climb (L.T.) with a slope of ~7 as the stress level is increased. The transition to the high stress region and power-law breakdown usually occurs at $\dot{\varepsilon}/D_\ell \simeq 10^{13}\, \mathrm{m}^{-2}$, where D_ℓ is the coefficient for lattice self-diffusion (Sherby and Burke, 1967).

Pure metals and solid solution alloys deform by similar creep mechanisms in the low and high stress regions so that, as in Fig. 1, the behavior is conveniently characterized in terms of the flow process at intermediate stresses. There are also several additional important characteristics of class M behavior at intermediate stresses, including an instantaneous strain on application of the load, a normal primary stage of creep in which the strain rate decreases to steady-state flow, and the formation of a substructure which remains constant in the steady-state stage and which consists of a regular array of subgrains. The change in substructure with change in deformation mechanism is illustrated in Fig. 2 for Harper-Dorn creep with n = 1 on the left (Yavari, Miller and Langdon, 1982) and class M behavior with n ≃ 5 on the right (Vastava and Langdon, 1980). In the Harper-Dorn region, the substructure consists of a random and reasonably uniform distribution of dislocations, with no evidence for subgrain formation: detailed measurements show that the average dislocation density is independent of stress in this region, the dislocations are predominantly close to edge orientation (Yavari, Miller and Langdon, 1982), and the behavior may be

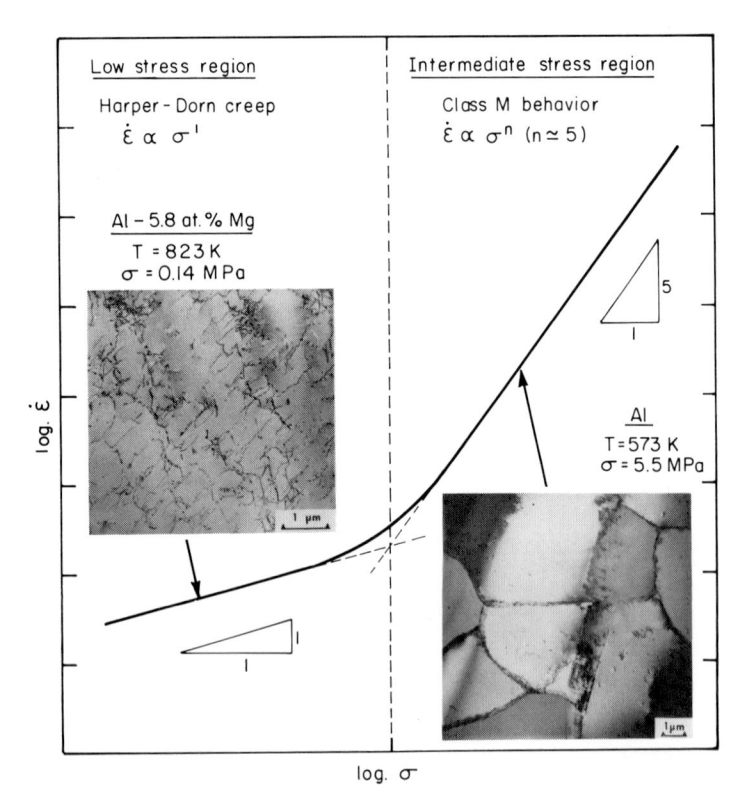

Fig. 2 The change in substructure from a random and reasonably uniform distribution of dislocations in Harper-Dorn creep with n = 1 on the left (Yavari, Miller and Langdon, 1982) to well-defined subgrains in class M behavior with n ≃ 5 on the right (Vastava and Langdon, 1980).

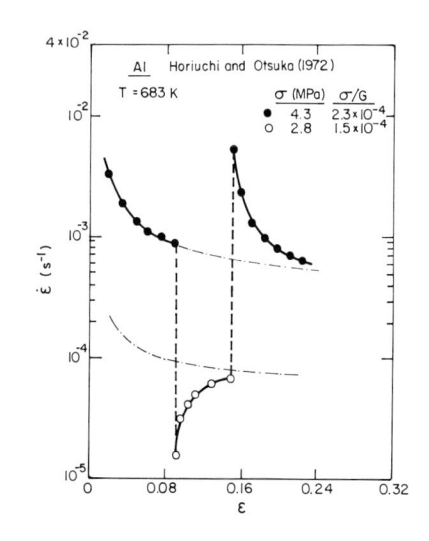

Fig. 3 Strain rate versus strain for pure Al at a temperature of 683 K, showing the effect of changes in the applied stress (Horiuchi and Otsuka, 1972).

interpreted in terms of the climb of edge dislocations under saturated conditions (Langdon and Yavari, 1982). In class M behavior, the steady-state substructure consists of well-defined subgrains; measurements from several materials show that the average subgrain size, λ, varies inversely with the applied stress, and the density of the dislocations within the subgrains is proportional to σ^2 (Bird, Mukherjee and Dorn, 1969).

An additional characteristic of class M behavior is the shape of the transient immediately following a change in the applied stress. This is illustrated in Fig. 3 in a plot of strain rate versus strain, ε, using data obtained on high purity Al tested at a temperature of 683 K (Horiuchi and Otsuka, 1972). The specimen was initially tested at a stress of 4.3 MPa (equivalent to $\sigma/G = 2.3 \times 10^{-4}$), the stress was abruptly dropped to 2.8 MPa ($\sigma/G = 1.5 \times 10^{-4}$) at a strain of ~9%, and then it was increased to the original value at a strain of ~15%. Following the stress reduction, there is a large inverted transient stage in which the strain rate increases towards steady-state flow. The two thin broken lines in Fig. 3 indicate the anticipated behavior at these two stress levels in the absence of a change in stress. Results of the type shown in Fig. 3 are often interpreted as indicative of an increase in strain rate following the stress reduction to a steady-state rate which is equal to the value which would have been attained at the same level of reduced stress in a constant stress experiment (Sherby, Klundt and Miller, 1977). However, close inspection shows that the experimental datum points in Fig. 3 are not strictly consistent with this claim because the stress level was subsequently increased at a strain of ~15% before a true steady-state condition was attained. It should be noted also that there is experimental evidence suggesting that the new steady-state strain rate following a stress reduction is slightly lower than the anticipated rate if the test is conducted only at the lower stress level (Langdon, Vastava and Yavari, 1979; Pontikis and Poirier, 1975).

Experiments show that the precise shape of the transient stage in a stress reduction test is not strictly definitive in class M behavior, because for large stress reductions there tends to be an initial decrease in strain rate and then a subsequent inverted transient (Blum, Hausselt and König, 1976; Blum and Pschenitzka, 1976; Langdon, Vastava and Yavari, 1979). The problem of anomalous transients is avoided in stress increment tests where there appears to be a normal transient of decreasing strain rate under all experimental conditions; an example is shown in Fig. 3 at $\varepsilon \simeq 0.15$. Furthermore, there is evidence that the new steady-state strain rate following an increase in stress is approximately equal to the anticipated value for the new stress level (Langdon, Vastava and Yavari, 1979).

DEFORMATION OF SOLID SOLUTION ALLOYS

Metallic solid solution alloys deform by the same mechanisms as pure metals in the low and high stress regions, but at intermediate stresses there are two distinct types of behavior. The first type is identical to class M, and it includes, as in pure metals, an instantaneous strain, a normal primary stage of creep, the formation of subgrains, and a normal transient after a stress increase. The second type, termed class A (Alloy type), occurs only in solid solution alloys, and it arises from the viscous glide of dislocations with solute atom atmospheres. As indicated in Table 1, class A behavior gives n ≃ 3, Q = Q_i, and p = 0. In principle, there is also the possibility of diffusion along the dislocation cores at low temperatures, giving a stress exponent of (n + 2) ≃ 5 and Q = Q_p; but in practice this low temperature behavior has not been reported experimentally, and henceforth the term viscous glide will be used to designate the standard class A behavior with n ≃ 3.

Class M - Class A Transition

Dislocation climb and viscous glide are sequential mechanisms, so that the observed creep rate is dictated by the slower of these two processes (Mohamed and Langdon, 1974). This leads to the possibility of a class M - class A transition, as illustrated schematically in Fig. 4. The behavior is similar to Fig. 1 in the low and high stress regions, but in the intermediate stress region there is a transition from n ≃ 5 (class M) to n ≃ 3 (class A) as the stress level is increased: viscous glide is therefore faster than climb in the n ≃ 5 region, and it is slower than climb in the n ≃ 3 region.

An experimental example of a class M - class A transition is shown in Fig. 5 for an Al-5.6 at. % Mg solid solution alloy tested at a temperature of 827 K (Yavari, Mohamed and Langdon, 1981). These tests were conducted using a shear configuration and the plot shows the steady-state shear strain rate, $\dot{\gamma}$, versus the shear stress, τ: the tests were performed using either a true constant shear stress in creep or a true constant shear strain rate on an Instron machine, and Fig. 5 shows the datum points obtained from both experimental procedures. In the intermediate stress region, there is a transition with increasing stress from high temperature climb (class M) with n = 4.4 to viscous glide (class A) with n = 3.1: since both of these processes are independent of grain size (p = 0 in Table 1), the datum points superimpose.

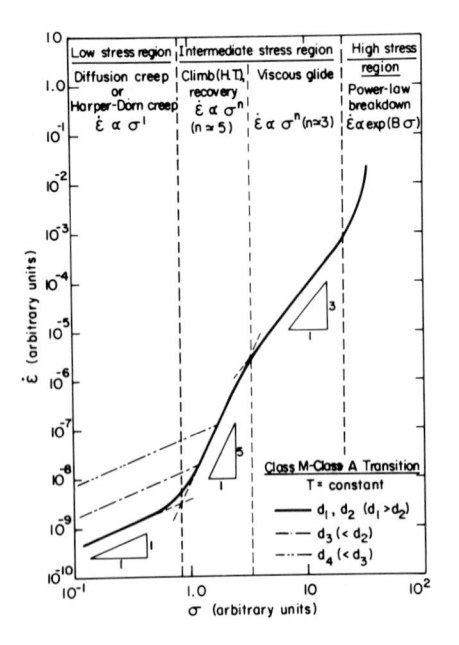

Fig. 4 Schematic illustration of a class M - class A transition in solid solution alloys, showing logarithmic $\dot{\varepsilon}$ versus logarithmic σ.

1110

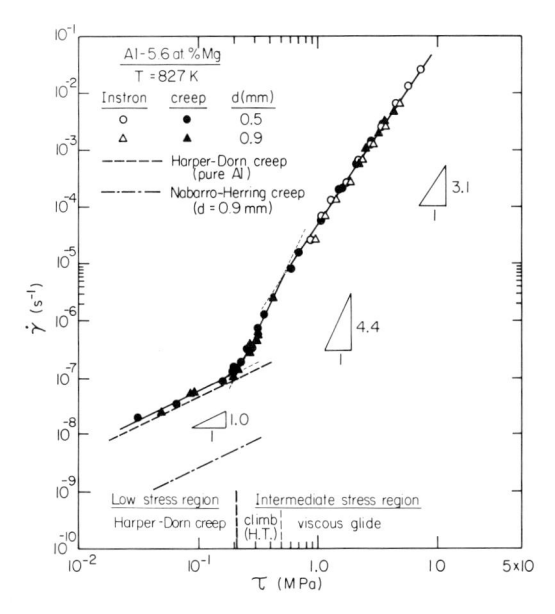

Fig. 5

Shear strain rate versus shear stress for Al-5.6 at. % Mg at a temperature of 827 K, showing an example of a class M - class A transition from n = 4.4 to n = 3.1: the behavior in the low stress region is due to Harper-Dorn creep with n = 1.0 (Yavari, Mohamed and Langdon, 1981).

for the two experimental grain sizes of 0.5 and 0.9 mm. It should be noted that the value of n = 4.4 obtained for class M behavior in this alloy is identical to the value of n reported from several sets of experiments on pure Al (Bird, Mukherjee and Dorn, 1969).

Figure 5 shows also that there is no dependence on grain size in the low stress region, and this is consistent with the occurrence of Harper-Dorn creep (see Table 1); this conclusion is further confirmed by the close agreement, shown by the upper broken line, with earlier data obtained on pure Al in this region (Harper and Dorn, 1957). In addition, the lack of agreement with diffusion creep is shown by the lower broken line estimated for Nabarro-Herring creep with a grain size of 0.9 mm. The experimental data in Fig. 5 were not extended to sufficiently high stresses to reveal the region of power-law breakdown.

A relationship may be derived to predict the transition from class M to class A behavior by equating the theoretical expressions for high temperature climb and viscous glide and fitting the resultant criterion to experimental results (Mohamed and Langdon, 1974). Using this procedure, and experimental data for an Al-3.3 at. % Mg alloy (Murty, Mohamed and Dorn, 1972), the limiting condition for class A behavior in solid solution alloys may be expressed in dimensionless form as

$$\frac{\sigma}{G} > \psi \left(\frac{kT}{eGb^3} \right) \left(\frac{1}{c} \right)^{1/2} \left(\frac{\tilde{D}}{D_\ell} \right)^{1/2} \left(\frac{Gb}{\Gamma} \right)^{3/2} \qquad (2)$$

where e is the solute-solvent size difference, c is the solute concentration, $\tilde{D}$ is the coefficient for interdiffusion of the solute atoms, Γ is the stacking fault energy of the alloy, and ψ is a dimensionless constant estimated from the experimental data as $\sim 3 \times 10^{-7}$: subsequent analyses, using different diffusion data for Al-Mg, have yielded values for ψ of $\sim 2 \times 10^{-7}$

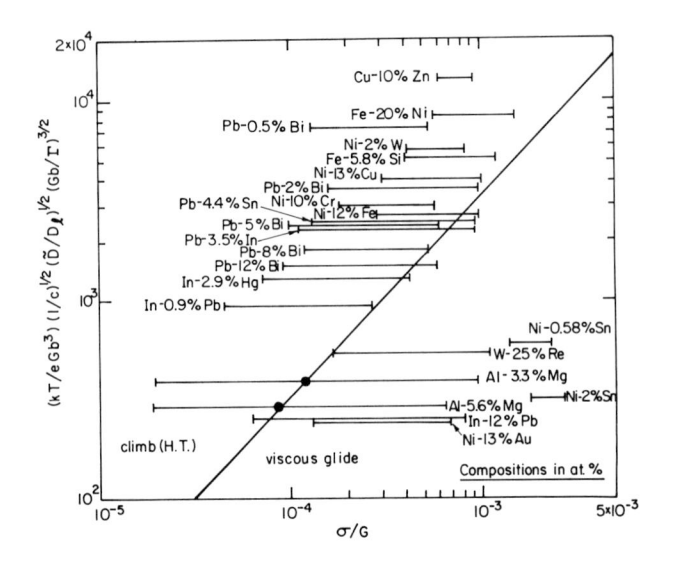

Fig. 6

The predicted transition from high temperature climb (class M) to viscous glide (class A) in solid solution alloys: the line at 45° separates the two classes of behavior.

(Oikawa, 1978) and ∿4 × 10^{-6} (Pahutová and Čadek, 1979), respectively.

Equation (2) is in excellent agreement with the available experimental data for creep of solid solution alloys. This agreement is illustrated in Fig. 6, where equation (2) with $\psi \simeq 3 \times 10^{-7}$ is given by the line at 45°, and this line separates high temperature climb (class M) at the lower stresses on the left from viscous glide (class A) at the higher stresses on the right. Each horizontal line in Fig. 6 represents a different alloy tested at a constant temperature, and the length of the line shows the range of normalized stresses covered experimentally. The position of the class M - class A transition line was placed using experimental data for Al-3.3 at. % Mg, as indicated by the upper solid circle; this line is in excellent agreement with the data for Al-5.6 at. % Mg given in Fig. 5, as shown by the lower solid circle marking the experimental point of transition from n = 4.4 to n = 3.1. All of the alloys on the left of the line in Fig. 6 were reported to exhibit high temperature climb (class M) and all of the alloys on the right of the line were reported to exhibit viscous glide (class A), thereby providing strong support for the normalized transition stress predicted by equation (2)

There are several important characteristics of class A behavior at intermediate stresses: there is little or no instantaneous strain on application of the load; there is either immediate steady-state behavior or a very brief primary which may be either normal, inverted or sigmoidal; and the steady-state substructure consists of an essentially uniform distribution of dislocations without any well-defined subgrains. When solid solution alloys exhibit a class M - class A transition, as in the Al-5.6 at. % Mg alloy shown in Fig. 5, there is a concomitant change in the steady-state substructure from the formation of subgrain boundaries in the high temperature climb region with n = 4.4 to a random distribution of dislocations in the viscous glide region with n = 3.1 (Yavari, Mohamed and Langdon, 1981).

There is also an important difference in class A behavior in stress change experiments, and this is illustrated in Fig. 7 for Al-3.0 at. % Mg tested at 683 K (Horiuchi and Otsuka, 1972); these experimental results may be

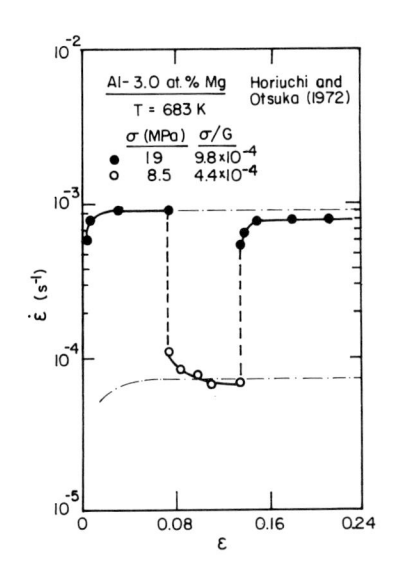

Fig. 7   Strain rate versus strain for Al-3.0 at. % Mg at a temperature of 683 K, showing the effect of changes in the applied stress (Horiuchi and Otsuka, 1972).

contrasted with the data shown in Fig. 3 for pure Al at the same testing temperature. In Fig. 7, the stress was decreased from 19 MPa ($\sigma/G = 9.8 \times 10^{-4}$) to 8.5 MPa ($\sigma/G = 4.4 \times 10^{-4}$) at a strain of ∼7%, and then increased to the original value at a strain of ∼14%. Two points should be noted with respect to the data. First, there is an initial very brief inverted primary, followed by a well-defined steady-state flow at a low total strain; this is typical of class A behavior and it contrasts with the long normal primary shown in Fig. 3 for class M behavior in pure Al. Second, the transients following a change in stress are very short and, contrary to class M, the transient is normal after a stress reduction and inverted after a stress increase. In practice, however, careful experiments indicate that the precise shape of the transient tends to be complex in stress reduction experiments (Pahutová, Čadek and Ryš, 1978), so that it is again preferable to characterize the behavior in terms of the transient following an increase in stress.

Class A - Class M Transition

The preceding section discusses the occurrence of a class M - class A transition in solid solution alloys with increasing stress, but there is also experimental evidence for an additional transition to a further regime of behavior at higher stress levels.

Two examples are shown in Fig. 8, in logarithmic plots of strain rate versus stress. In Fig. 8(a), there is a transition from n = 3.1 to n = 5.1 in an Al-2.2 at. % Mg alloy tested at 573 K, but there is no corresponding transition in an Al-5.5 at. % Mg alloy tested at the same temperature over a similar range of stresses (Oikawa, Sugawara and Karashima, 1976). Figure 8(b) shows data for Al-2.0 at. % Mg single crystals tested in <001> orientation over a range of temperatures, and in this case the bulk of the datum points have a slope of n ≃ 3.5 but there is an apparent increase in the value of n at both high and low stress levels (Kikuchi, Motoyama and Adachi, 1980).

Although it may be anticipated initially that the increase in slope at high stresses is due to power-law breakdown, close inspection of the available data indicates that this is unlikely (Yavari and Langdon, 1982). First, the datum points obtained at high stress levels in these and other experiments tend to exhibit a well-defined linear region rather than an exponential dependence on stress. Second, calculations show that the deviation from viscous glide behavior and n ≃ 3 occurs at normalized strain rates which are often about two orders of magnitude lower than the usual value of $\dot{\varepsilon}/D_\ell \simeq 10^{13}$ m^{-2} associated with power-law breakdown. Third, it is clear from Fig. 8(a) that the deviation from viscous glide at high stresses depends critically on the solute concentra-

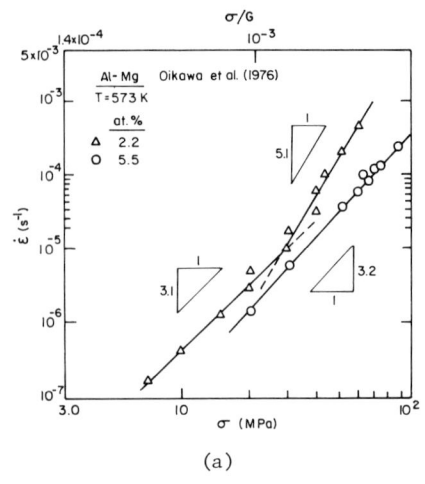

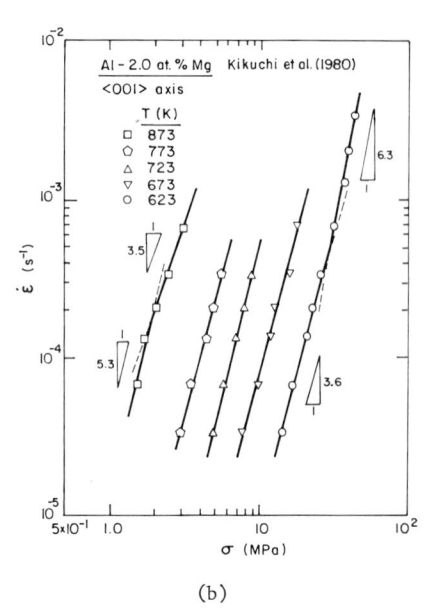

Fig. 8 Strain rate versus stress showing the transition to a higher value of n at high stress levels in (a) Al-2.2 at. % Mg (Oikawa, Sugawara and Karashima, 1976) and (b) Al-2.0 at. % Mg single crystals (Kikuchi, Motoyama and Adachi, 1980).

tion of the alloy.

The results in Fig. 8, and other similar data, may be explained by noting that the gliding dislocations break away from their solute atom atmospheres at a critical and well-defined breaking stress, σ_b (Friedel, 1964). Using this concept, the upper limiting stress for class A behavior in solid solution alloys may be expressed in dimensionless form as (Yavari and Langdon, 1982)

$$\frac{\sigma}{G} = \frac{W_m^2 c}{5Gb^3 kT} \qquad (3)$$

where W_m is the binding energy between the solute atom and the dislocation. Thus, the glide of dislocations is no longer retarded by the dragging of solute atom atmospheres at normalized stresses above the breakaway condition given in equation (3), and this suggests a transition to some modified form of high temperature climb (class M).

Equation (3) is plotted in Fig. 9, and the line at 45° separates viscous glide (class A) at low stress levels on the left from some form of high temperature climb (class M) at stress levels above the breakaway condition on the right (Yavari and Langdon, 1982). Each horizontal line represents a different alloy tested over a range of stresses at a constant temperature, and the solid circles denote the experimental stress levels where there are clear deviations from viscous glide behavior with n ≃ 3. For three solid solution alloys, shown near the upper edge of Fig. 9, there is no experimental deviation from class A behavior, and the lines are shown without solid circles.

Two important conclusions may be drawn from Fig. 9. First, the experimental

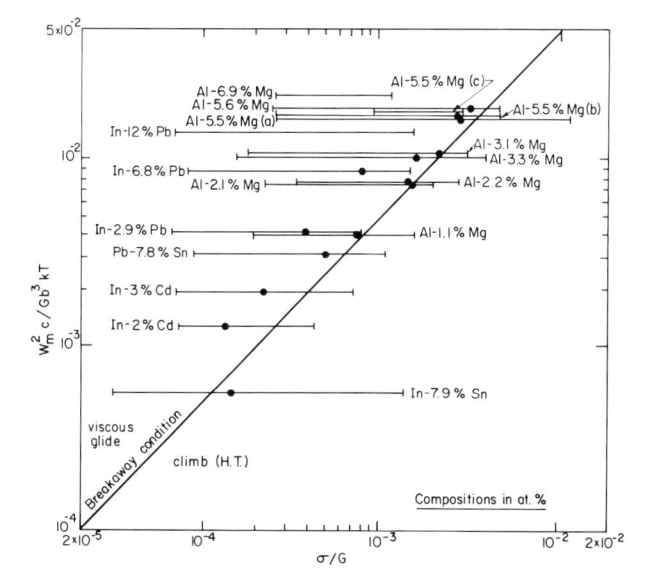

Fig. 9

The predicted breakaway condition marking the transition from viscous glide (class A) to some form of high temperature climb (class M): the line at 45° separates the two classes of behavior, and the solid circles show the experimental transition points (Yavari and Langdon, 1982).

transition points, shown by the solid circles, are in excellent agreement with the breakaway stresses predicted by equation (3); this agreement is to within a factor of three for all of the alloys used to construct Fig. 9, and for most of the alloys the agreement is to within a factor of two. Furthermore, the three alloys not exhibiting a deviation from n ≃ 3 behavior lie correctly within the region for viscous glide over the entire stress range examined experimentally. Second, this procedure explains the apparent dichotomy in behavior exhibited by the two Al-Mg alloys shown in Fig. 8(a), since the stress range for the Al-2.2 at. % Mg alloy crosses the breakaway line in Fig. 9 whereas the stress range for the Al-5.5 at. % Mg alloy [labelled Al-5.5% Mg (c) in Fig. 9] lies exclusively within the region of viscous glide.

It should be noted that, because of the uncertainties in the precise numerical factors occurring in the theoretical expressions for high temprrature climb and viscous glide, it was necessary in the preceding section to obtain the transition stress from class M to class A by fitting equation (2) to experimental data to provide a reasonable value for the constant ψ: the result is shown in Fig. 6. By contrast, equation (3) is derived directly from the breaking stress, σ_b, and, since it contains no unknown terms, the excellent agreement in Fig. 9 provides strong support for the concept of a breakaway condition. In addition, this concept is consistent with the experimental evidence, clearly visible from inspection of several alloys in Fig. 9, that the transition from n ≃ 3 behavior is displaced to higher stress levels as the solute concentration, c, is increased.

The transition from class A behavior at the lower stress levels to some modified form of class M behavior at the higher stress levels is accompanied by a change in the substructural features during steady-state flow, as illustrated in Fig. 10: there is no subgrain formation in class A behavior with n ≃ 3, and the substructure consists of an essentially uniform distribution of dislocations (Horiuchi and Otsuka, 1972; Matsuno and Oikawa, 1981; Orlová and Čadek, 1979; Yavari, Mohamed and Langdon, 1981); whereas at higher stress levels, in the class M region with n ≃ 5, the substructure is heterogeneous and consists of large areas which are essentially free of dislocations surrounded by tangles of dislocations in the form of loose-knit subgrain boundaries (Yavari and Langdon, 1982). A similar change in substructure, from random dislocations at the lower stresses to subgrain formation at the higher stresses, was observed also in several different Al-Mg alloys (Matsuno, Oikawa and Karashima, 1974) and in a Mg-0.8 wt. % Al solid solution alloy (Vagarali and Langdon, 1982): for the latter material, the subgrain structure was well-defined and it was shown that the average subgrain size, λ, was of the anticipated magnitude for class M behavior.

The concept of a breakaway from solute atmospheres at high stresses is supported by other experimental observations (Yavari and Langdon, 1982), including a change in the shape of the creep curves with increasing stress (Kuchařová and Čadek, 1972) and a change in the value of the effective stress exponent for the dislocation velocity (Oikawa, Kariya and Karashima, 1974). Thus, the available evidence indicates that, for some solid solution alloys, Fig. 4 is incomplete and there is the possibility of a double transition in the intermediate stress region to give an additional regime with n $\simeq$ 5 between viscous glide and power-law breakdown. This class M – class A – class M

Fig. 10 The change in substructure from an essentially uniform distribution of dislocations in class A behavior with n $\simeq$ 3 on the left (Yavari, Mohamed and Langdon, 1981) to loose-knit subgrain boundaries in class M behavior with n $\simeq$ 5 on the right (Yavari and Langdon, 1982): the change in n from 3 to 5 corresponds to the breakaway condition in Fig. 9.

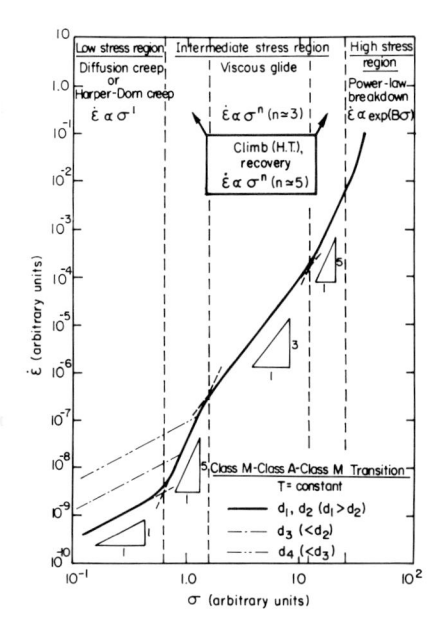

Fig. 11 Schematic illustration of a class M - class A - class M transition in a solid solution alloy, showing logarithmic $\dot{\varepsilon}$ versus logarithmic σ.

transition is illustrated schematically in Fig. 11: experimental examples of this behavior include the Al-2.0 at. % Mg single crystals shown in Fig. 8(b) (Kikuchi, Motoyama and Adachi, 1980) and polycrystalline Al-2.2 at. % Mg (Oikawa, Sugawara and Karashima, 1978) and Al-5.5 at. % Mg (Kuchařová and Čadek, 1972).

In the intermediate stress region, the class M - class A transition at the lower stresses is given by equation (2), and it represents the point at which viscous glide with solute atmospheres becomes slower than high temperature climb. As indicated in Fig. 6, this transition occurs at lower stress levels, so that class A behavior is favored, when both the solute-solvent size difference and the solute concentration are increased. The class A - class M transition at the higher stresses is given by equation (3), and it represents the point at which the dislocations break away from their solute atmospheres. As indicated in Fig. 9, this transition occurs at higher stress levels, so that class A behavior is favored, when the solute concentration is increased. If tests are conducted with high solute concentrations, the upper regime of class M behavior with $n \simeq 5$ is likely to be obscured by the normal transition to power-law breakdown.

From equations (2) and (3), it follows that class A behavior with $n \simeq 3$ is restricted in solid solution alloys to a limited range of normalized stresses which is given by

$$\psi \left(\frac{kT}{eGb^3} \right) \left(\frac{1}{c} \right)^{1/2} \left(\frac{\tilde{D}}{D_\ell} \right)^{1/2} \left(\frac{Gb}{\Gamma} \right)^{3/2} < \frac{\sigma}{G} < \frac{W_m^2 c}{5Gb^3 kT} \qquad (4)$$

where $\psi \simeq 3 \times 10^{-7}$.

DISCUSSION

Pure Metals

As indicated in Fig. 1, the steady-state creep of pure metals at high temperatures exhibits transitions from a low stress region with $n = 1$ to an intermediate stress region with $n \simeq 5-7$ and to a high stress region with an exponential dependence on stress.

In the low stress region, two of the possible deformation mechanisms depend on grain size (Nabarro-Herring and Coble creep) and the other mechanism is independent of grain size (Harper-Dorn creep). Thus, in tests with single crystals, the experimental datum points lie along the line for Harper-Dorn creep (Harper and Dorn, 1957). In polycrystalline materials, the limiting grain sizes marking the transitions from Harper-Dorn creep at large grain size to diffusion creep at small grain sizes are given by 1.5×10^6 b and 1.4×10^4 $b^{2/3}$ $(\delta D_{gb}/D_\ell)^{1/3}$ for Nabarro-Herring and Coble creep, respectively, where δ i the width of the grain boundary and D_{gb} is the coefficient for grain boundary diffusion (Langdon and Yavari, 1982).

At intermediate stresses, several theories of creep are available based on dislocation climb (Weertman 1968, 1975), although these theories are generally not sufficiently detailed to account for all of the observations. For example there is evidence that the steady-state creep rate in class M varies with $(\Gamma/Gb)^3$, and this leads to the inclusion of a stacking fault energy term in equation (2) (Mohamed and Langdon, 1974): attempts have been made to interpret this relationship by invoking the climb of jogs in extended edge dislocations (Argon and Moffatt, 1981) and by using a dislocation network model (Burton, 1982). There is also the possibility that additional mechanisms, not included in Fig. 1, may become important under some experimental conditions or in some crystal structures. For example, experiments on h.c.p. Mg reveal a transition at high temperatures to a region where deformation is controlled by the cross-slip of dislocations from the basal to the prismatic planes (Vagarali and Langdon, 1981).

An important problem in class M behavior is the stability of the subgrain structure in stress reduction experiments, since there is evidence indicating both an increase in subgrain size during the inverted transient following a stress reduction in pure Al (Ferreira and Stang, 1979) and a negligible change in subgrain size in pure Al (Langdon, Vastava and Yavari, 1979) and pure Cu (Parker and Wilshire, 1976). This dichotomy is evident also in stress reduction experiments on non-metallic single crystals, where there are reports of an increase in subgrain size in NaCl (Eggeler and Blum, 1981) and no change in subgrain size in AgCl (Pontikis and Poirier, 1975). The latter result of subgrain stability is observed also in olivine (Ross, Avé Lallemant and Carter, 1980), thereby suggesting an important limitation on the use of subgrain measurements as an indirect indicator of tectonic stresses in geological materials (Twiss, 1977).

At high stresses, very extensive data are available for pure Al to show that the exponential dependence on stress extends for up to 15 orders of magnitude of normalized strain rate beyond the point of power-law breakdown (Luthy, Miller and Sherby, 1980). In addition, pipe diffusion tends to dominate under these conditions, so that the activation energy is closer to Q_p than Q_ℓ.

Solid Solution Alloys

The steady-state creep of solid solution alloys is similar to pure metals in the low and high stress regions. At low stresses, the similarity is demonstrated by the datum points for Al-5.6 at. % Mg and the line for pure Al in the region of n = 1 in Fig. 5. At high stresses, the similarity is shown by the observation that alloys without subgrains exhibit power-law breakdown at the same value of $\dot{\varepsilon}/D_\ell$ ($\sim 10^{13}$ m^{-2}) as for pure metals (Kassner, 1982).

The critical difference in the creep of solid solution alloys occurs at intermediate stresses where there is the possibility of a class M - class A and/or a class A - class M transition with increasing stress: these transitions are illustrated schematically in Fig. 11. The basic and significant character-

TABLE 2   Characteristics of Class M and Class A Behavior

	Class M	Class A
1.	$n \simeq 5$ (or $\simeq 7$)	$n \simeq 3$
2.	Instantaneous strain	Little or no instantaneous strain
3.	Normal primary	No primary, or very short normal, inverted or sigmoidal primary
4.	Regular array of subgrains	Essentially uniform distribution of dislocations
5.	Normal transient after stress increase	Short inverted transient after stress increase

istics of class M and class A behavior are summarized in Table 2, and these may be used, together with the criteria illustrated in Figs. 6 and 9, to identify the two types of behavior. In practice, the second and third characteristics listed in Table 2 tend to be qualitative, and it is experimentally easier to check the shape of the transient after an increase in stress at constant temperature; as noted earlier, the shape of the transient is not always definitive after a stress reduction. Having determined the transitions in behavior depicted in Fig. 11, it is a simple procedure to display the mechanisms in a solid solution alloy in the form of a deformation map of grain size versus stress at constant temperature (Mohamed and Langdon, 1975).

SUMMARY AND CONCLUSIONS

1. The steady-state creep of pure metals may be divided into three distinct regions: a low stress region with $n = 1$, an intermediate stress region due to dislocation climb with $n \simeq 5$-7 (class M behavior), and a high stress region with an exponential dependence on stress.

2. The steady-state creep of solid solution alloys is similar to pure metals in the low and high stress regions, but at intermediate stresses there may be a region of viscous glide with $n \simeq 3$ (class A behavior). This introduces the possibility of a class M - class A and/or a class A - class M transition with increasing stress. Equations are available to predict these transitions in solid solution alloys.

ACKNOWLEDGEMENT

This work was supported by the United States Department of Energy under Contract DE-AM03-76SF00113 PA-DE-AT03-76ER10408.

REFERENCES

Argon, A.S., and W.C. Moffatt (1981). *Acta Met.*, 29, 293-99.
Bird, J.E., A.K. Mukherjee and J.E. Dorn (1969). In D.G. Brandon and A. Rosen (Eds.), *Quantitative Relation Between Properties and Microstructure.* Israel Universities Press, Jerusalem. pp. 255-342.
Blum, W., J. Hausselt and G. König (1976). *Acta Met.*, 24, 293-97.
Blum, W., and F. Pschenitzka (1976). *Z. Metallk.*, 67, 62-65.
Burton, B. (1982). *Acta Met.*, 30, 905-10.

Coble, R.L. (1963). *J. Appl. Phys.*, **34**, 1679-82.
Eggeler, G., and W. Blum (1981). *Phil. Mag.*, **44A**, 1065-84.
Ferreira, I., and R.G. Stang (1979). *Mater. Sci. Eng.*, **38**, 169-74.
Friedel, J. (1964). *Dislocations*. Pergamon, Oxford.
Harper, J., and J.E. Dorn (1957). *Acta Met.*, **5**, 654-65.
Herring, C. (1950). *J. Appl. Phys.*, **21**, 437-45.
Horiuchi, R., and M. Otsuka (1972). *Trans. Japan Inst. Metals*, **13**, 284-93.
Kassner, M.E. (1982). *Scripta Met.*, **16**, 265-66.
Kikuchi, S., Y. Motoyama and M. Adachi (1980). *J. Japan Inst. Light Metals*, **30**, 449-55.
Kuchařová, K., and J. Čadek (1972). *Kovové Materiály*, **10**, 240-48.
Langdon, T.G. (1975). In R.C. Bradt and R.E. Tressler (Eds.), *Deformation of Ceramic Materials*. Plenum, New York. pp. 101-26.
Langdon, T.G., R.B. Vastava and P. Yavari (1979). In P. Haasen, V. Gerold and G. Kostorz (Eds.), *Proc. ICSMA 5*. Pergamon, Oxford. v. 1, pp. 271-76.
Langdon, T.G., and P. Yavari (1982). *Acta Met.*, **30**, 881-87.
Luthy, H., A.K. Miller and O.D. Sherby (1980). *Acta Met.*, **28**, 169-78.
Matsuno, N., and H. Oikawa (1981). *Scripta Met.*, **15**, 319-22.
Matsuno, N., H. Oikawa and S. Karashima (1974). *J. Japan Inst. Metals*, **38**, 1071-76.
Mohamed, F.A., and T.G. Langdon (1974). *Acta Met.*, **22**, 779-88.
Mohamed, F.A., and T.G. Langdon (1975). *Scripta Met.*, **9**, 137-40.
Murty, K.L., F.A. Mohamed and J.E. Dorn (1972). *Acta Met.*, **20**, 1009-18.
Nabarro, F.R.N. (1948). In *Report of a Conference on Strength of Solids*. The Physical Society, London. pp. 75-90.
Oikawa, H. (1978). *Scripta Met.*, **12**, 283-85.
Oikawa, H., J. Kariya and S. Karashima (1974). *Metal Sci.*, **8**, 106-11.
Oikawa, H., K. Sugawara and S. Karashima (1976). *Scripta Met.*, **10**, 885-88.
Oikawa, H., K. Sugawara and S. Karashima (1978). *Trans. Japan Inst. Metals*, **19**, 611-16.
Orlová, A., and J. Čadek (1979). *Mater. Sci. Eng.*, **38**, 139-44.
Pahutová, M., and J. Čadek (1979). *Phys. Stat. Sol. (a)*, **56**, 305-13.
Pahutová, M., J. Čadek and P. Ryš (1978). *Mater. Sci. Eng.*, **36**, 185-91.
Parker, J.D., and B. Wilshire (1976). *Phil. Mag.*, **34**, 485-89.
Pontikis, V., and J.P. Poirier (1975). *Phil. Mag.*, **32**, 577-92.
Robinson, S.L., and O.D. Sherby (1969). *Acta Met.*, **17**, 109-25.
Ross, J.V., H.G. Avé Lallemant and N.L. Carter (1980). *Tectonophys.*, **70**, 39-61.
Sherby, O.D., and P.M. Burke (1967). *Prog. Mater. Sci.*, **13**, 325-90.
Sherby, O.D., R.H. Klundt and A.K. Miller (1977). *Met. Trans.*, **8A**, 843-50.
Twiss, R.J. (1977). *Pure Appl. Geophys.*, **115**, 227-44.
Vagarali, S.S., and T.G. Langdon (1981). *Acta Met.*, **29**, 1969-82.
Vagarali, S.S., and T.G. Langdon (1982). *Acta Met.*, **30**, 1157-70.
Vastava, R.B., and T.G. Langdon (1980). In I. Le May (Ed.), *Advances in Materials Technology in the Americas - 1980*. ASME, New York. v. 2, pp. 55-60.
Weertman, J. (1968). *Trans. ASM*, **61**, 681-94.
Weertman, J. (1975). In J.C.M. Li and A.K. Mukherjee (Eds.), *Rate Processes in Plastic Deformation of Materials*. ASM, Metals Park, Ohio. pp. 315-36.
Yavari, P., and T.G. Langdon (1982). *Acta Met.* (in press).
Yavari, P., D.A. Miller and T.G. Langdon (1982). *Acta Met.*, **30**, 871-79.
Yavari, P., F.A. Mohamed and T.G. Langdon (1981). *Acta Met.*, **29**, 1495-1507.

Strengthening from the Melt: Castings and Weldments

G. J. Davies

Department of Metallurgy, University of Sheffield, Mappin Street, Sheffield, UK

ABSTRACT

Traditionally castings and weldments have been seen as having inferior properties to wrought alloys. However, the performance of cast materials depends to a significant degree on the service conditions to which they are to be subjected. In some cases sound performance depends upon avoiding premature failures, such as those resulting from intergranular fracture, and this can require careful control of conditions with a view to the elimination or minimisation of defects. In other cases controlled processing, e.g. making use of very rapid or of directional solidification, can give rise to enhanced properties. Enhancement can also be achieved through the utilisation of phase relations to produce in situ composites. Additionally a range of new processes has been developed which can be used to produce materials with substantially improved properties. With fusion weldments the position is more complex and strengthening more usually results from careful structural control. Various ways in which improved properties can be produced in structures developed during solidification (castings and weldments) are reviewed and directions for future developments are discussed.

KEYWORDS

Solidification; fusion welding; casting defects; continuous casting; rapidly quenched metals; in situ composites; rheocasting.

INTRODUCTION

There are two principal ways in which the production of metals and alloys with enhanced properties directly from the melt can be approached. The first is primarily concerned with quality control. Often the properties of as-cast structures are seriously influenced by the existence of microscopic or macroscopic defects such as segregates, inclusions, gas and/or shrinkage porosity, hot tears etc. During the working processes involved in the production of wrought materials many (if not all) of these defects are eliminated and this has led to wrought products having their reputation for being, in general, superior to as-cast products. However, when the nature and origin of the casting defects are well understood steps can be taken to eliminate them or to minimise their influence. This forms the basis of the first approach. The second approach involves the development and use of new processes to yield as-cast materials which have properties which are intrinsically superior to normal castings or even to competing wrought materials. Here enhanced strength in its own right is the goal. The directness of the solidification process is taken advantage of so that often quite complicated artefacts can be manufactured in a single processing stage. A number of process routes are available e.g. directional growth of single-phase and polyphase (composite) alloys, rapid solidification processing, rheocasting and thixocasting, which need to be examined in the present context. These all fall broadly into the category of the production of strong microstructures from the melt. The basic concepts were reviewed in an earlier publication (Davies, 1971) which also reviewed the then state-of-the-art. There have been many developments in the intervening years which will be considered in the present survey.

The development of strong microstructures in fusion welds is more complex since the nature of the welding process makes more difficult close microstructural control of the solidified fusion zone. Solidification structures and properties of fusion welds have been reviewed by Davies and Garland (1975) but again there have been many subsequent developments which warrant attention. In this paper consideration will be given to a number of these developments with a view to identifying points of significance and openings for future work.

ENHANCEMENT OF THE PROPERTIES OF CASTINGS

In this Section we will be concerned with the methods available for generally improving the quality of normal cast materials and castings. Attention will be given not only to batch processes but also to the continuous casting of strands and slabs which are to be subjected to subsequent hot working stages.

Avoiding Premature Failure

Design philosophies based on the use of the principles of fracture mechanics are now finding greater application in the foundry industry (Jackson and Wright, 1977). The aim is to supplement conventional design factors with a more quantitative measure of the fracture resistance of the cast material. However, this can only be achieved satisfactorily provided the mechanism of failure under the prevailing service conditions can be predicted and the fracture toughness data used are representative of this mechanism. Temperature is known to have an effect on the properties of cast materials, particularly steels, with many alloys showing a transition from fibrous to cleavage fracture over a comparatively narrow temperature range. This change in behaviour can be taken into account when the materials are specified since the effects of alloying, heat treatment etc. on the transition temperature and mechanical property data are usually documented. Matters become complicated, however, when an unexpected mechanism intervenes and produces embrittlement under conditions where high toughness should be obtained. This can be the case with intergranular embrittlement such as occurs when aluminium nitride precipitates in steels. This has frequently been cited as a cause of premature failure in cast steels (Briant and Banerji, 1978; Lorig and Elsea, 1947; Woodfine and Quarrell, 1960; Wright and Quarrell, 1962). Most commonly it has been assumed that aluminium nitride precipitation and other embrittling precipitation phenomena occur in the solid state and thus a means of control is offered through heat treatment after solidification is complete. Although this can be useful in some cases, recent work (Croft, Entwisle and Davies, 1982) has shown that because of the significantly greater quantities of precipitate formers that can occur in cast metals the solubility temperature is often very high and precipitates can form while liquid is still present. This is particularly the case with aluminium nitride since frequently concentrations of up to 0.10%Al are obtained as a consequence of the deoxidation practice. The precipitates formed are dendritic in character and then the relative grain boundary/matrix strengths play an important role and the variation of fracture strength with temperature shows three regimes (Fig.1).

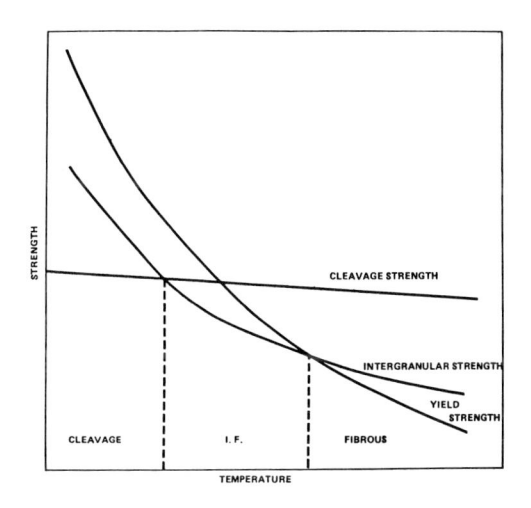

Fig. 1 Schematic diagram showing the variation of cleavage, intergranular and yield strength with temperature which can be used to explain premature failure.

Because of the high concentrations a high temperature solution treatment is not a practical possibility because the precipitates will coarsen rather than dissolve (Croft, Entwisle and Davies, 1982; Gladman and Pickering, 1967). Then the only resort is avoidance of precipitation by elimination of nitrogen and a change in deoxidation practice to one in which ladle additions are used, e.g. Zr, Ti, which do not lead to deleterious precipitates.

Elimination or Minimisation of Defects

Microsegregation is a short-range phenomenon which extends over distances of the order of the cast grain size or less. The dendrite arm spacings (primary and secondary) are important variables (Flemings, 1974a) since they are fundamental in determining the effective "wavelength" of the compositional variations which need to be eliminated by heat treatment. Although it is well known that both coarsening effects (Dann, Hogan and

Eady, 1979; Kattamis, Loughlin and Flemings, 1967; Reeves and Kattamis, 1971; Spittle and Lloyd 1977; Young and Kirkwood, 1975) and migration effects (Allen and Hunt, 1976) can affect the dendrite arm spacing, it is generally acknowledged that these arm spacings are dependent on the local cooling rate (although not in as simple a way as originally proposed by Flemings (1974b)). This was recognised by Flemings (1974a) who saw freezing not as a fortuitous event but as a process to be controlled to attain closely defined metallurgical objectives, his so-called "Premium Quality" casting. Here the aim was the production of castings with guaranteed mechanical properties that were much higher than those of conventional castings. That this was achievable was illustrated by data for an aluminium alloy casting in which chills were used judiciously for producing fine dendrite arm spacings (Table 1).

TABLE 1 Mechanical Properties of a Cast Al-Mg-Si Age Hardenable Alloy (after Flemings (1974a))

	Tensile strength (MN_m^{-2})	Yield strength (MN_m^{-2})	Elongation (%)
Guaranteed properties in premium quality castings	345	275	5
Guaranteed properties in conventional castings	155	100	<1

Surprisingly this approach does not seem to have been exploited to the extent that might have been warranted except in its ultimate form through powder metallurgy or rapid solidification processing (see below).

Macrosegregation

This is a longer range phenomenon extending over distances approaching the casting dimensions and as such is not sensibly responsive to heat treatment. In its most extreme form it is manifest as the substantial compositional inhomogeneities (channel segregates) that appear in large static cast ingots. Since they are not able to be eliminated by heat treatment they must be controlled by inhibiting their formation during the initial solidification stages. Methods of control through the use of controlled heat flow and related factors have been outlined by Flemings (1976). Furthermore, since it has been shown that the formation of channel segregates (and a related phonomemon "freckle" (Giamei and Kear, 1970)) is the result of fast flowing liquid jets in the partly solidified zone which are driven by density inversions produced by thermal and solute transport effects (Copley and coworkers, 1970; Mehrabian, Keane and Flemings, 1970) control can also be exerted by influencing the density changes. One approach is to balance the temperature-density change with a composition-density change (Burden, Hebditch and Hunt, 1974). Another is to eliminate the gravitational convection by addition of small quantities of alloying elements (McCartney and Hunt, 1981; Bridge and Beech, 1982). This latter approach offers a great deal of promise for improving quality and performance in both as-cast materials and wrought products. It is not well researched and clearly warrants further attention.

The gains to be made from the elimination of long range segregates in ingots or the related "ghost bands" in castings (Jackson, 1979) are substantial.

Porosity and Shrinkage

Cavities can be created in castings either because of the evolution of gas in the liquid or as a consequence of the volume contraction that occurs on solidification. In either case these cavities can be small and widely dispersed (microporosity or microshrinkage) or more massive. The latter can be dealt with comparatively easily through careful control of macroscopic heat flow and the judicious use of feeders and risers (Beeley, 1979). In contrast, research on the origin and control of microporosity is still relatively scarce despite the fact that its presence can degrade the properties of the cast structure.

A number of models exist which account for the nucleation and growth of gas bubbles during casting, e.g. Campbell (1968), Burns and Beech (1974), Fredriksson and Svensson (1976), and in principle gas-induced defects can be avoided by melt degassing techniques. On the other hand dispersed shrinkage arises from constraints on fluid flow through the residual passages that exist in the dendrite array in the final stages of solidification (Piwonka and Flemings, 1966; Streat and Weinberg, 1976). Thus microshrinkage pores can form without gas being present. The structure sensitivity is variable but certainly the mechanical properties, including the fracture and fatigue strengths are affected deleteriously (Bachelet and Lesoult, 1978; Pook and coworkers, 1980; Selby and Hough, 1980). Further work is needed on the mechanics and control of flow through porous dendrite arrays. Notwithstanding that, however, it must be accepted that in fine-grained castings with small dendrite arm spacings a degree of microshrinkage is inevitable. Then attention must be given to improvement of properties by subsequent treatments. One of the most promising routes for the removal of casting porosity involves the use of hydrostatic pressure during heat treatment near the melting temperature (Coble and Flemings, 1971). In particular hot isostatic pressing (HIPing) has been proposed as a means of healing pores but to date very little work on this topic has been reported, e.g. Basaran and coworkers, (1973), Stevens and Flewitt, (1980), despite the manifest rewards that should result.

Inclusions

The reduction and/or elimination of inclusions in cast metals must also lead to the enhancement of properties. Exogeneous inclusions e.g. slag, sand etc. can be easily controlled by good foundry practice (Sharman, Sims and Ashton 1980). Endogenous inclusions associated, for example, with steelmaking or ironmaking practice, have been the subject of much study and their origins and control are well understood (see, for instance, Kiessling 1980; Kiessling and Lange, 1978). This understanding makes it possible to optimise the technical and economical factors related to the occurrence of inclusions.

Surface Defects

When a liquid is cast against a mould wall the normal expectation is that the cast surface will merely assume the smoothness of that wall. In practice this is rarely the case and the surface quality has been found to depend to a considerable extent on the casting conditions. However, until recently the formation of surface irregularities, such as ripples on ingots and reciprocation marks on continuously cast strands, had not been extensively studied. The reason for this apparent neglect seems to have been that they could be relatively easily dealt with by scarfing or by other subsequent processing. The effects of the surface features on the formation of defects such as transverse cracks and on operational problems associated in particular with breakouts in continuous casting where high casting speeds are required for steels prone to surface unevenness was not, however, accounted for and needed further clarification. Work relating to surface defects was normally associated with specific cases or systems (see Saucedo, Beech and Davies, (1982a) for a review of data and proposed mechanisms). This created the general idea that each process produced a different type of defect. Any connection such as between say ripples on ingots or electroslag remelted strands and reciprocation marks in DC casting of aluminium or continuous casting of steel was normally dismissed as non-existent. However, since all of these processes involved a melt-mould interaction, the possibility of a connection in the formation of surface ripples in different processes and conditions could not be ignored. This possibility was examined by Saucedo, Beech and Davies (1980,1982a) and independently by Wray (1981) and Tomono, Kurz and Heinemann (1981). The casting processes where surface rippling occurs are listed in Table 2.

TABLE 2 Casting Processes where Surface Rippling Occurs

Process	Mould	Strand	Casting Speed
DC casting of Al and its alloys	Static	Continuous withdrawal	Constant
DC casting of steel billets (experimental)	Static	Continuous withdrawal	Constant
Continuous casting of non-ferrous alloys	Static	Continuous withdrawal	Constant
Ingot casting	Static	Generally constant	Function of melt steel, generally constant
Electroslag Remelting	Static or mobile	Static or mobile	Function of strand or mould speed
Semi-continuous casting of non-ferrous alloy	Static	Intermittent withdrawal	Function of withdrawal
Continuous casting of steel billets & blooms	Reciprocating	Continuous withdrawal	Function of reciprocation
Continuous casting of steel slabs	Reciprocating	Continuous withdrawal	Function of reciprocation

The conclusion from the various studies was that the surface defects were attributable to a meniscus freezing mechanism which could be applied to all processes. This was subsequently confirmed by a transient heat flow analysis (Saucedo, Beech and Davies, 1982b). As a consequence it was proposed that a reduction in the severity of the defect and enhancement of properties could be achieved by lowering the rate of heat extraction in the meniscus region. Reductions in heat extraction can be obtained through increasing casting speed, changing the lubrication practice, using higher frequencies of mould oscillation, etc. The use of a mould with reduced heat extraction at the top is another possible solution for the case of continuous casting.

Improving the Internal Quality of Continuously-Cast Strands

Continuously cast strands normally have a considerable columnar zone at the outer surface with a comparatively small central equiaxed zone. The columnar structure favours heavy centreline segregation (Flemings, 1974; Weinberg, 1975) and increases the probability of internal crack formation (Brimacombe and Sorimachi, 1977). Although some improvement can be achieved by reducing the casting temperature this represents something of a "knife-edge" between quality improvement and loss of output due to casting stoppages. A more positive approach is to utilise electro-magnetic stirring either in or below the mould (Diserens and coworkers, 1981; Iwata and coworkers, 1976; Hurtuk and Travaras, 1977; Melford and coworkers, 1982; Schrewe 1981; Widdowson and Marr, 1979). This causes both crystal multiplication and more efficient packing of the equiaxed crystals (Melford, 1980). The influence of temperature and electromagnetic stirring on the extent of the equiaxed zone is shown in Fig.2.

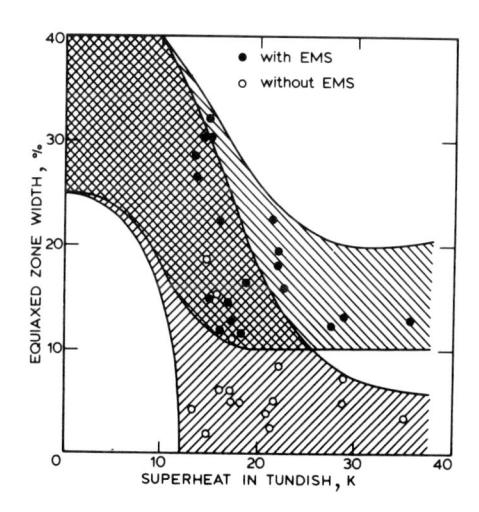

Fig. 2 Influence of temperature and electromagnetic stirring on equiaxed solidification

An alternative but less controllable procedure is to induce a preferred flow pattern by the use of specially designed submerged nozzles (Offerman, 1981). However, this approach is likely to be prone to inconsistencies due to nozzle blocking and changes in flow during ladle changeover.

Developments in Foundry Technology

In addition to enhancements of the mechanical properties of castings arising from fundamental knowledge of the causes of defects and their avoidance, there are benefits that arise because of technological developments in foundry practice. The more straightforward of these are either associated with improved moulding methods e.g. the vacuum sealed moulding process (Schneider, 1974) or freeze moulding (Clegg, 1979), or improved melt treatment especially through procedures leading to enhanced grain refinement, e.g. Campbell (1981). There have also been developed a number of alternative casting processes and some of these are considered in more detail in a later section of this paper.

CONTROLLED PROCESSING

Directional Solidification

Very significant improvements in the creep behaviour of alloys for gas-turbine blades have been made by the use of controlled directional solidification to produce fully columnar or single crystal structures (Kear and Piearcey, 1967; Versnyder and Shank, 1970). In the normal as-cast form the alloys usually failed in creep by intergranular void linkage along grain boundaries oriented transverse to the major stress axis. A fully columnar structure which is free of transverse grain boundaries has a longer creep life but the most marked change results from complete elimination of grain boundaries. The single crystal blades have both an increased life and a decreased creep rate. They also show a markedly increased resistance to thermal shock and an improved impact strength. Process control was traditionally difficult but the advent of liquid metal cooling (Giamei and Tschinkel, 1976) provided a practical solution to many of the problems and opened the way to production under a wider range of conditions. There is ample opportunity for further examination and exploitation of this technique.

Rapid Solidification

Rapid solidification processing by a variety of techniques e.g. melt spinning, melt extraction, planar flow casting etc., offers a range of opportunities for the production of material directly from the melt with enhanced properties. Principally rapid solidification is used for,

(i) the production of amorphous alloys,
(ii) the production of hypersaturated microcrystalline alloys,
(iii) the production of compactible material from alloys which are difficult to work.

The process has many advantages among which are the reduction of segregation, the formation of ultrafine dendrites and grains, the formation of "new" phases and the attainment of large increases in terminal solubility. Aspects of the production and properties of some rapidly solidified alloys are dealt with elsewhere in this Conference. The study of rapidly solidified alloys is very much an activity in its own right as evinced by the two most recent Conferences dedicated to the topic (Cantor, 1978; Masumoto and Suzuki, 1982) at which were presented 127 and 416 papers, respectively. As such it cannot be given justice in the present paper.

One facet which is, however, of more special interest is laser glazing (laser surface melting) which is used to surface treat bulk cast materials. This is considered further below.

GROWTH OF IN-SITU COMPOSITES

The production and properties of in-situ composites by directional solidification has been widely studied. Initially the use of controlled eutectic growth was seen to be most attractive because it offered the possibility of growth in complex shapes or with specified fibre layouts particularly since both fibrous and lamellar geometries were feasible. After early exploratory work (see Davies (1971, 1974) attention has largely concentrated on either,

(a)   the γ/γ -δ eutectic systems (Lemkey and Thompson, 1973; Pearson and Lemkey, 1979; Thompson and Lemkey, 1969), or

(b)   the γ/γ' -MC eutectic systems (Lemkey and Thompson, 1971).

Both systems were very successful in so far as the ambient and high temperature properties were concerned. Other related systems have subsequently been developed (Quested, Miles and McLean, 1980; Van dem Boomgaard and Wolff, 1972) and additionally off-eutectic compositions have been utilised (Boettinger, Biancaniello and Coriell, 1981; Frydman and Courtney, 1982; Holder and Oliver, 1974) following a suggestion by Mollard and Flemings (1967a,b). Clearly this is a very well-established technology in which the effects of process variables (growth rate, temperature gradients) are well documented. Even the effects of changes in the growth conditions have been carefully examined (Farag, Matera and Flemings, 1979).

It was anticipated initially that because eutectic composites have preferred low energy interphase interfaces (Chadwick, 1963) these should impart excellent stability at elevated temperatures even under service loads. In general this view is supported by experimental data obtained under isothermal conditions even when there are stress variations e.g. Johnson and Stoloff (1980), Thompson and Lemkey (1969). However, in service, uniform thermal conditions are not common and artefacts, such as turbine blades, are subject to both temperature gradients and thermal cycling. Then a different picture emerges and considerable microstructural instability occurs. This can, in turn, impair the strengthening effects.

Temperature Gradient Stability

In the presence of a temperature gradient it is expected that particles (or fibres) will migrate up the gradient when the solubility of the dispersed phase increases with increasing temperature. This migration may be aided or retarded by thermal diffusion effects depending on the sign of the Soret coefficient. Experimental evidence indicates that migration and coarsening does occur (Davies, Courtney and Przystupa, 1980; McLean, 1982) under these conditions, albeit slowly. For both lamellar and rod-like eutectic systems it is possible to formulate a set of principles that should be satisfied if an alloy is to be capable of service under non-isothermal conditions. Where breakdown is largely a consequence of diffusion in the bulk phases, structural stability may result if there are low diffusivities and solid solubilities. In many systems the degeneration process seems to depend on interfacial diffusion. In this case since little can be done to achieve low diffusivities then structural breakdown can best be prevented by selecting alloys having extremely sluggish interface kinetics. Before passing on to consider thermal cycling effects it is necessary to point out that the limited number of data available for temperature gradient ageing have been obtained at mean temperatures of the order of 0.96 T_E. This is much in excess of anticipated operating conditions for this class of alloy.

Thermal Cycling Effects

This is an important analogue of in-service behaviour and relatively complete results are available for the principal eutectic systems referred to above. For the γ/γ'-MC alloys serious degradation of strength occurs after thermal cycling (Billingham and Cooper, 1981; Cooper and Billingham, 1980; Dunlevey and Wallace, 1974). In contrast the γ/γ'-δ alloys seemed much less affected although some coarsening of the γ'-precipitates was reported. The damage produced by thermal cycling of directionally-grown in-situ composites is thus a recognised source of great concern and must be taken into consideration in all studies of potentially useful alloys.

NEW PROCESSES

Strengthening by Fractional Melting

Goodwin, Davami and Flemings (1980) developed a process in which an alloy is heated above its solidus temperature into a semi-solid state and compressed against a filter in order to remove liquid. The process starts with the alloy ingot somewhat richer in solute than the final required composition and during heating some solute-rich and impurity-rich lower melting point liquid is removed from the interdendritic spaces while at the same time some homogenisation of the residual primary phase occurs. Work was successfully carried out on complex aluminium alloys, especially Al - 8.0Zn - 4.0Mg - 2.1Cu - 0.21Cr which was reduced to Al - 4.2/6.5Zn - 2.0/3.0Mg - 0.7/2.4Cu - 0.19/0.25Cr after "squeezing". At the same time the amounts of some deleterious impurities with k<1, e.g. Fe,Si, were reduced. Quite substantial improvements in properties after working relative to conventionally produced alloys with the same final alloy content were reported. This approach could well be extended to other aluminium alloys and other alloy systems.

Stircasting (Rheocasting) Processes

Rheocasting. Conventional metal forming processes make use of either fully solid or fully liquid changes. It is not usual to employ semi-solid metals because the dendritic structure which forms during solidification cannot be deformed without cracking. However, it has been shown that if the cooling liquid is vigorously agitated during the initial stages of solidification a non-dendritic structure is developed which can be successfully processed in the semi-solid state (Spencer, Mehrabian and Flemings, 1972; Flemings, Riek and Young,1976). When the highly fluid semi-solid slurry is cast directly into shape the process is known as rheocasting. Although there is still much work needed to establish completely the fundamental mechanisms of slurry formation there is a considerable range of actual and potential applications for the processed material. In the main these involve preprocessing of the charge for subsequent use taking advantage of the thixotropic properties of the slurry. Depending upon the degree of remelting the resulting processes are known as thixocasting or thixoforging. One direct process with considerable promise is continuous (or semi-continuous) casting in which a strand is produced directly from the treated semi-solid. This offers the opportunity of high casting rates because of the removal of a substantial proportion of the latent heat before casting. It has been applied successfully to the continuous casting of aluminium alloys (Nurthen, 1981) and a model study has been described of continuous strip production (Matsumiya and Flemings, 1981).

Thixocasting

Much more progress has been made in the use of reheated rheocast material as a charge for diecasting. Both non-ferrous alloys (Backman, Mehrabian and Flemings, 1977; Fescetta and coworkers, 1973; Mehrabian and Flemings, 1972; Young and coworkers, 1976) and ferrous alloys (Law, Hostetler and Schulmeister, 1979; Oblak and Rand, 1976; Young, Riek and Flemings, 1979) have been successfully processed using diecasting machines. The resultant castings have a high degree of soundness which is in part attributable to the partial solidification which reduces the degree of shrinkage during final processing. There is little gas entrapment because of the lower temperatures than with liquid charges, die temperatures and thermal shock are much reduced (Backman, Mehrabian and Flemings, 1977). These advantages are of considerable importance in the diecasting of steels and since there is some evidence that the heat treatment response is improved with the thixocast structure there would appear to be potential for enhanced strengthening.

Thixoforging

This differs from thixocasting in that lower temperatures and higher solid fractions are made use of. The liquid-metal equivalent of thixoforging is squeeze casting (see below). Both processes use a pair of shaped dies for component production. Thixoforging could, however, have a distinct advantage because the partially-solidified charge is inclined to homogeneous deformation. Data have been published for thixoforged aluminium alloys (Chen and coworkers, 1979; Ramati and coworkers, 1978) although there is clearly much room for further development. Again good mechanical properties and enhanced heat treatment response have been reported.

Compocasting

In this process particulate or fibrous material is added to the semi-solid slurry prior to casting (Mehrabian, Riek and Flemings, 1974; Sato and Mehrabian, 1976). Particles and fibres of a variety of materials, including SiC, Al_2O_3 and glass, have been incorporated into alloys. The major gain seems to be in enhanced wear resistance (Hosking and coworkers, 1982). An important requirement is that the particles should be wetted by the liquid. This can be achieved by an interface interaction (Hosking and coworkers, 1982) or by coating the particles. This latter procedure has been used when introducing graphite particles (Pai and Rohatgi, 1978; Surappa and Rohatgi, 1978) although more recent work has shown that alloying additions to the matrix can induce wetting, allowing incorporation of even very fine uncoated particles ($\sim 5\mu m$ dia.) (Banerji and Rohatgi, 1982; Krishnan, Surappa and Rohatgi, 1981). Compocast materials have a low energy content and as such provide the opportunity of property enhancement at low cost.

Squeeze Casting

Squeeze casting involves the solidification under pressure of liquid metal in heated re-usable dies. As such it is effectively a combination of gravity die casting and closed die forging. The process has been the subject of recent reviews (Das and Chatterjee, 1981; Williams and Fisher, 1981) and since it is reported to improve strength and ductility it justifies inclusion as a means of strengthening from the melt. One important aspect of the process is the elimination of porosity. Additionally the manufactured components have excellent dimensional reproducibility. Of particular note is the report that the high strength 7000 series aluminium alloys can be successfully processed by squeeze casting (Williams and Fisher, 1981).

Laser Glazing

This is not so much a new process as a new solidification-related treatment that can be used to produce improvements in strength and performance. In its normal form (Kear, Breinan and Greenwald, 1979) the laser beam is focussed on the surface of a metal and scanned, producing rapid surface melting. Since there is intimate liquid-substrate contact, very rapid quenching can result. Depending on the rate of cooling (up to $10^8 Ks^{-1}$) quite different structures can result from fine dendritic forms, through homogeneous crystalline structures to amorphous structures. These can lead to superior surface mechanical properties as well as improved corrosion resistance. Aluminium-bronze (Draper, 1981), 304 stainless steel (Anthony and Cline, 1978), alloy steels (Kear, Brienan and Greenwald, 1979; Strutt and coworkers, 1978) and nickel-base superalloys (Kear, Brienan and Greenwald, 1979) are among the materials that have been effectively treated. However, there are still many basic data needed before the full capabilities of the process are defined.

ENHANCEMENT OF THE PROPERTIES OF WELDMENTS

In their review of solidification structures and properties of fusion welds, Davies and Garland (1975) pointed out that in the main the high level of understanding of the fundamental aspects of casting and solidification that had been built up in the previous twenty years, had not been widely directed to the study of weld-pool solidification. Then, as now, it was clear that the control of weld-pool solidification to produce weldments with enhanced properties was a desirable target. They defined methods whereby the fusion zone structure (and the associated properties) could be controlled. These methods involved (i) control by inoculants, (ii) control by stimulated surface nucleation, (iii) control by dynamic grain refinement, or (iv) grain refinement by arc modulation. The data available then were sparse (see, for example, Garland (1974); Tseng and Savage (1971)) and progress in the intervening period has been insubstantial. It must be acknowledged, on the other hand, that there is greater understanding of fusion zone behaviour even if real control of the as-solidified structure has rarely been achieved.

Grain Structure Control

Arc oscillation and pulsing of the welding current have been used effectively to grain refine tungsten inert gas (TIG) welds in tantalum (Grill, 1981a,b; Sharir, Pelleg and Grill, 1978). Additionally a significant increase in elongation and a slight increase in both yield and tensile strength were obtained. Similar changes in grain structure have also been reported for carbon steels (Ganaha and Kerr, 1978) and aluminium alloys (Ganaha, Pearce and Kerr, 1980) although in both these cases the grain refinement was a consequence of changes in the process parameters (heat input, welding speed) rather than because of the introduction of perturbations. Unfortunately no mechanical property data were presented for either of these latter two groups of alloys.

Fusion Zone Geometry

Apart from the internal structure the macroscopic geometry of the fusion zone and weld pool are known to have an effect on fusion zone properties especially the likelihood of solidification cracking (Davies and Garland, 1975). Thermal control of geometry has been made use of but it is surprising that more attention has not been given to minor element effects. While it is recognised that minor elements in the weld pool can change the weld pool shape through interaction with the arc (Metcalfe and Quigley, 1977; Savage, Nippes and Goodwin, 1977), changes in the liquid-vapour or solid-liquid interfacial energy (Bradstreet, 1968; Heiple and Roper, 1982; Roper and Olson, 1978), and alterations of fluid flow patterns in the weld pool (Woods and Milner, 1971), there does not seem to have been any systematic study of how the changes affect the properties and whether strength improvements can be achieved. It is feasible that enhancement could result since microscopic segregation patterns are also dependent on the fusion zone behaviour.

There are other aspects of weld solidification structures that could be dwelt upon if space had allowed. Not least of these is the solidification behaviour of stainless steel weldments, particularly since the solidification mode is known to affect hot tearing behaviour (for recent work see, for example, Cieslak, Ritter and Savage (1982); Gooch and Honeycombe, 1980; Lippold and Savage 1979,1980). However, this topic alone could sensibly occupy a Conference session.

REFERENCES

Allen, D.J. and Hunt, J.D. (1976) Metall. Trans.7A, 767-770.
Anthony, T.R. and Cline, H.E. (1978) J.Appl.Phys. 49, 1248-1255.
Bachelet, E. and Lesoult, G. (1978). In D.Coutsaradis and others (Eds.) "High Temperature Alloys for Gas Turbines", Applied Science, London. pp.665-691.
Backman, D.G., Mehrabian, R. and Flemings, M.C. (1977) Metall. Trans. 8B, 471-477.
Banerji, A. and Rohatgi, P.K. (1982) J.Mat.Sci.17, 335-342.
Basaran, M., Kattamis, T.Z., Mehrabian,R. and Flemings, M.C. (1973) Metall. Trans. 4, 2429-2434.
Beeley, P.R. (1979). In "Solidification and Casting of Metals", The Metals Society, London. pp.319-324.
Billingham, J. and Cooper, S.P. (1981) Metal Science 15, 311-316.
Boettinger, W.J., Biancaniello, F.S. and Coriell, S.R. (1981) Metall. Trans. 12A, 321-327.

Bradstreet, B.J. (1968) Welding J. 47, 314s-322s.
Briant, C.L. and Banerji, S.K. (1978). Int.Met.Reviews 23, 164-199.
Bridge, M.R. and Beech, J. (1982). In "Solidification Technology in Foundry and Casthouse", The Metals Society, London, in press.
Brimacombe, J.K. and Sorimachi, K. (1977) Metall. Trans. 8B, 489-505.
Burden, M.H., Hebditch, D.J. and Hunt, J.D. (1974) J.Crystal Growth 20, 121-127.
Burns, D. and Beech, J. (1974) Ironmaking and Steelmaking 1, 239-250.
Campbell, J. (1968). In "The Solidification of Metals" I.S.I., London. pp. 18-26.
Campbell, J. (1981) Int.Met.Reviews 26, 71-108.
Cantor, B. (1978) (Ed.) "Rapidly Quenched Metals III" The Metals Society, London.
Chadwick, G.A. (1973). Prog. Materials Science 12, 97-182.
Cieslak, M.J., Ritter, A.M. and Savage, W.F. (1982). Welding J. 60, 1s-8s.
Clegg, A.J. (1979) Metall. and Mat. Technologist 11, 85-93.
Coble, R.L. and Flemings, M.C. (1971) Metall. Trans. 2, 409-415.
Cooper, S.P. and Billingham, J. (1980) Metal Science 14, 225-229.
Copley, S.M., Giamei, A.F., Johnson, S.M. and Hornbecker, M.F. (1970). Metall. Trans. 1, 2193-2204.
Croft, N.H., Entwisle, A.R., and Davies, G.J. In Advances in the Physical Metallurgy of Steels, The Metals Society, London. 1983, in press.
Dann, P.C., Hogan, L.M. and Eady, J.A. (1979). Metals Forum 2, 212-219.
Das, A.A. and Chatterjee, S. (1981). Met.and Mat.Tech.13, (3) 137-142.
Davies, G.J. (1971). In A.Kelly and R.B.Nicholson (Ed.), Strengthening Methods in Crystals, Elsevier, London. Chap.9, pp.485-533.
Davies,G.J. (1974). In "Practical Metallic Composites", Inst.Metallurgists, London. pp.D1-D14.
Davies,G.J. and Garland, J.G. (1975). Int.Met.Reviews 20, 83-106.
Davies, J.R., Courtney, T.H. and Przystupa, M.A. (1980). Metall. Trans. 11A, 323-332.
Diserens, M., Hafonen, T., Ristimaki, E. and Tukiainen,M. (1981) Scand. J. Met. 10, 19-23.
Draper, C.W. (1981) J.Mat.Sci.16, 2774-80.
Dunlevey, F.M. and Wallace, J.F. (1974). Metall. Trans. 5, 1351-1356.
Farag, M.M., Matera, R. and Flemings, M.C. (1979) Metall. Trans. 10B, 381-388.
Fascetta, E.F., Riek, R.G., Mehrabian, R. and Flemings, M.C. (1973) Trans. AFS, 95-100.
Flemings, M.C. (1974a). Metall. Trans. 5, 2121-2134.
Flemings, M.C. (1974b). Solidification Processing, McGraw-Hill, New York, p.83.
Flemings, M.C. (1976). Scand. J. Met. 5, 1-15.
Flemings, M.C., Riek, R.G. and Young, K.P. Mat.Sci.Eng. 25, 103-117.
Fredriksson, H. and Svensson, I. (1976) Metall. Trans. 7B, 599-606.
Frydman, S.S. and Courtney, T.H. (1982) Metall. Trans. 13A, 967-973.
Garland, J.G. (1974) Metal Construction 6, 121-127.
Giamei, A.F. and Kear, B.H. (1970) Metall. Trans. 1, 2185-2192.
Giamei, A.F. and Tschinkel, J.G. (1976) Metall. Trans. 7A, 1427-1434.
Gladman, T and Pickering, F.B. (1967). J.I.S.I. 205, 653-664.
Gooch, T.G. and Honeycombe, J. (1980) Welding J. 59, 233s-241s.
Goodwin, F.E., Davami, P. and Flemings, M.C. (1980) Metall. Trans. 11A, 1777-1787.
Grill, A (1981a) Metall. Trans. 12B, 187-192.
Grill, A. (1981b) Metall. Trans. 12B, 667-674.
Heiple, C.R. and Roper, J.R. (1982) Welding J. 61, 97s-102s.
Hosking, F.M., Folgar Portillo, F. Wunderlin, R. and Mehrabian, R. (1982) J.Mat.Sci. 17, 477-498.

Hurtuk, D.J. and Travaras, A.A. (1977) Metall. Trans. 8B, 243-251.
Iwata, H., Yamada, K., Fujita, T. and Hayashi, K. (1976)
   Trans. I.S.I. Japan 16, 37-47.
Jackson, W.J. (1979) J.Research SCRATA No.47, 26-30.
Jackson, W.J. and Wright, J.C. (1977) Metals Tech. 4, 425-433.
Johnson, W.A. and Stoloff, N.S. (1980) Metall. Trans. 11A, 307-317.
Kattamis, T.Z., Coughlin, J.C. and Flemings, M.C. (1967)
   Trans. Met. Soc. AIME 239, 1504-1511.
Kear, B.H., Breinan, E.M. and Greenwald, L.E.(1979) Metals Tech.6, 121-129.
Kear, B.H. and Piearcey, B.J. (1967) Trans. Met. Soc. AIME 239, 1209-1215.
Kiessling, R. (1980) Metal Science 14, 161-172.
Kiessling, R and Lange, N. (1978). "Non-Metallic Inclusions in Steel",
   The Metals Society, London.
Krishnan, B.P., Surappa, M.K. and Rohatgi, P.K. (1981) J.Mat.Sci. 16,
   1209-1216.
Law, C.C., Hostetler, J.D. and Schulmeister, L.F. (1979) Mat.Sci.Eng.38,
   123-137.
Lemkey, F.D. and Thompson, E.R. (1971) Metall. Trans. 2, 1537-1544.
Lemkey, F.D. and Thompson, E.R. (1973). In "The Microstructure and Design
   of Alloys" Vol.1, The Metals Society, London. pp.271-275.
Lippold, J.C. and Savage, W.F. (1979) Welding J. 58, 362s-374s.
Lippold, J.C. and Savage, W.F. (1980) Welding J. 59, 48s-58s.
Lorig, C.H. and Elsea, A.R. (1947). Trans. A.F.S. 55, 160-174.
McCartney, D.G. and Hunt, J.D. (1981) Acta Met. 29, 1851-1863.
Masumoto, T. and Suzuki, K. (1982) (Eds.) "Rapidly Quenched Metals IV"
   Japan Inst. Metals, Sendai. Vols.1 and 2.
Matsumiya, T. and Flemings, M.C. (1981) Metall. Trans. 12B, 17-31.
Mehrabian, R. and Flemings, M.C. (1972) Trans. AFS 80, 173-182.
Mehrabian, R., Keane, M. and Flemings, M.C. (1970) Metall. Trans. 1,
   1209-1220.
Mehrabian, R., Riek, R.G. and Flemings, M.C. (1974) Metall. Trans. 5,
   1899-1905.
Melford, D.A. (1980) Ironmaking and Steelmaking 7, 89-92.
Melford, D.A., Whittington, K.R., Funnell, G.D. and Armstrong, G.R. (1982).
   In "Continuous Casting" The Metals Society, London, Paper B3
Metcalfe, J.C. and Quigley, M.B.C. (1977) Welding J. 56, 133s-139s.
Mollard, F.R. and Flemings, M.C. (1967a) Trans. Met. Soc. AIME 239,
   1526-1533.
Mollard, F.R. and Flemings, M.C. (1967b) Trans. Met. Soc. AIME 239,
   1534-1546.
Nurthen, P.D. (1981) Metals and Materials, (Oct.) 25-26.
Oblak, J.M. and Rand, W.H. (1976) Metall. Trans. 7B, 699-703 and 705-709.
Offerman, C. (1981) Scand. J. Met. 10, 25-28.
Pai, B.C. and Rohatgi, P.K. (1978) J.Mat.Sci. 13, 329-335.
Pearson, D.D. and Lemkey, F.D. (1979). In "Solidification and Casting of
   Metals" The Metals Society, London. pp.526-532.
Piwonka, T.S. and Flemings, M.C. (1966) Trans.Met.Soc.AIME 236, 1157-1165.
Pook, L.P., Greenan, A.F., Found, M.S. and Jackson, W.J. (1980)
   J.Research SCRATA No.49, 28-35.
Quested, P.N., Miles, D.E. and McLean, M. (1980). Metals Tech. 7, 433-439.
Reeves, J.J. and Kattamis, T.Z. (1971). Scripta Met. 5, 223-229.
Roper, J.R. and Olson, D.C. (1978) Welding J. 57, 104s-107s.
Sato, A. and Mehrabian, R. (1976) Metall. Trans. 7B, 443-451.
Saucedo, I.G., Beech, J. and Davies, G.J. (1980). In "Special Melting"
   (Proc. 6th Int. Vacuum Metallurgy Conf.) Amer. Vacuum Soc., New York.
   pp.885-904.
Saucedo, I.G., Beech, J. and Davies, G.J. (1982a). In "Solidification Tech-
   nology in Foundry and Casthouse", The Metals Society, London. in press.

Saucedo, I.G., Beech, J. and Davies, G.J. (1982b) Metals Tech.9,in press.
Savage, W.F., Nippes, E.F. and Goodwin, G.M.,(1977) Welding J. 56, 126s-132s.
Schneider, P. (1974) Foundry Trade J. 136, 723-733.
Schrewe, H. (1981) Ironmaking and Steelmaking 8, 85-90.
Selby, K. and Hough, M.R. (1980) J.Research SCRATA No.51, 17-23.
Sharir, Y., Pelleg, J. and Grill, A. (1978) Metals Tech. 5, 190-196.
Sharman, S.G., Sims, B.J. and Ashton, M.C. (1980). J.Research SCRATA No.48, 16-19.
Spittle, J.A. and Lloyd, D.M. (1977). In "Solidification and Casting of Metals", The Metals Society, London, pp.15-20.
Spencer, D.B., Mehrabian, R. and Flemings, M.C. (1972) Metall. Trans.3, 1925-1932.
Stevens, R.A. and Flewitt, P.E.J. (1980) Metal Science 14, 81-88.
Streat, N. and Weinberg, F. (1976) Metall. Trans. 7B, 417-423.
Strutt, P.R., Nowotny,H., Tuli, M. and Kear, B.H. (1978) Mat.Sci.Eng.36, 217-222.
Surappa, M.K. and Rohatgi, P.K. (1978) Metals Tech. 5, 358-361.
Thompson, E.R. and Lemkey, F.D. (1969) Trans ASM 62, 140-154.
Tomono, H., Kurz,W. and Heinemann, W. (1981) Metall.Trans.12B, 409-411.
Tseng, T.F. and Savage, W.F. (1971) Welding J. 50, 777-786.
Van den Boomgaard, J. and Wolff, L.R. (1972) J.Crystal Growth 15, 11-15.
Ver Snyder, F.L. and Shank, M.E. (1970) Mat.Sci.Eng. 6, 213-247.
Weinberg, F. (1975). Metall. Trans. 6A, 1971-1985.
Widdowson, R. and Marr, H.S. (1979). In "Solidification and Casting of Metals" The Metals Society, London. pp.547-552.
Williams, G. and Fisher, K.M. (1981) Metals Tech. 8, 263-267.
Woodfine, B.C. and Quarrell, A.G. (1960) J.I.S.I. 195, 409-414.
Woods, R.A. and Milner, D.R. (1971) Welding J. 50, 163s-173s.
Wray, P.J. (1981) Metall. Trans. 12B, 167-176.
Wright, J.A. and Quarrell, A.G. (1962) J.I.S.I. 200, 299-307.
Young, K.P. and Kirkwood, D.H. (1975) Metall. Trans. 6A, 197-205.
Young, K.P., Riek, R.G. and Flemings, M.C. (1979) Metals Tech. 6, 130-137.
Young, K.P., Riek, R.G., Boylan, J.F., Bye, R.L., Bond, B.E. and Flemings, M.C. (1976) Trans AFS 85, 169-174.

ADDITIONAL REFERENCES

Chen, C.Y., Sekhar,J.A., Backman, D.G. and Mehrabian, R. (1979) Mat.Sci.Eng. 40, 265-272.
Ganaha, T. and Kerr, H.W. (1978) Metals Tech.5, 62-69.
Ganaha, T., Pearce, B.P. and Kerr, H.W. (1980) Metall.Trans.11A, 1351-1359.
Holder, J.D. and Oliver, B.F. (1974) Metall.Trans.5, 2423-2437.
McLean, M. (1982) Metal Science 16, 31-36.
Ramati, S.D.E., Abbaschian, G.J., Backman, D.G. and Mehrabian, R. (1978) Metall.Trans.9B, 279-286.

Deformation Problems in Minerals and Rocks

M. S. Paterson

Research School of Earth Sciences, Australian National University, Canberra 2600, Australia

ABSTRACT

The nature of the deformation processes in minerals and rocks is briefly reviewed with special reference to similarities and differences relative to metals. Illustrative examples are given of calcite, quartz and olivine-rich rocks and the problem of extrapolation to geological conditions is discussed.

KEYWORDS

Plastic deformation, minerals, rocks, marble, quartz, dunite, brittle-ductile transition, deformation mechanism map, geological extrapolation.

INTRODUCTION

Rocks are polycrystalline bodies, made up of grains that comprise one, or more often, several types of minerals. An interest in the mechanical properties of rocks therefore leads one to study both the behaviour of the single crystals of the individual minerals and the behaviour of their polycrystalline aggregates, the rocks themselves, in direct parallel with the study of metal single crystals and their polycrystalline aggregates in metallurgy. However, compared with metals, minerals and rocks involve additional complexities and introduce a number of novel and interesting aspects to the study of mechanical behaviour.

In this brief survey we shall first consider the nature of minerals and rocks from a materials point of view, attempting to identify the principal similarities and differences compared with metals. Next, the brittle-ductile transition as a function of confining pressure will be touched upon, which introduces a number of properties that are of particular interest in rocks deformed under compressive conditions. Finally, we shall consider briefly some typical examples of plastic deformation of minerals and rocks and discuss the problem of extrapolating to geological conditions.

THE NATURE OF PLASTICITY IN MINERALS AND ROCKS

The same basic processes of crystal plasticity, slip and twinning, are observed in minerals as in metals (they were, in fact, discovered first in minerals), but important differences in the dynamics of these processes arise from difference in crystal structure and bonding. It is sufficient here to point to three of these differences relative to typical metals:
(1) From a purely geometrical point of view, the unit cells of minerals in general contain a number of atoms and these are of several kinds; consequently the Burgers vectors for slip are relatively large and the diffusion processes complex.
(2) The bonding between the atoms is normally of mixed ionic and covalent character, tending to be more directional than in metals.
(3) Owing to there being more than one kind of atom, a complex defect chemistry is potentially involved, including electrically charged defects.

Special attention may also be drawn to the silicon-oxygen bond which is an important factor in determining the properties of quartz and the silicates, the chief family of rock-forming minerals. This bond is strong and directional and is evidently not easily disrupted in the passage of dislocations; some analogy with the behaviour of the covalently bonded germanium and silicon may therefore be expected. The other families of minerals that are of interest for their deformation properties are the carbonates, sulphides and halides; these in general tend to behave mechanically more like the archetypal ionic crystal, sodium chloride.

The deformation phenomena in rocks are, again, generally similar to those in polycrystalline metals but the factors just mentioned, which tend to make the constituent grains more difficult to deform than in the case of metals, are even more effective in promoting brittleness in the rocks, even at high temperatures. The following are particularly important factors in the study of the plastic deformation of rocks:

(1) The deformation is normally carried out under a confining pressure. The reason for doing so is usually more to promote ductility than to simulate the triaxial nature of geological stress states. The differential stress or stress difference, that is, the stress component superimposed on the confining pressure to give the nonhydrostatic conditions for plastic flow, may be compressive or tensile but the overall stress state is generally compressive in that all three principal stress components are normally non-zero and compressive. The use of a confining pressure may also be necessary at times to ensure that the experiments are conducted within the desired fields of stability of the phases present in the rock.

(2) Many minerals are of relatively low crystallographic symmetry and so such potential slip systems as they may possess tend to be of low multiplicity. Therefore, as grains in a polycrystal, minerals tend to lack sufficient slip or twinning systems to meet intergranular compatibility requirements, as expressed in the von Mises criterion of five independent slip systems for homogeneous deformation or in any slightly less stringent criterion that may apply when grain-to-grain inhomogeneity of deformation is allowed for. This factor, on the one hand, is an important one in leading to brittleness of rocks and, on the other hand, lends a greater potential importance to alternative or supplementary deformation processes such as diffusional flow, grain-boundary sliding and even microcracking itself (as a "cataclastic" component in ductile behaviour). Another possible consequence of low symmetry in the grains is to force the activity of several non-equivalent slip systems, the relative strengths of which depend

differently on temperature and strain rate so that a complexity of texture transitions with change in deformation conditions can appear in the study of crystallographic preferred orientations (Lister and Paterson, 1979).

(3) Grain boundary processes are potentially of greater relative importance in rocks than in metals, partly because of the inadequacy of intracrystalline mechanisms just pointed to but partly also because of there being many situations where a fluid phase is present. Rocks tend to have a significant amount of porosity and permeability which, in natural situations, leads to the presence and movement of fluid phases, usually aqueous; in other cases, partial melting may be of importance. Even minute amounts of water present in a rock may have an important influence on its high temperature rheology, as will be illustrated later. Also in fine-grained rocks, a mechanism of relative grain movement analogous to that in superplasticity in metals can be demonstrated.

(4) Recrystallization commonly accompanies high temperature deformation in rocks. However, this is found to involve not only the classical nucleation and growth process but also a process, not normally seen in metals, of progressive relative rotation of subgrains to develop high angle grain boundaries in situ (Hobbs, 1968). Another recrystallization phenomenon, of much interest to structural geologists concerned with metamorphic rocks, is the growth of new mineral phases in a rock that is initially not in mineralogical equilibrium under the conditions of deformation; little progress has yet been made in this area experimentally because of severe kinetic difficulties.

We shall now move on to consider some specific topics in experimental rock deformation.

THE BRITTLE-DUCTILE TRANSITION

The transition from brittle to ductile behaviour with increasing confining pressure is of special interest in the study of rock deformation and introduces some interesting features associated with predominantly compressive states of stress. Thus, even in the brittle field, significant departure from elastic behaviour is usually detectable in a rock under uniaxial compressive stresses above about one-third to two-thirds of the macroscopic compressive fracture stress. This departure from elastic behaviour is accompanied by inelastic dilatation of the specimen, increase in the rate of acoustic emission, changes in elastic wave velocities, and other changes in properties (see review by Paterson, 1978, Chapter 7). All these changes can be attributed to the proliferation of microcracks throughout the specimen. Virgin rock already contains a population of microcracks and this population grows and is modified and augmented under stress. Thus the microcrack plays a role as a fundamental entity underlying inelastic behaviour in the brittle or near-brittle field, somewhat analogous to the role of the dislocation in crystal plasticity.

The development of a macroscopic fracture (normally a shear fracture in the case of predominantly compressive conditions) can be viewed as arising out of an instability in the pattern of proliferation of microcracking whereby the population of microcracks becomes locally concentrated and eventually leads to a complete disruption of cohesion. However, increase in the hydrostatic component of the stress tends to inhibit the onset of this instability leading to two consequences for ductility. On the one hand, the continued proliferation of microcracking increases the possibilities of relative movement of fracture fragments and thereby the contribution of the so-called cataclastic component of deformation (which is rather analogous to the deformation mechanism of a granular material such as sand); this is the first factor contributing to an increase in macroscopic ductility. At the same time, higher stress differences are required to continue the microfracturing processes under higher confining pressures, thereby increasing the chance of reaching the plastic yield stress of the grains; this is the second and generally more important factor contributing to increasing ductility with pressure. Such a progression with increasing pressure from brittle fracture, through intermediate stages involving substantial amounts of cataclastic flow, to a ductile regime involving only intragranular plasticity is well illustrated in the behaviour of marble at room temperature (Paterson, 1978, Chapter 8).

However, in silicate rocks, increase in pressure alone is generally inadequate at room temperature to induce ductility and an increase in temperature is required as well. Thus for geological application, the interest mainly lies in deformation behaviour under simultaneous high pressure and high temperature, conditions which also more closely simulate the geological environment. We now consider some high pressure-high temperature studies that illustrate some of the simpler problems that arise in this field of rock deformation.

SOME PARTICULAR DEFORMATION PROBLEMS

Marble and Limestone. The plastic deformation of marble and limestone, rocks consisting essentially of polycrystalline calcite ($CaCO_3$), is readily achieved at relatively low confining pressures and temperatures. It represents a rather straight-forward case of plastic deformation involving both slip and twinning in the calcite grains, which behave as typical ionic crystals. However, from the rheological measurements and microscopical observations (Heard, 1963; Heard and Raleigh, 1972; Rutter, 1974; Schmid et al., 1977, 1980), several rheological regimes have been recognized which appear to reflect changes in detail in the underlying mechanisms, as follows:

(1) At relatively low temperatures ($<\sim 600°C$) and high differential stresses (> 100 to 200 MPa), there is work hardening behaviour involving both twinning and slip which can approach a steady state creep with exponential stress dependence in the upper part of this temperature range.

(2) With increase in temperature and decrease in stress, there is a change to a power creep law and twinning is no longer an important deformation mechanism. In the coarser-grained marbles, the stress exponent is at first high ($n \approx 7-8$) and there is an obvious inhomogeneity or "core and mantle" distribution of deformation within the grains. This gives way at lower differential stresses ($<\sim 20$ MPa) to flow with a lower stress exponent ($n \approx 4$) and more nearly homogeneous deformation within the grains, which also involves marked development of subgrains as well as the in situ recrystallization by subgrain rotation, mentioned earlier; the size of subgrains and recrystallized grains is inversely proportional to the differential stress.

(3) In very fine-grained limestone (~ 4 μm), the power-law regime with intermediate stress exponent (~ 4) borders immediately on the exponential creep regime but in turn gives way at stresses of a few hundred MPa and lower to a regime of low stress exponent ($n \approx 2$). Microscopical evidence

indicates that in the latter regime grain boundary sliding makes a major contribution to the strain and that the mode of deformation therefore corresponds to that known as superplasticity in metals. This represents the first experimental observation of such a deformation mode in rocks, although it had already been proposed as occurring in nature (Boullier and Gueguen, 1975).

Thus, apart from details such as the mechanism of recrystallization, the deformational behaviour of marble and limestone shows much that is parallel to that of typical metals. Moreover, similarities in microstructure indicate that the mechanisms involved in such rocks in nature are of the same type, and extrapolation of the flow laws suggests stress levels for the natural deformation that would seem to be realistic (Schmid, 1975; Schmid et al., 1977). However, when we move on to quartz and silicate rocks, the situation is much less clear.

Quartz. There is widespread evidence that rocks consisting predominantly of quartz (SiO_2) commonly undergo large plastic deformations under geological conditions. However, for a long time, attempts to deform quartz rocks or single crystals in the laboratory either failed or involved flow stresses which were comparable to the "theoretical strength" and were too high to be geologically realistic. Thus great importance attached to the discovery of "hydrolytic weakening" in quartz by Griggs and Blacic (1964, 1965), which appeared to resolve the apparent conflict by demonstrating that quartz can be readily deformed in the presence of water, which is, of course, ubiquitous in nature; it also raises the possibility that a similar effect may be important in all silicates where deformation involves the breaking of Si - O bonds.

The original observation on hydrolytic weakening was that, after a natural quartz crystal had been heated for a number of hours in the presence of water from dehydrating talc at around 900°C under 1500 MPa confining pressure in a solid-medium high pressure apparatus, the quartz could be plastically deformed at relatively low differential stresses, of the order of 200 MPa. Soon after making this observation, Griggs and Blacic discovered that certain synthetic quartz crystals were also readily deformable provided the temperature was above a critical value, T_c; these crystals showed a broad infrared absorption band near 3 µm, corresponding to the presence of OH bonds, and T_c was shown to be a function of the "water" content as determined by the integrated absorption in the 3 µm band. Many studies have since been done on the deformation properties of "wet" synthetic quartz crystals, from which it would appear that the hydrolytic weakening effect involves both the lowering of the Peierls stress and the facilitating of the climb of dislocations and hence of recovery processes.

In spite of the studies just referred to, there is still much that is not clear in the phenomenology of hydrolytic weakening in quartz, including the dependence on pressure and temperature of the solubility and rates of diffusion of the water-related species, the proper characterization of this species as a crystal defect, and the question of whether there are any other factors that are also essential to the hydrolytic weakening effect. In view of this situation, attempts to identify the mechanism of hydrolytic weakening are necessarily somewhat speculative. However, it seems appropriate to mention two recent proposals because they are also of interest in indicating the sort of factors that may need to be taken into account in discussing the deformation of minerals.

It has been noted that there is some similarity in yielding behaviour between hydrolytically weakened quartz and the semiconductor crystals Ge and Si (Hobbs et al., 1972). Hirsch (1981) has pressed this comparison further and proposed that the role of the water-related defect in quartz may be analogous to that of electrically active dopants in semiconductor crystals; he envisages an effect such as a lowering of the Fermi level which would lead to an increase in the equilibrium concentration of negatively charged kinks on the dislocations and hence to increased dislocation mobility, with some similar influence on diffusion and hence on climb and recovery rates. The proposal of Hobbs (1981) also points to the possibility that the role of "water" is an electronic one; he discusses the defect chemistry of quartz in some detail, suggesting that the "water" gives rise to charged defects that perturb the Fermi level and thereby influence the concentration of other charged species, especially oxygen vacancies, which in turn influences the diffusion and hence climb and recovery rates.

In order to pursue these ideas, we require, firstly, a clearer picture of the hydrolytic weakening phenomenon, especially in respect of its kinetics and its dependence on pressure, temperature, and possibly other impurities, and, secondly, a better characterization of the weakening species, including its influence on the electronic structure of quartz. An interesting current development in the latter connection is a study by McLaren and coworkers (to be published) on the physics of the precipitation of water in "wet" quartz crystals, involving measurements on the kinetics of the precipitation, on the changes in bulk density and in infrared absorption, and on the growth of dislocation loops associated with the growing bubbles. These observations are very neatly rationalized by a model of the precipitation which assumes that the water-related entity in solid solution is the $(4H)_{Si}$ defect (a substitution of four hydrogen atoms for a silicon atom, as in hydrogarnets) or an association of these defects.

Dunite. Although a relatively rare rock in nature, dunite has assumed considerable importance in experimental studies and will be discussed briefly here for several reasons:
(1) it serves to typify the problems that arise in the deformation of silicate rocks;
(2) it enables these problems to be studied in relatively simple circumstances because of its predominantly monomineralic nature (it consists mainly of olivine, $(Mg,Fe)_2SiO_4$);
(3) its rheology is of considerable geophysical interest because olivine is believed to be the principal constituent of the Earth's upper mantle.

The main observations on approximately steady-state flow of dunite in the laboratory are those of Carter and Ave'Lallemant (1970), Kirby and Raleigh (1973) and Post (1977), all obtained with solid-medium apparatus, and those of Chopra and Paterson (1981) obtained with gas apparatus. Although the experimental variations are considerable, there is a measure of agreement between the various studies, especially under "wet" conditions; in the latter case, if a power law is fitted the apparent activation energy for steady-state creep is found to be fairly high, around 400-500 kJ mol^{-1}, while the stress exponent n lies in the normal range 3-5 at higher stresses but falls below 3 at lower stresses (below 100-200 MPa). However, more refined work on two different dunites in the temperature range 1100-1300°C under "wet" conditions, as defined by the dehydration of small amounts of hydrous minerals initially present, reveals that there is an inverse dependence of flow stress on grain size (Chopra and Paterson, 1981). This observation strongly suggests that there is a weakening effect associated with the grain boundaries. In confirmation of the grain boundary weakening, when the rocks are carefully pre-dried by heating above 1000°C at an oxygen fugacity within the olivine stability field the strength is markedly increased and the inverse grain size dependence disappears.

Optical and electron microscope studies are still in progress on the "wet" dunite specimens but preliminary studies (FitzGerald, Chopra and Paterson, to be published) reveal the presence of an amorphous phase in the grain boundaries, which was presumably molten during the deformation. Similar observations have been made on polycrystalline quartz deformed under wet conditions (Mainprice and Paterson, to be published). These observations are thought to be very significant for experimental rock deformation studies for the following reasons:
(1) Grain boundary effects can strongly influence the observed rheological behaviour of rocks; moreover, since the effects are very sensitive to the actual composition and structure of the grain boundary region, it is possible that through them the rheological behaviour of the rock may be rather sensitive to minor bulk compositional or structural variations.
(2) Since minor degrees of alteration commonly occur in the grain boundary regions in rocks during weathering or minor retrograde metamorphism (for example, minor serpentinization in dunite), the deformation behaviour measured in the laboratory on as-received rock may not truly represent the behaviour of the parent rock as it existed under the original geological conditions.
(3) The effect of water-rich phases in the grain boundaries may obscure, or make difficult to establish, any hydrolytic weakening within the grains of the rock.

It is therefore evident that much more detailed studies are required on the structure and behaviour of grain boundaries in rocks.

EXTRAPOLATION TO GEOLOGICAL CONDITIONS

Apart from the materials science interest, one of the principal aims of rock deformation studies is to throw light on geological parameters, for example, on the stresses associated with natural deformations. However, relating the laboratory observations to geological conditions involves a large extrapolation in strain rate, from the laboratory range of around 10^{-3} to 10^{-7} s^{-1} (for substantial strains) to the geological range of 10^{-10} to 10^{-14} s^{-1}, an extrapolation of at least as many orders of magnitude in strain rate as those over which measurements have been made. Clearly no great precision can be expected. Moreover, such extrapolation is only valid if made within the same deformation regime in respect of mechanism and flow law.

In the previous sections we have seen that the same rock can show several distinct types of deformation behaviour under different conditions. Therefore extrapolation from a particular regime for which measurements have been obtained in the laboratory is not necessarily relevant to given geological conditions; it must first be demonstrated that the same mechanism and flow law is applicable under the geological conditions as under the laboratory conditions from which extrapolation is being done. This requirement makes it very important to study the microscopical evidence relating to the mechanism of deformation both in the laboratory specimens and in the rocks in the field in order to establish similarity of mechanism. Conversely it poses the problem of selecting laboratory conditions that will enable one to study deformation processes of the same type as occur geologically. The situation is most conveniently depicted and appreciated with reference to a deformation mechanism map, such as proposed by Ashby (1972), where fields for different mechanisms are depicted in a stress-temperature plot with constant strain rate contours, or, perhaps more conveniently for the present purpose, such a map projected on the strain rate-temperature plane with constant stress contours (an example is shown in Fig. 1).

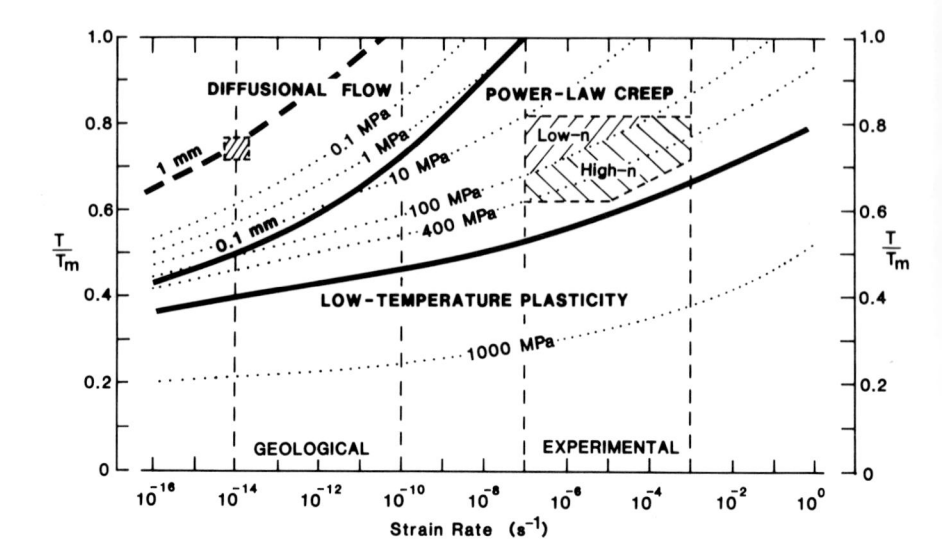

Fig. 1. Flow regime map for polycrystalline olivine, derived from Stocker and Ashby (1973) and Ashby and Verrall (1978). Heavy lines delineate flow regime boundaries (alternative boundaries between diffusional flow and power-law creep are shown for 1mm and 0.1mm grain sizes). The dotted lines represent constant stress contours. The shaded area for experimental strain rates represents roughly the experimental range so far explored, with the two fields for high-n and low-n power laws found by Chopra and Paterson (1981) shaded differently. The small shaded area at 10^{-14} strain rate indicates approximately the conditions of interest for upper mantle flow in the earth; however, it should be noted that the increased hydrostatic pressure at depth in the earth can be expected to raise the stress contours significantly from those shown (Ashby and Verrall, 1978).

The data needed for plotting deformation mechanism maps for rocks are so far very sparse and any maps calculated are necessarily rather speculative. Moreover, severe difficulties are likely to become evident in relating laboratory and geological situations in practice where either the geological conditions fall in a strain rate-temperature field that does not extend to sufficiently large strain rates to permit large-strain experiments in the laboratory or the potential extension of the field to large strain rates only covers a high temperature range in which extraneous factors such as partial melting or phase transformations prevent relevant experimentation; raising the stress in order to achieve laboratory strain rates at lower temperatures in such cases would simply take one into an irrelevant mechanism field. Thus, while fairly straightforward extrapolation from laboratory to geological conditions may be possible in some cases, such as in calcite rocks where geological deformation tends to occur at relatively low temperatures and the rocks are stable in the laboratory to much higher temperatures, in other cases indirect means may have to be found for relating laboratory and geological studies.

REFERENCES

Ashby, M.F. (1972). Acta Met., 20, 887-897.
Ashby, M.F. and Verrall, R.A. (1977). Phil. Trans. R. Soc. Lond., A288, 59-95.
Carter, N.L. and Ave'Lallemant, H.G. (1970). Bull. Geol. Soc. Amer., 81, 2181-2202.
Chopra, P.N. and Paterson, M.S. (1981). Tectonophysics, 78, 453-473.
Griggs, D.T. and Blacic, J.D. (1964). EOS Trans. Amer. Geophys. Union, 45, 102 (abs.).
Griggs, D.T. and Blacic, J.D. (1965). Science, 147, 292-295.
Boullier, A.M. and Gueguen, Y. (1975). Contrib. Mineral. Petrol., 50, 93-104.
Heard, H.C. (1963). J. Geol., 71, 162-195.
Heard, H.C. and Raleigh, C.B. (1972). Bull. Geol. Soc. Amer., 83, 935-956.
Hirsch, P.B. (1981). J. de Phys., 42, Colloqu.C3, Suppl. No. 6, C3-149 to C3-159.
Hobbs, B.E. (1968). Tectonophysics, 6, 353-401.
Hobbs, B.E. (1981). Tectonophysics, 78, 335-383.
Hobbs, B.E., McLaren, A.C. and Paterson, M.S. (1972). In Flow and Fracture of Rocks (Ed. H.C. Heard and others), Amer. Geophys. Union, Monogr. 16, 29-53.
Kirby, S.H. and Raleigh, C.B. (1973). Tectonophysics, 19, 165-194.
Lister, G.S. and Paterson, M.S. (1979). J. Structural Geol., 1, 99-115.
Paterson, M.S. (1978). Experimental Rock Deformation: The Brittle Field, Springer-Verlag, Heidelberg.
Post, R.L. (1977). Tectonophysics, 42, 75-110.
Rutter, E.H. (1974). Tectonophysics, 22, 311-334.
Schmid, S.M. (1975). Eclogae Geol. Helv., 68, 251-284.
Schmid, S.M., Boland, J.N. and Paterson, M.S. (1977). Tectonophysics, 43, 257-291.
Schmid, S.M., Paterson, M.S. and Boland, J.N. (1980). Tectonophysics, 65, 245-280.
Stocker, R.L. and Ashby, M.F. (1973). Rev. Geophys. Space Phys., 11, 391-426.

Design Against Variable Amplitude Fatigue — an approach through Cyclic Stress-Strain Response

Campbell Laird, Fernando Lorenzo and Alex S. Cheng

Department of Materials Science and Engineering, University of Pennsylvania, Philadelphia, PA 19104, USA

ABSTRACT

Some of the unsolved problems in variable amplitude fatigue are described, and we present a method of attacking them through cyclic stress-strain response. Thus we review cyclic stress-strain behavior and its structure sensitivity at low amplitudes, taking account of the effects of variable loading. The role of mean stress, and how to handle it in variable amplitude fatigue, is explored, and also the load interaction mechanism on fracture initiation. The method is appropriate to "smooth" specimens and components in which crack propagation kinetics are less important for determining life.

KEYWORDS

Variable amplitude fatigue, cyclic stress-strain response, dislocation structure, persistent slip bands, cyclic creep, mean stress effects, crack initiation, load interaction effects, damage summation.

INTRODUCTION

The organizers of this 6th International Conference requested us to contribute a paper which covered a practical aspect of the fatigue problem and which also emphasized the microstructure-sensitive aspect. One of the most difficult unsolved practical problems in fatigue concerns life prediction under variable loading, particularly when dealing with "smooth" specimens or components. Certainly the problem of predicting crack propagation rate under variable loads has received considerable attention and seems to be quite well in hand. The fracture mechanics approach is not appropriate, however, to the design of many small components such as engine parts, axles, gears, bearings, and components in such structures as helicopter rotor heads, machine tools, and energy converting machines. We feel, therefore, that there is room to improve significantly the traditional methods of designing against fatigue, and our purpose is to describe results which we hope will be helpful in this direction.

Some of the problems of coping with fatigue in variable loading can be focused with reference to Fig. 1, which shows schematically a typical segment of a load- (or strain-) time recording in a fatigued component. Five peaks and valleys in this recording segment have been labelled A to E. Because of the well-known difficulty of defining a cycle within such a recording, we shall follow here the now-established method of counting "reversals" for summing fatigue damage. A reversal is counted whenever the direction of loading changes, i.e. there are five reversals in the segment shown in Fig. 1.

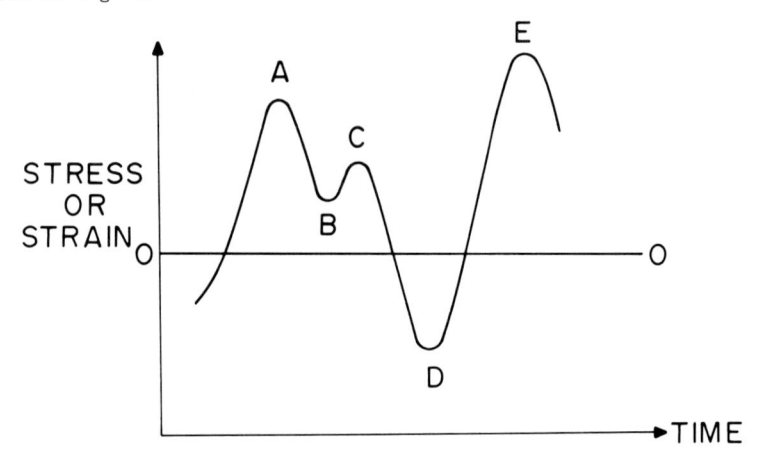

Fig. 1. A typical segment of a variable load (strain)-time recording.

One of the problems inherent in summing the fatigue damage of variable loads is deciding which reversals to include or exclude in the count. For example, the sequence A, B, C, D, could be counted as three reversals AB, BC and CD or simply as one reversal AD, which might be more damaging than the alternative three reversals because of its much larger range. Advances in counting have been considerable in recent years, culminating in the "rainflow counting" or "pagoda roof" method which has gained some acceptance (Socie, 1975; Mitchell, 1979). Such methods depend on the cyclic stress-strain response of the material, and implicitly contain a limited "load-interaction" effect (What effect has a large excursion on an immediately-subsequent small excursion?) through such phenomena as cyclic hardening or softening. Accurate counting and the inclusion of load interaction effects must depend for their success on a thorough knowledge of cyclic response under variable loads. In the next section, we describe recent results on such response, although we are still far from obtaining the understanding necessary.

In subsequent sections, we treat two other problems, as follows: a) Note that, in Fig. 1, the reversals AB, BC and DE have tensile mean stresses, while reversal CD has a zero mean stress. How can one account for non-zero mean stresses in the damage summation? b) How can one also account for the load-interaction effect on <u>fracture mechanisms</u>, which presumably is different from that in cyclic response? This problem may turn out to be one of the most complicated and difficult. Within each section we indicate, where possible, the microstructure-sensitive facets of the behavior. We emphasize that our purpose is not to improve a method of summing damage, but to point out the underlying science which might be helpful for improvements to be developed later, especially for methods based on cyclic stress-strain response such as that of Wetzel (1971).

CYCLIC STRESS-STRAIN RESPONSE UNDER VARIABLE LOADING -- PURE METALS OR SIMPLE ALLOYS

Cyclic stress-strain response is the term given to cover the hardening which occurs in most soft ductile metals when subjected to cyclic strain, or the softening which can occur in a metal hardened by some means (cold work, alloying, or by cycling previously at high strain). Most pure metals and simple alloys (e.g. solid solutions) undergo large degrees of hardening under cyclic plastic strains even if the strain amplitude, usually maintained constant, is quite small (for review, see Laird, 1981). Bcc metals, in many circumstances, are unique in showing little or no hardening at constant strain amplitudes less than 10^{-3}. Depending on strain rate and interstitial alloy content, bcc metals can differ significantly in cyclic response behavior from that of fcc metals. Lack of space here prevents discussion of bcc behavior (for review, see Laird, 1981, or Mughrabi, Herz and Stark, 1981). However, in many respects the cyclic response of commercial bcc alloys is similar to that of fcc metals, which are used here for examples almost exclusively.

In both single crystals and polycrystalline metals, the cyclic hardening rate is rapid early in life, and for a wide range of monocrystalline orientation, the rate is not affected by orientation provided the strain amplitude is low ($<2 \times 10^{-3}$). As the applied (constant) strain amplitude is increased, the rate of hardening is lower for orientations which are definitely single slip, but the overall hardening rate increases with increasing strain amplitude in both mono- and polycrystals (Laird, 1981).

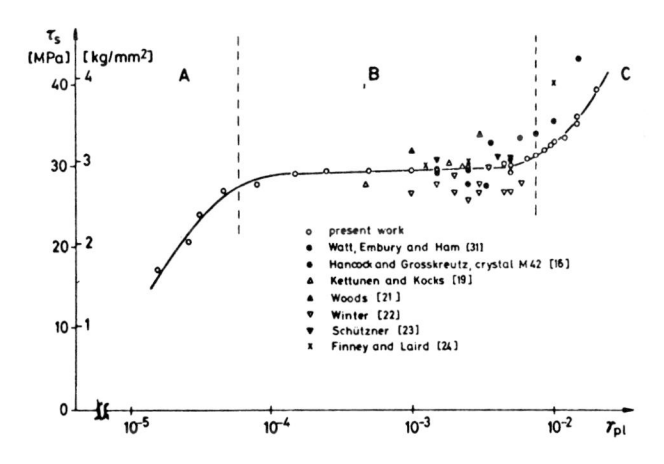

Fig. 2. The cyclic stress-strain curve for copper single crystals: resolved shear stress versus plastic resolved shear strain amplitude (courtesy of Mughrabi, 1978).

One of the most typical features of cyclic hardening is that the hardening rate falls to zero at large cumulative strains (the strain from one reversal to the next is summed without respect to sign) and the material is then described as being in saturation. A plot of the saturation stress versus the applied plastic strain amplitude is termed the cyclic stress-strain curve and it is, of course, most easily defined for tests at constant strain amplitude. The issue of whether such a curve can apply for cycling under variable amplitudes has frequently been addressed (Wetzel, 1971; Laird, 1977; Koibuchi and Kotani, 1973) and we examine it here again in the light of recent advances obtained from results on monocrystals. Since saturation is reached early in fatigue life and the dislocation micro-structure associated with it dominates the fatigue mechanisms, especially in long life fatigue, we assume here that hardening transients have little effect on fracture behavior in variable loading fatigue. We therefore focus on saturation behavior.

The cyclic stress-strain curve for copper single crystals in single slip orientation and for constant applied strains is shown in Fig. 2. It will be noted that the curve contains a plateau for which the stress is constant over the range of applied strain 6×10^{-5} to 7.5×10^{-3}. Within this range occur persistent slip bands (PSB's), the major source of damage in long life fatigue. Below this range, no PSB's occur but just dislocation loop patches, the cyclic deformation is homogeneously and finely distributed, and a fatigue limit applies (Laird, 1976). Above this range the dislocation structures differ from those at lower amplitudes, consisting of fatigue-type cells in which there is little or no misorientation across the cell walls, i.e. the walls are highly dipolar in structure in spite of the fact that the cell walls must contain dislocations of at least three different Burgers vectors. However, any wall may not contain all three. Although Fig. 2 applies to single slip orientation, it is interesting that the shape of the cyclic stress-strain curve is little or no different for orientations close to (and probably including) multi-slip orientations (Cheng and Laird, 1981).

The discovery of the plateau in the monocrystalline cyclic stress-strain curve attracted much interest. It is explained by the fact that plastic strain is localized in the PSB's (which are excited by a constant stress) and within the plateau range, increasing strain serves only to increase the volume fraction of PSB's. At the high strain end of the plateau, essentially the whole specimen becomes a super-PSB. The magnitude of the localized strain averages to ~0.01, but individual PSB's can show major variations from this figure (see Laird, 1981, for review). The dislocation structure of the PSB consists of uniformly separated dipolar walls about $0.1 \mu m$ thick, in which primary dislocations are dominant.

The suggestion that the connection between the cyclic response of mono- and polycrystalline material might be much simpler than that in monotonic deformation was first made when stages I and II/III hardening mechanisms were identified with low and high strain fatigue, respectively (Feltner and Laird, 1967). Bhat and Laird (1978) extended the connection when they claimed, on the basis of Lukáš and Klesnil's data, that the cyclic stress-strain curve of polycrystalline copper contained a plateau at low strain amplitudes and that the plateau levels in the two kinds of material could be related by the Taylor factor. Since their claim, much work on cyclic response and dislocation structures has been completed for low amplitudes in polycrystalline metal (Rasmussen and Pedersen, 1980, a and b; Mughrabi and Wang, 1980 and 1981; Figueroa and co-workers, 1981, a and b; Cheng and co-workers, 1981; Kettunen and Tiainen, 1981; Winter and co-workers, 1981), and the following facts emerge clearly: Provided the grain size is small (typically 100μm in diameter), the cyclic stress-strain curve obtained under strain control does not contain a plateau. A typical example of such a curve is shown in Fig. 3. It is interesting that the dislocation structures observed in polycrystals correspond to those of monocrystals, i.e. at plastic strains less than 2×10^{-5} dislocation loop patches occur; PSB's are first observed at this strain and increase in volume fraction up to about 7×10^{-4}; at still higher strains, cell structures occur. PSB's are most easily observed in near-surface grains but they occur throughout the bulk also. It will be noted from Fig. 3 that a plateau occurs in the cyclic stress-strain curve for tests conducted under stress control. The reason for the difference of this behavior from that in strain control is connected with the cells introduced by the large strain amplitudes caused by the initial load cycles. On the other hand, if the grain size is large (~1mm), a plateau will occur in the cyclic stress-strain curve (Kettunen and Tiainen, 1981). Since these workers used a large specimen, their result is a true polycrystalline result, not a multi-crystalline result. We believe that stress gradients associated with grain incompatibility act to suppress the plateau when the grain size is small. The connection between the flow stresses of mono- and polycrystals is an unresolved issue; a reader interested in the current state is directed to the references from the Riso Conference cited above.

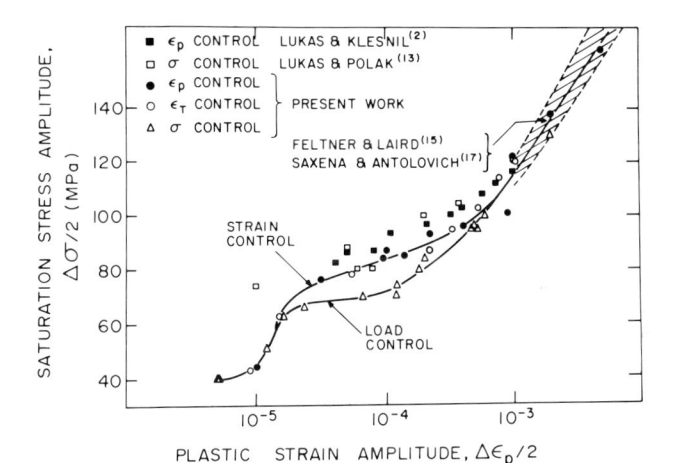

Fig. 3. The cyclic stress-strain curve of small-grained polycrystalline copper over a large range of strains and for tests under various forms of control, compared to the pioneering results of Lukáš and Klesnil (1973) at low strain and other workers at high strains. Taken from Figueroa and co-workers (1981a).

Before about 1979, studies of the effect of variable loading on cyclic response have usually been confined to high strain fatigue. Improved testing capability has permitted cyclic response to be studied recently at increasingly lower strains, and to date such studies for variable amplitudes have been carried out on both mono- and polycrystals. Typical results for polycrystals are shown in Fig. 4, in the form of cyclic stress-strain curves.

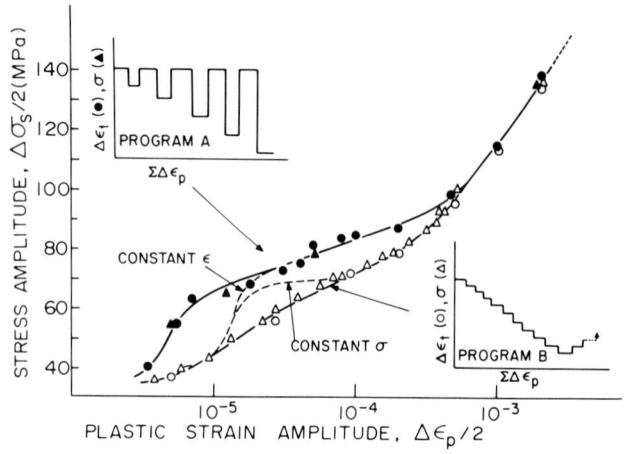

Fig. 4. Cyclic stress-strain curves for copper associated with two kinds of decremental step tests, illustrated in the inserts, compared with the cyclic stress-strain curves measured in constant amplitude tests run in strain control and stress control, and shown in detail in Fig. 3. Taken from Figueroa and Laird (1981).

These curves were obtained for two kinds of decremental step tests: a) A "large step" test in which cycling was first carried out to saturation at a strain giving a uniform dislocation cell structure, after which cycling was continued at progressively lower strains. However, between steps at lower strains, the initial cell structure was regenerated at the highest level chosen. b) A "small step" test in which cycling was initiated at the same high strain as used in the large step test, but the steps were decreased by small amounts seriatim without excursions to higher strain. In both these kinds of tests, cycling was maintained long enough at each step to reach saturation, and both stress and strain were used for control purposes.

It will be noted that the cyclic stress-strain curve for the large step test is congruent with that for constant _strain_ cycling except at the very lowest strains in which cyclic softening cannot apparently be completed. The cyclic stress-strain curve for the small step test is congruent with that for constant _stress_ cycling down to the beginning of the plateau, but it then falls to lower stress values. Figueroa and Laird (1981) concluded that these two curves represented extremes for _cyclic_ history effects, and we feel that they should adequately describe the response for more complex loading spectra. Since most structures are subject to load cycling rather than strain cycling (except perhaps for material at notches) the cyclic stress-strain curve for the small step test will yield larger strains for a given load, correspondingly shorter lives, and more conservative design.

Smoothly-varying spectra should also be expected to fit this curve for design purposes. Single periodic overloads of large magnitude may well subsequently (and temporarily) yield the rather less damaging stresses/strains for the cyclic stress-strain curve of the large step test. In this respect, the behavior is similar to that of retardation in crack propagation caused by periodic overloads, but for entirely different reasons, of course. It is important to point out that the decremental small step test has a strong resemblance to the "engineering" test known as the "incremental" test (a gradually increasing and then decreasing envelope of cyclic strains), which was found quite successful for life prediction under variable loading (Wetzel, 1971) although measured at quite high strains. However, these considerations are quite speculative, and much more work is required to establish a sound approach to designing against variable fatigue loads.

Tests for large and small decremental tests have now been completed on copper monocrystals with results which correspond to those observed for polycrystals (Cheng and Laird, 1982).

CYCLIC STRESS-STRAIN RESPONSE UNDER VARIABLE LOADING -- ALLOYS CONTAINING SHEARABLE PRECIPITATES

While the above description of cyclic response applies to pure metal, almost identically parallel behavior is shown by complex alloys, for example, Al-4% Cu alloy containing a dense population of coherent shearable θ" precipitates. The cyclic stress-strain curve for the monocrystalline form of this alloy is shown in Fig. 5, and obviously it contains a pronounced plateau but shifted to lower strains than that observed for pure metals because the PSB's in such an alloy show a much larger localized strain, even as high as 0.6. The strict definition of the cyclic stress-strain curve does not apply to this alloy because it undergoes hardening to a peak and then cyclic softening due to degradation of the precipitates. The cyclic stress-strain curve shown in Fig. 5 is based on the peak stress.

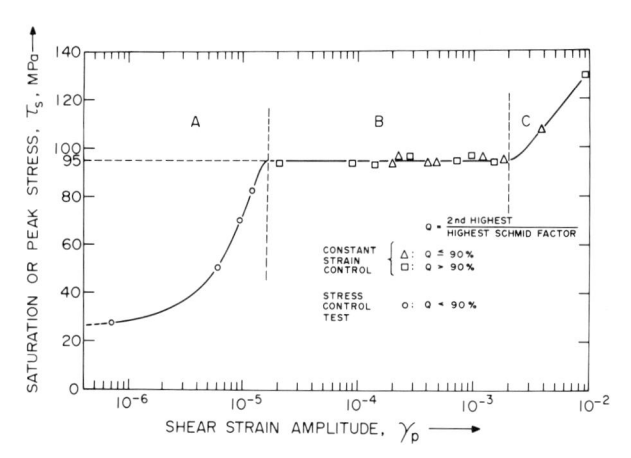

Fig. 5. Cyclic stress-strain curve for Al-Cu single crystals containing θ" precipitates. The specimens were of "random" orientation and many would not commonly be regarded as of single slip orientation--nevertheless, they behaved in a single slip mode. The orientation is indicated by the parameter Q, defined as the percentage ratio of the stress acting on the second most highly stressed system to that on the primary. Taken from Lee and Laird (1982).

Again like copper, the cyclic stress-strain curve of the polycrystal also contains a plateau if the grain size is large (>1mm), as shown in Fig. 6. This plateau can be suppressed, as in copper, by reducing the grain size (Horibe and Laird, 1981). Because this alloy localizes strain strongly in its PSB's, decrementing the strain after first cycling at a strain high enough to form PSB's causes a lowering in the level of the cyclic stress-strain curve. Typical examples of cyclic stress-strain curves for various loading programs and for monocrystals are shown in Fig. 7, and the behavior corresponds to that of copper observed for the small step test, with the exception that flow stresses of decremental or decremental-incremental tests tend to be higher at the lowest strains. However, like copper, a variable test begun within the plateau (rather than above it) shows no significant differences in cyclic response from that of the regular cyclic stress-strain curve. Note that results on the polycrystalline form of this alloy correspond to these shown here for single crystals (Horibe and Laird, 1981). Not shown in Fig. 7 is the result for the large step test -- different from the result in copper, it agrees closely with those of the small step tests.

These results indicate that the speculations concerning pure metal in variable amplitude fatigue noted above should apply with minor variation for

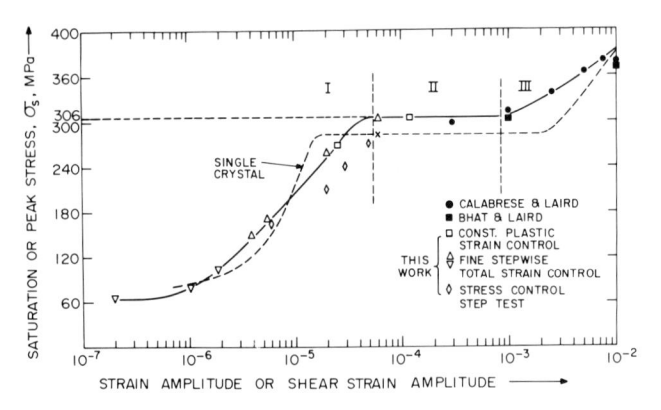

Fig. 6. The cyclic stress-strain curve for polycrystalline Al-4%Cu alloy, of large grain size, containing θ" precipitates compared with that for the same alloy in monocrystalline form. Note that the ordinate scale should be divided by three to apply to the monocrystal; the abscissa applies to both forms of material. Courtesy of Lee (1980).

complex alloys, provided strain localization occurs. However, if the microstructure contains large particles which homogenize the cyclic deformation, then different behavior can be expected. For example, Renard (1981) noted that no plateau in cyclic response was observed for 7075-T651 alloy containing constituent particles. We would anticipate, therefore, that there would be little difference in cyclic response for tests conducted at constant amplitude from those in variable amplitudes. Further work, however, is needed for alloys of this class, particularly in variable amplitude fatigue at low amplitudes.

VARIABLE AMPLITUDE FATIGUE -- THE EFFECT OF MEAN LOADS ON CYCLIC RESPONSE

In the approach to design against variable amplitude fatigue for which cyclic stress-strain response was employed (e.g. Wetzel, 1971), the problem of mean stress was handled by a "stress-strain function" explored by Smith

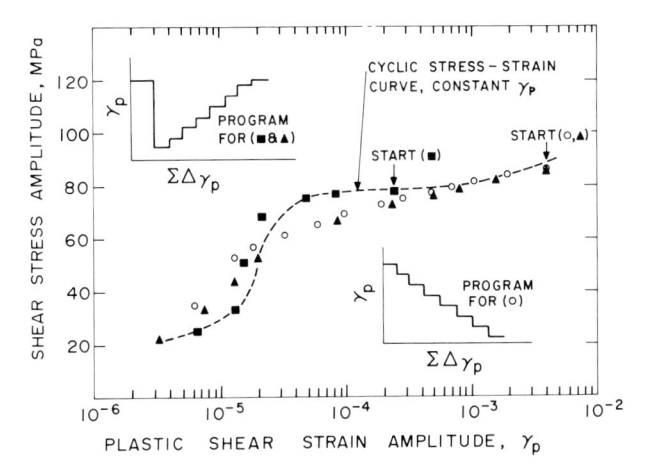

Fig. 7. Cyclic stress-strain curves for Al-4w/oCu alloy monocrystals containing θ" precipitates, associated with decremental step tests illustrated in the inserts, and compared with the cyclic stress strain curve measured in constant amplitude tests. All tests conducted in strain control. Courtesy of Horibe.

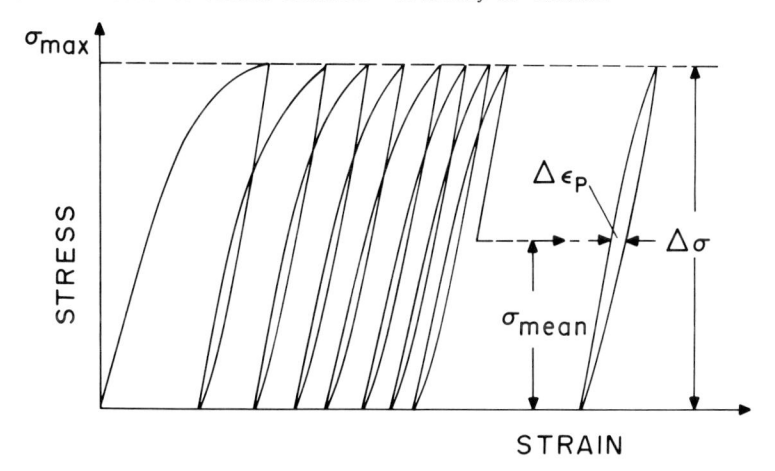

Fig. 8. Schematic stress-strain response in cyclic creep with 100% unloading showing the initial hardening behavior, the development of a "saturated loop", and the parameters used to define the cyclic stress-strain curve for cyclic creep.

and co-workers (1970) and Landgraf (1970). These workers found that a plot of this function, $\sqrt{\sigma_{max} \varepsilon_a E}$ (having units of stress), where σ_{max} is the maximum tensile stress, ε_a the strain amplitude and E the modulus of elasticity, against the life, yielded a Wohler curve which was <u>independent of mean stress</u>. Since it is easy to identify these parameters in a loading spectrum, the stress-strain function can be easily employed for linear damage summation or other kinds of summation (Wetzel used linear summation) where mean stresses are involved. We have recently explored cyclic stress-strain behavior during cyclic creep (Lorenzo, 1981) and believe we can shed some light on the astonishing result of Smith and his co-workers (1970).

Figure 8 shows schematically typical stress-strain behavior for copper cycled in pulsating tension, i.e. cyclic creep with 100% unloading. After the large strains of the first few cycles, the material work hardens and eventually reaches a saturated loop shown on the right of Fig. 8. If the maximum stress is high enough for creep to persist, this saturated loop is not quite closed and marches to higher strains. At intermediate stresses, the creep strain may fall to such a small value that the closure discrepancy cannot be detected in a single loop. If the stress is low enough for the elapsed strain to cease, the loop will become both closed and stationary. Whatever the nature of the loop, we can define a cyclic stress-strain curve for cyclic creep in terms of the semi-range of stress and the half-width of the loop at the mean stress. These parameters are shown in Fig. 8. The cyclic stress-strain curve for cyclic creep of copper, based on numerous tests, both constant amplitude and incremental step tests (which yield similar results) is shown in Fig. 9 and compared with the cyclic stress-strain curves of Figueroa and co-workers (1981a) in fully-reversed cycling. It is extremely interesting that the different kinds of cyclic stress-strain curves roughly agree over the whole of the low strain regime, in spite of the fact that the cyclically-crept material (containing cells) differs markedly in dislocation structure from the copper cycled in push-pull.

Now let us consider the variable of mean stress in greater detail. For a given maximum stress, one can conduct pulsating tension tests with different degrees of unloading, indicated by the mean of the maximum and minimum stresses. One can also define a cyclic stress-strain curve for such conditions. In Fig. 10, the width of the saturated loop is plotted against the mean stress for cyclic creep tests with partial unloading; a family of parallel curves for different values of maximum stress is shown. It is fascinating that these curves extrapolate to the plastic strain amplitudes

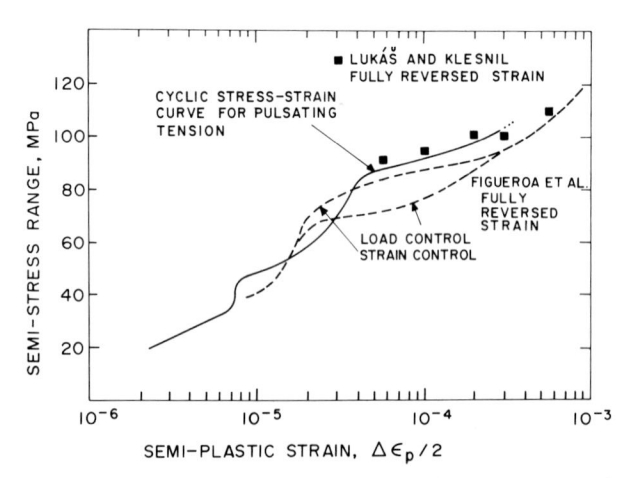

Fig. 9. The cyclic stress-strain curve for cyclic creep in polycrystalline copper compared with the cyclic stress-strain curves for fully reversed cycling, taken from Fig. 3. The specimens were all of roughly similar grain size.

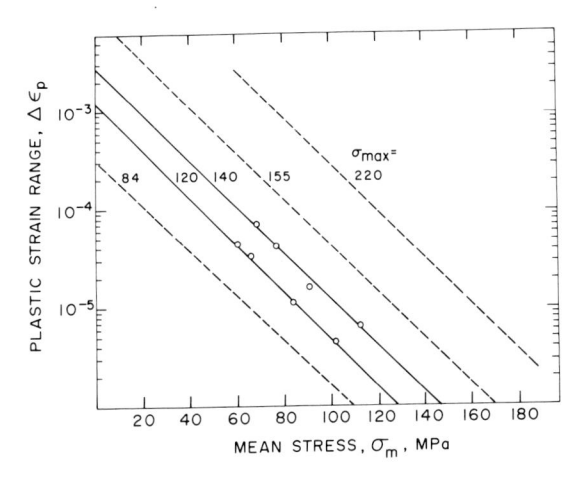

Fig. 10. The width of the saturated hysteresis loop measured in cyclic creep of polycrystalline copper with partial unloading, plotted against the mean stress.

for push-pull loading (i.e. mean stress is zero) and the given maximum tensile stress. Figures 9 and 10 show the intimate connection between cyclic stress and strain for fully reversed cycling and cycling with mean stress, in line with the stress-strain function of Smith and co-workers (1970) and Landgraf (1970).

Since these investigators made a connection of stress and strain to fracture through Wohler curves, it would be interesting to examine whether or not a connection to fracture could be established through a Coffin-Manson curve under conditions in which a mean stress applies. This would entail measuring fracture lives for different mean stresses and all the appropriate cyclic deformation parameters for the same material, ideally in the same investigation. Since no one appears to have looked at the mean stress problem in the manner we now use, such data are quite rare and we haven't made a complete set of measurements ourselves. However, there exists a report by Pokluda and Stanek (1978) from which the necessary data can be culled. Their investigation was made on low carbon steels and we justify using their results here on the grounds that iron is a metal of wavy slip mode like copper, and the presence of alloying additions in a commercial steel would cause a similarity in cyclic response to that of f.c.c. metal. The plastic strain amplitudes extracted from Pokluda and Stanek's results are shown in Fig. 11 plotted against life in the manner of Coffin and Manson, for both fully reversed cycling and cycling with mean stress. While there is a tendency for the fatigue lives in cyclic creep to lie on the low side, probably because the tests with tensile mean stress have increased the stress intensity during crack propagation (an effect especially pronounced at short lives because of the greater fraction of life spent in crack propagation), the agreement between the two sets of measurements is remarkably close. Apparently, the magnitude of plastic strain in a cycle has a more controlling influence on life than the mean stress, and this conclusion again underlines the success of the stress-strain function in predicting life. Furthermore, the fact that Fig. 11 applies to steel while earlier figures apply to copper indicates that the present considerations should apply to materials of widely-varying microstructure.

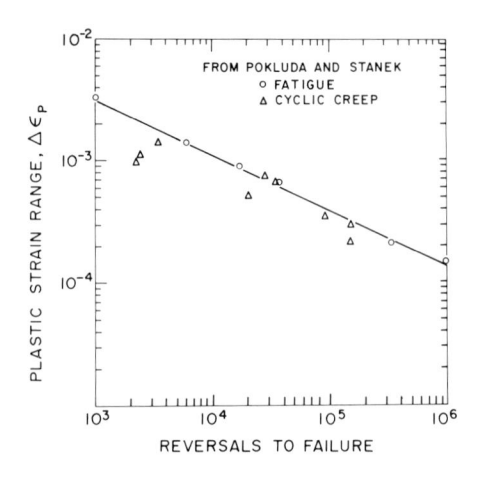

Fig. 11. Coffin-Manson plot for low carbon steel in cyclic creep and fully-reversed cycling. Data extracted from the results of Pokluda and Stanek (1978).

THE LOAD INTERACTION EFFECT ON FRACTURE MECHANISM IN VARIABLE AMPLITUDE FATIGUE

Although the load interaction problem is widely regarded as significant for fatigue under variable amplitudes, investigators of the problem are quite rare. Wood was one of the most persevering investigators in this direction and his book contains a summary of his views (1971). He felt that, in a high-low sequence, the high strain being in the "H" regime (in his terminology) and the low strain in his "F" regime, the disorientation produced by the high initial strain would prevent the slip-zone concentrations under subsequent "F" amplitudes. Nine and Wood (1967) using copper single crystals, found that first cycling at $300°C$, for thousands of cycles, caused large increases of life when cycling was subsequently continued at ambient temperature. In the light of modern research, we now understand that the cycling at $300°C$ produced cell structures, whereas the same strain applied at room temperature, absent prior cycling, would have produced PSB's. However, the cells prevented slip localization on subsequent cycling at room temperature and thus caused the life to increase. In low-high sequences, Wood (1971) expected (and showed, in some circumstances) that the subsequent "H" cycles would distort and open-up the "slip-zones" prematurely. This implies that the Miner summation would be decreased but we normally find Miner summations to be greater than unity in low-high sequences. The foregoing means that the nature of damage accumulation depends very sensitively on the state of damage at which load changes occur.

The load interaction problem has recently been investigated in our laboratory (Figueroa and Laird, 1981c) by means of high-low and low-high stress tests, conducted in strain control. The two amplitudes chosen were 10^{-3} and 10^{-4}, such as to give cells and PSB's/loop patches respectively (see Fig. 3). Since polycrystalline specimens were investigated, crack nucleation at grain boundaries was quite prevalent. Such nucleation at high strains is well-known (see Laird, 1981, for review) but its wide extent at low strains was found surprising. The mechanism of nucleation at low

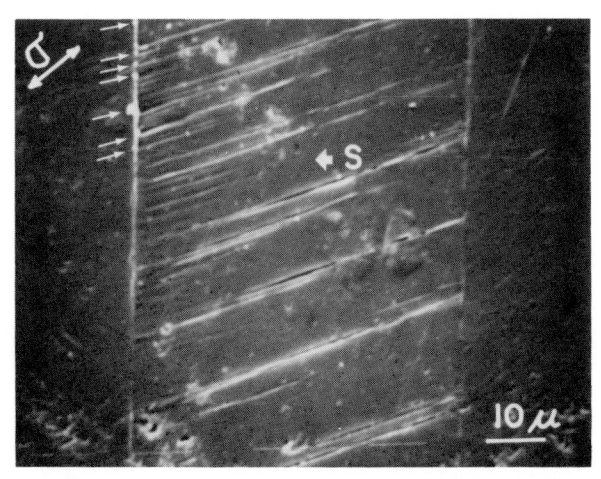

Fig. 12. Interaction of PSB's with grain boundaries in fatigued pure copper. Plastic strain amplitude = 10^{-4}. Note that cracks have opened in the PSB's, but the main crack formed in the boundary to the left, marked with arrows. The double-headed arrow indicates the push-pull axis. Courtesy of Figueroa (1979).

strains was PSB impingement on the grain boundaries (see Fig. 12) at which the kinetics of nucleation and early growth were greater than in the PSB's themselves. This mechanism was found to be a true mechanical effect and not due to oxide contamination of the specimens, for example.

Thus, in high-low sequences, the grain boundary cracks nucleated in the initial high strain cycling were simply inherited and propagated in the subsequent low strain cycling. Because of their higher density, they linked-up more easily and caused reduced Miner summations. Typical sequences of damage evolution in a high-low test are shown in Fig. 13. Note that the life to failure at the high level was about 12,000 cycles (24,000 reversals), so that the damage introduced at the high level corresponded to half of the life before the step to low level occurred. The life at the low level lay in the scatter range 150,000 to 450,000 cycles.

On the other hand, the initial damage in a low-high sequence was observed to be effectively ignored by the subsequent cycling at high strain over a considerable range of life fractions at the low level. This then led to Miner summations greater than unity in line with the most usual experience for low-high tests. The evolution of crack nuclei and their density in the high and low levels is shown schematically in Fig. 14,* as well as their interaction in step tests. It will be noted that our results differ considerably from those of Wood, noted above, and the load interaction mechanism can be expected to vary widely depending on the specifics of the system (temperature, environment, stress-state, stress levels) and of the material. Clearly much more research is needed in this area before we have adequate guidelines for dealing with classes of complex materials.

ACKNOWLEDGEMENTS

The results described in this paper were derived from the following projects: 1) cyclic response of copper, National Science Foundation, Grant

---

*Figure 14 appears in References section

No. DMR80-19914; 2) cyclic response of Al-4%Cu, Army Research Office, Grant No. DAAG29-78-C-0039; 3) load interaction effects in copper, National Science Foundation MRL Grant No. DMR79-23647; and 4) cyclic creep, the Dept. of Energy, Grant No. DE-AC02-80ER10570. All of the mechanical tests and microscopic observations for these projects were conducted in the Mechanical Testing and Electron-microscope central facilities of the Laboratory for Research on the Structure of Matter, University of Pennsylvania. We are grateful for this support, and to the authors who have kindly given us permission to reproduce the figures referenced herein.

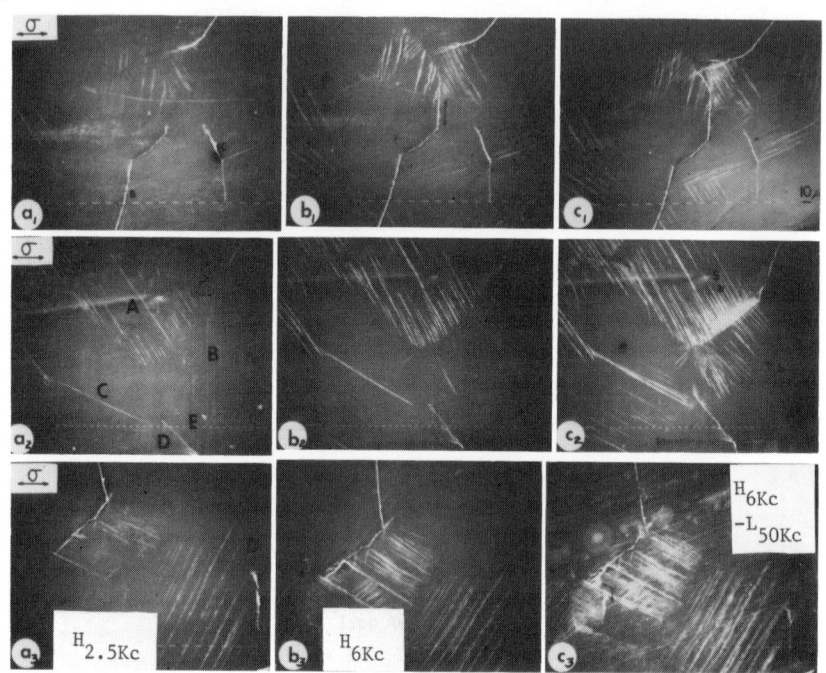

Fig. 13. Three sequences of surface damage evolution observed by means of replicas in a specimen first cycled 6000 cycles at high level (10^{-3}), and then to failure at low level (10^{-4}). (1) Three microcracks A, B, and C are present after 2500 cycles at the high level; A and B coalesce after the step forming a well-defined crack; crack C did not extend significantly after 2500 cycles because it was sheltered by crack B. 2) A crack is formed after the step at grain boundary A/B; S denotes the additional slip activation caused by the stress intensification of the crack. 3) Propagation at the low level of an intergranular crack previously well developed at the high level. Taken from Figueroa and Laird (1981c).

REFERENCES

Bhat, S.P. and Laird, C. (1978). The Cyclic Stress-Strain Curves in Monocrystalline and Polycrystalline Metals, Scipta Met., 12, 687-692.

Cheng, A.S. and Laird, C. (1981). Mechanisms of Fatigue Hardening in Copper

Single Crystals: the Effects of Strain Amplitude and Orientation, Mat. Sci. Eng., 51, 111-121.

Cheng, A.S., Figueroa, J.C., Laird, C. and Lee, J.K. (1981). The Cyclic Deformation of Polycrystalline Pure Metals and Precipitation Hardened Alloys, in Deformation of Polycrystals, Riso, Denmark, 405-416.

Cheng, A.S. and Laird, C. (1982). Unpublished work. University of Pennsylvania, Philadelphia.

Feltner, C.E. and Laird, C. (1967). Cyclic Stress-Strain Response of F.c.c. Metals and Alloys, Parts I & II, Acta. Met. 15, 1621-1632 and 1633-1654.

Figueroa, J.C. (1979). Cumulative Fatigue Damage in Copper Polycrystals, Ph.D. Thesis, University of Pennsylvania, Philadelphia.

Figueroa, J.C., Bhat, S.P., de la Veaux, R., Murzenski, S. and Laird, C. (1981a). The Cyclic Stress-Strain Response of Copper at Low Strains, Part I, Constant Amplitude Testing, Acta. Met. 29, 1667-1678.

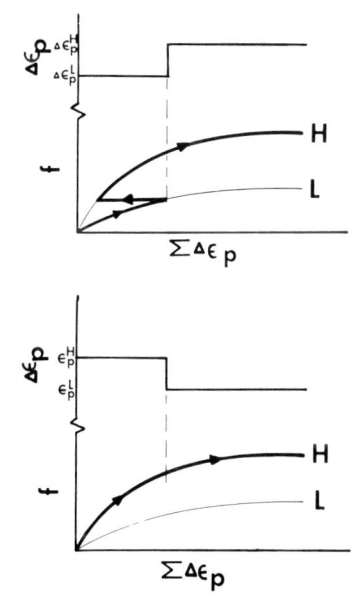

Fig. 14. Schematic evolution of crack nuclei and their density, embraced in a damage parameter F, in step tests (heavy lines) (a) L-H, (b) H-L, as related to the evolution in constant amplitude tests --light lines. Taken from Figueroa and Laird (1981c).

Figueroa, J.C. and Laird, C. (1981b). The Cyclic Stress-Strain Response of Copper at Low Strains, Part II, Variable Amplitude Testing, Acta. Met. 29, 1679-1684.

Figueroa, J.C. and Laird, C. (1981c). Cumulative Damage in Copper Polycrystals Cycled Under Constant Strain Amplitudes and in Step Tests, Part I, Crack Initiation Mechanisms, Part II, Crack Propagation Mechanisms, University of Pennsylvania, Philadelphia.

Horibe, S. and Laird, C. (1981). Unpublished Work. University of Pennsylvania, Philadelphia.

Kettunen, P. and Tiainen, T. (1981). Strain Hardening in Monotonic and Cyclic Deformation of Polycrystalline F.c.c. Metals, in Deformation of Polycrystals, Hansen, N., Horsewell, A., Leffers, T. and Lilholt, H. (Eds.), Riso, Denmark, 437-444.

Koibuchi, K. and Kotani, S. (1973). The Role of Cyclic Stress-Strain Behavior on Fatigue Damage Under Varying Load, Am. Soc. Testing Mat., Spec. Tech. Pub. 519, 229-245.

Laird, C. (1976). The Fatigue Limits of Metals, Mat. Sci. Eng. 22, 231-236.

Laird, C. (1977). The General Cyclic Stress-Strain Response of Aluminum Alloys, in Am. Soc. Test. Mat. Special Tech. Publ. 637, 3-35.

Laird, C. (1981). Cyclic Deformation, Fatigue Crack Nucleation and Propagation in Metals and Alloys, in J.K. Tien and J.F. Elliott (Eds.), Metallurgical Treatises, AIME, New York, 505-528.

Landgraf, R.W. (1970). The Resistance of Metals to Cyclic Deformation, Am. Soc. Testing Mat., Spec. Tech. Pub. 467, 3-36.

Lee, J.K. (1980). Cyclic Deformation and Strain Localization in Al-4w/oCu Alloy Containing θ'' Precipitates, Ph.D. Thesis, University of Pennsylvania, Philadelphia.

Lee, J.K. and Laird, C. (1982). The Cyclic Stress-Strain Response of Al-4w/o Cu Alloy Single Crystals Containing θ'' precipitates, Part I, Cyclic Stress-Strain Response, Mat. Sci. Eng., 54, 39-52.

Lorenzo, F. (1981). Unpublished work on cyclic creep, University of Pennsylvania, Philadelphia.

Lukáš, P. and Klesnil, M. (1973). Cyclic Stress-Strain Response and Fatigue Life of Metals in Low Amplitude Region, Mat. Sci. Eng. 11, 345-351.

Mitchell, M.R. (1979). Fundamentals of Modern Fatigue Analysis for Design in Fatigue and Microstructure, Am. Soc. Met., Metals Park, Ohio, 385-438.

Mughrabi, H. (1978). The Cyclic Hardening and Saturation Behavior of Copper Single Crystals, Mat. Sci. Eng. 33, 207-220.

Mughrabi, H. and Wang, R. (1980). Cyclic Strain Localization and Fatigue Crack Initiation in PSB's in F.c.c. Metals and Single Phase Alloys, Int. Sym. Defects and Fracture, Tuczno, Poland, Sijthoff and Noordhoff: (1981) Cyclic Deformation of F.c.c. Polycrystals: A Comparison with Observations on Single Crystals, in Deformation of Polycrystals, Riso, Denmark, 87-98.

Mughrabi, H., Herz, K. and Stark, X. (1981). Cyclic Deformation and Fatigue Behavior of α-iron Mono- and Polycrystals, Int. J. Fracture 17, 193-220.

Nine, H. and Wood, W.A. (1967). An Improvement of Fatigue Life by Dispersal of Cyclic Strain, J. Inst. Met. 95, 252-254.

Pokluda, J. and Stanek, P. (1978). Cyklicky Creep Oceli 12010, 14331 a Vliv Asymetrie Cyklu na Životnost v oblasti Časované Únavové Pevnosti, Kovové Materiály, 5(16), 583-599.

Rasmussen, K.V. and Pedersen, O.B. (1980a). Cyclic Deformation in Polycrystals, 5th Int. Conf. on Strength of Metals and Alloys, Haasen, P., Gerold, V. and Kostorz, G. (Eds.), Pergamon Press, Oxford, 1219-1223: (1980b). Fatigue of Copper Polycrystals at Low Plastic Strain Amplitudes, Acta. Met. 28, 1467-1478.

Renard, A. (1981). Cyclic Response at Low Amplitudes of Al-Zn-Mg Alloy and Commercial Alloys Based on this System, M.Sc. Thesis, University of Pennsylvania, Philadelphia.

Smith, K.N., Watson, P. and Topper, T.H. (1970). A Stress-Strain Function for the Fatigue of Metals, Journal of Materials 5, 767-778.

Socie, D.F. (1975). Fatigue Life Prediction Using Local Stress-Strain Concept, SESA Spring Meeting, Chicago, Ill., 11-16 May.

Wetzel, R.M. (1971). A Method of Fatigue Damage Analysis, Ford Motor Co., Scientific Research Staff, Dearborn, Michigan.

Winter, A.T., Pedersen, O.B. and Rasmussen, K.V. (1981). Dislocation Microstructures in Fatigued Copper Polycrystals, Acta. Met. 29, 735-748.

Wood, W.A. (1971). The Study of Metal Structures and their Mechanical Properties, Pergamon Press, New York, 274-275.

Note:

After preparing this paper, we found a reference by Manson and Halford (addendum to: Manson, S.S. and Halford, G.R. (1981). "Practical Implementation of the Double Linear Damage Rule and Damage Curve Approach for Treating Cumulative Fatigue Damage", Int. J. of Fracture, 17, 169-192) which bears upon the question of the mean stress effect. These authors tested three specimens of 316 stainless steel at room temperature, for all of which the stress range was 43.2×10^3 lb/in^2 (297 MPa). However one test was conducted in push-pull, one in pulsating tension and one in pulsating compression. They report that, for all three cases, the widths of the hysteresis loops were nearly the same. This finding supports our conclusion that the cyclic stress-strain curve is unaffected by mean stress.

Deformation for Manufacture: Forming and Shaping

T. Furukawa

Fundamental Research Laboratories, Nippon Steel Corporation, Nakahara-ku, Kawasaki City, Kanagawa 211, Japan

ABSTRACT

An overview of current research is made in terms of alloy and microstructural designs for cold-rolled deep-drawing steels and dual-phase steels, with emphasis on current problems in cold-pressing of high-strength steels.

KEYWORDS

Deformation; fracture; plastic strain ratio; work-hardening exponent; continuous annealing; texture control; ageing; dual-phase steel.

INTRODUCTION

The title "Deformation for Manufacture" encompasses a very wide technological area covering hot and cold working of metals and alloys. In technical terms, forgeing, swageing, extrusion, wire-drawing, pressing, ironing and many others involving a variety of materials may be mentioned. Theories relevant to the present title may also cover almost all the areas of physical and mechanical metallurgy and plasticity. A comprehensive discussion on the basis of this title is, therefore, far beyond the author's reach. Since it will be necessary to restrict the scope of this paper, the major topics will be limited to only a few specific cases relating to press-formable sheet steels for cold forming. After a short description of deformation characteristics required for optimum press formability, it will be shown how steel compositions and microstructures can be modified to provide these characteristics, citing some recent research papers written for the practical and industrial benefits. Hence, metallurgical efforts to develop sheet steels characterized by improved formability will be paid more attention in this paper, rather than dealing with deformation processing itself. Materials to be dealt with in this paper are deep-drawing steels and recently developed dual-phase steels.

DEFORMATION CHARACTERISTICS

It would be worthwhile to look over the deformation behaviour of sheet steels, from the viewpoints of both practical and fundamental aspects.

Yielding Behaviour and Work Hardening

Discontinuous yielding (Lüders strain) should be suppressed to improve the surface quality of cold-formed products, especially of shallow stampings. For smooth yielding behaviour, it is considered that abundant mobile dislocations should exist in the material before straining and should move easily under external stress. Temper-rolling after annealing (to introduce mobile dislocations) and some means such as steel composition and annealing heat cycle designs to minimize interstitial solute amounts (to suppress a return of discontinuous yielding at room temperature) are required in the deep-drawing steel production.

As steel is cold-deformed, flow stress increases in a manner of linear relationship with square root of the dislocation density (Keh and Weissmann, 1963). Dislocation cell structure formation starts at several percent deformation (in uni-axial tension), resulting in increased fraction of immobile dislocation density. Further deformation behaviour may be controlled by the balance between work hardening and dynamic recovery in which cross slip plays an important role. Considering dislocation-impurity interactions, it may be surmised that a cleaner ferrite (with respect to interstitial solutes, fine precipitates and others) is less prone to cross slip, leading to an improved ductility as a result of prolonged high level of work hardening at high strains.

Factors Affecting Large Strain Behaviour

At large strains fracture behaviour becomes important. In interstitial-free deep-drawing steels, brittle intergranular fracture sometimes occurs on heavy deformation, presumably due to weak grain-boundary strength (McMahon, 1966). Trace amount of boron addition is reported to prevent this embrittlement, probably because of the grain boundary strengthening effect owing to boron atoms segregated at the boundaries (Takahashi and others, 1982a).

In steels having a second phase such as dual-phase steels, decohesion of ferrite/second phase interface or cracking of the second phase takes place during heavy plastic straining. These "voids" grow upon further plastic straining until the ductile fracture of ferrite by localized shear is completed (Speich and Miller, 1979). Therefore such factors as mechanical properties and morphology of the second phase, cohesive strength of interface as well as the nature of ferrite matrix are considered to be quite influential on total strain to fracture. This kind of effect on ductility due to the presence of "foreign phases" seems especially important in such deformations as heavy bending, stretch flangeing (Nishimoto and others, 1981b) and cold forgeing (Nagumo and others, 1972) in which the morphology control of carbides and/or non-metallic inclusions is significant.

Another factor contributing to ductility is reported to be the strain rate dependence of deformation stress (Mizunuma and others, 1980; Stevenson, 1980) which merits further study to improve this parameter through composition and structure of steel.

Plastic Strain Ratio and Work-Hardening Exponent

It is well known that r value, or plastic strain ratio ($r = \varepsilon_w/\varepsilon_t$, where ε_w: true strain in width, and ε_t: true strain in thickness) and n value (n as in the well known empirical equation $\sigma = k\varepsilon^n$, where σ: true flow stress, ε: true strain, and k: constant) are the simplified, still very important parameters to estimate press-formability of a sheet steel. Both parameters can be gathered by conducting a uni-axial tensile test. Commonly accepted view is that r and n values indicate deep-drawability and stretchability respectively, although not all the press performance results can be predicted by these parameters alone since the stress-strain history that an actual sheet material undergoes may be extremely complicated. Roughly speaking, r value can be improved by controlling of crystallographic textures, and n value is increased as the steel is made softer.

In this connexion, Abe and others (1979) examined the crystal orientation dependencies of deep-drawability and stretchability using high-purity iron single crystals. After preparation of disk specimens with various planar orientations, they conducted deep-drawing and stretching tests by means of a mini press machine. The experimental results are shown in Fig. 1 and Fig. 2. After some analytical discussions they obtained the results as shown in Fig. 3,

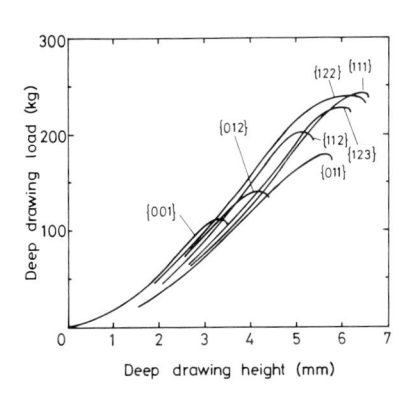

Fig. 1. Drawing test results (Abe and others, 1979).

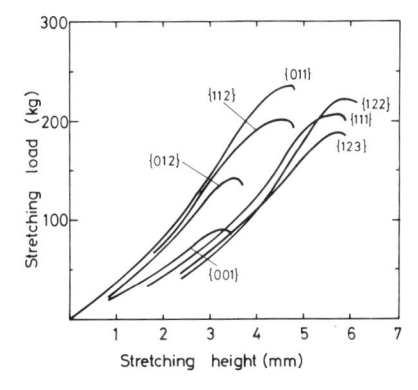

Fig. 2. Stretching test results (Abe and others, 1979).

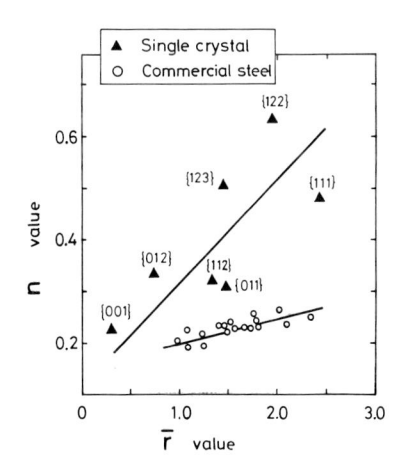

Fig. 3. Relation between n and r (Abe and others, 1979).

and concluded that: (a) Both deep-drawability and stretchability are high for planar orientations {111}, {122} and {123} having high values of n as well as r, (b) Variation of deep-drawability with planar orientations in the single crystals is directly related to the texture dependence of deep-drawability in commercial steels, and (c) Texture dependence of stretchability in commercial steels, however, is not obvious presumably because the exponent n depends largely upon the grain size of steels. Thus, value of n practically obeys the following relation

$$n = 5/(10 + d^{-\frac{1}{2}}), \quad (1)$$

where d is the grain diameter in mm, as Morrison (1966) suggested.

In Japan, recent production technology of highly formable cold-rolled sheet steels employs continuous annealing process which gives rise to different problems in controlling textures, softness and strain-ageing characteristics of sheet steels compared to the classical batch-annealing process. This topic will be described in the next section.

METALLURGICAL PROPOSAL TO OBTAIN DEEP-DRAWING STEELS THROUGH CONTINUOUS ANNEALING

Texture Control

It was suggested by Adams and Bevan (1966) that a vacuum degassed, titanium added steel would be a candidate for a deep-drawing steel adaptable to continuous annealing. The scavengeing effect of titanium exerted on carbon and nitrogen was considered to result in "clean" ferrite matrix leading to a soft, non-ageing steel. However, information regarding texture control was not available in their paper. Recently, Takahashi and others (1982b) thought this kind of scavengeing could be applied to the design of steel compositions for continuously-annealed deep-drawing steels, and have shown experimentally the generalized effect of scavengeing due to various elements — carbide, nitride, oxide, or sulphide forming elements — on improvement of r value. They used steels and processings as shown in Table 1 in which carbon, nitrogen, sulphur and oxygen were regarded to be impurities and manganese, aluminium, boron and titanium were intentionally added scavengers in each steel. They assumed the interaction between the scavenger X and impurity Y to form precipitate XmYn as denoted by

$$mX + nY = XmYn \qquad (2)$$

(e.g. for $Mn + S = MnS$, X: Mn, Y: S, m = 1, n = 1). An index $f(X,Y)$ was then introduced;

$$f(X,Y) = (wt.\% \ X) / \frac{m}{n} \cdot \frac{p}{q} (wt.\% \ Y) \ , \qquad (3)$$

where p and q were atomic weight of X and Y respectively. If $f(X,Y) = 1$, the content ratio X/Y should correspond to the stoichiometry of XmYn. Takahashi and others considered possible reactions and $f(X,Y)$ in each steel as in Table 2. After tensile testing of the specimens, results as shown in Fig. 4 and Fig. 5 were obtained. The $f(X,Y)$ range of about 1 to 3 gives a peak of r value in each series of steels.[1] Also, both r value and total elongation increase as the grain size increases. It seems from these results that minimizing solute amounts of both the scavengers and impurities favours the grain growth of favourably oriented recrystallization nuclei during the short-time annealing, as they pointed out. However, different levels of r value can be seen in Fig. 5, depending on the steel series, for a given grain size. This would mean that recrystallization textures depend upon oriented nucleation behaviour of recrystallization as well as grain growth.

With regard to recrystallization behaviour of cold-rolled pure iron, initial grain boundaries before cold rolling are reported to be selective nucleation sites of {111} grains (Inagaki, 1976; Matsuo and others, 1971), owing to the

---

[1] It should be noted that practically optimum value of $f(X,Y)$ would be influenced by such factors as diffusivities of X and Y (the coiling temperatures are, in general, kept at rather high temperatures to ensure the long distance diffusion of the scavengers), solubility limits of X and Y, exact composition and precipitation kinetics of XmYn. Inasmuch as the detailed information regarding these factors is not always available, practically optimum $f(X,Y)$ will not necessarily be at $f(X,Y) = 1$ (Takahashi and others, 1982b).

TABLE 1 Chemical Composition and Processing of Steels
(Takahashi and others, 1982b)

Steel	Chemical Composition (wt. %)								Hot Coiling Temp. (°C)	Heating Cycle	
	C	Mn	S	O	N	Al	B	Ti		Annealing (°C - s)	Over-Ageing (°C - s)
Capped Steel	0.05 ∫ 0.06	0.18 ∫ 0.26	0.009 ∫ 0.028	0.037 ∫ 0.073	0.0021 ∫ 0.0026	-	-	-	690 ∫ 710	700 - 60	400 - 180
Aluminium-Killed Steel	0.04 ∫ 0.05	0.15 ∫ 0.28	0.010 ∫ 0.027	0.0026 ∫ 0.0051	0.0043 ∫ 0.0049	0.018 ∫ 0.101	-	-	740 ∫ 750	850 - 40	400 - 180
Boron Steel	0.02 ∫ 0.03	0.17 ∫ 0.22	0.008 ∫ 0.010	N.D.	0.0031 ∫ 0.0032	0.031 ∫ 0.035	0.0003 ∫ 0.0081	-	600 ∫ 650	800 - 40	400 - 180
Titanium Steel	0.003 ∫ 0.004	0.11 ∫ 0.14	0.005 ∫ 0.012	N.D.	0.0023 ∫ 0.0035	0.038 ∫ 0.051	-	0.031 ∫ 0.058	670 ∫ 680	800 - 30	-

P : 0.013 - 0.018 wt. %,   Si : 0.01 - 0.02 wt. %,   Cold Rolling : 70 - 77 %.

TABLE 2 Possible Reactions and Stoichiometry Index f(X,Y),
(Takahashi and others, 1982b)

Steel	Reaction	f(X,Y)	percent in wt.
Capped Steel	Mn + O = MnO Mn + S = MnS	}	$f(Mn,O,S) = (\%\,Mn)/[\frac{55}{16}(\%\,O) + \frac{55}{32}(\%\,S)]$
Aluminium-Killed Steel	Mn + S = MnS 2Al + 3O = Al$_2$O$_3$ Al + N = AlN	} }	$f(Mn,S) = (\%\,Mn)/\frac{55}{32}(\%\,S)$ $f(Al,O,N) = (\%\,Al)/[\frac{2}{3}\cdot\frac{27}{16}(\%\,O) + \frac{27}{14}(\%\,N)]$
Boron Steel*	B + N = BN		$f(B,N) = (\%\,B)/\frac{11}{14}(\%\,N)$
Titanium Steel**	Ti + C = TiC Ti + N = TiN	}	$f(Ti,C,N) = (\%Ti)/[\frac{48}{12}(\%\,C) + \frac{48}{14}(\%\,N)]$

* In boron steel, boron-oxygen reaction is not taken into account because of preceding deoxidation by aluminium in steel making. Aluminium-nitrogen reaction is not taken into account either, owing to the low temperature hot coiling of this steel (cf. Table 1).
**In titanium steel, titanium-oxygen reaction is neglected because of preceding deoxidation by aluminium in steel making. Aluminium-nitrogen reaction is also neglected considering the strong affinity of titanium for nitrogen.

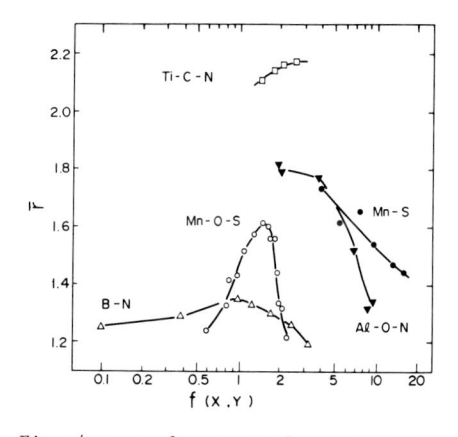

Fig. 4. r values as a function of stoichiometry index f(X,Y) (Takahashi and others, 1982b).

Fig. 5. (Right) Mechanical properties as a function of grain size (Takahashi and others, 1982b).

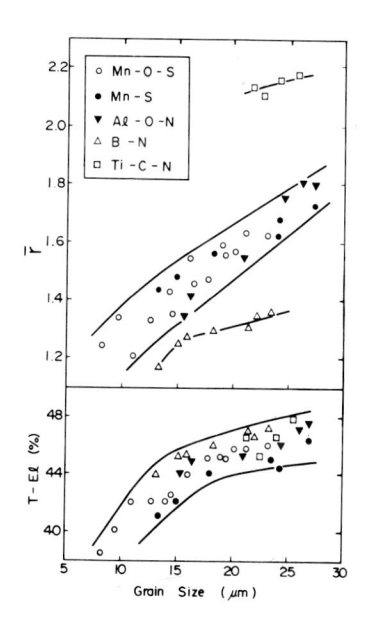

local formation of {111} regions with high dislocation densities adjacent to the initial grain boundaries, as a result of multiple slip induced by cold rolling (Abe and others, 1980). The fact that titanium steel shows the highest level of r value for a given grain size, as seen in Fig. 5, seems to indicate the effectiveness of selective nucleation mechanism observed in pure iron mentioned above, since the interstitial-free titanium steel is of the closest composition to pure iron among the steels given in Table 1. A similar steel containing niobium as a scavenger has also been developed (Ohashi and others, 1981),

It is also surmised that fractions of nucleation and growth of unfavourably oriented grains may be influenced by such factors as the interstitial and substitutional solute contents, and the amounts, size and distribution of precipitates, which can vary depending on the steel composition and $f(X,Y)$. Further studies on these factors in reference to $f(X,Y)$ would be important.

Effect of Carbon on Ductility

Besides grain size, interstitial solutes and fine precipitates formed at low temperatures are quite influential upon ductility. For a given chemical composition, solute carbon amount and fine carbide distribution are controlled by the combined processes of cooling and over-ageing. It was found (Abe and others, 1981; Takahashi and others, 1982b) that solute carbon mainly deteriorates post-uniform elongation by necking localization due probably to dynamic strain ageing, whereas a high density of fine carbide distribution within the ferrite matrix mainly decreases uniform elongation i.e. the work-hardening exponent. It is therefore important to reduce both carbide distribution density and solute carbon content by means of the controls of cooling rate and over-ageing conditions.

DUAL-PHASE STEELS

Since a new class of continuously-annealed, high-strength sheet steels characterized by a microstructure consisting of a dispersion of martensite particles in a soft, ferrite matrix was introduced in the name of "dual-phase steels" (Hayami and Furukawa, 1975; Rashid, 1976), the intense interest in these steels has been aroused because of their unique properties such as continuous yielding behaviour, a high work-hardening rate, a low yield strength and a high tensile strength combined with a high elongation (e.g. Bucher and Hamburg, 1977). Fig. 6 dramatically shows how the stress-elongation characteristics can be changed for a given steel composition, if the structure is changed from ordinary ferrite-pearlite into ferrite-martensite such as shown in Fig. 7, through a proper heat treatment. As a result, dual-phase steels have ductility superior to that of precipitation hardened or solid-solution hardened high-strength steels, as shown in Fig. 8. Dual-phase steels are also characterized by a high strain-ageing response which is adaptable to auto-body panel manufacture, in spite of their practically non-ageing property before deformation (Araki and others, 1977; Hayami and Furukawa, 1975; Nishida and others, 1978). Recent studies on physical metallurgy of dual-phase steels to understand the property mechanisms and to improve the production technology are voluminous. Many of these research papers have been collected in three volumes of the TMS-AIME symposium proceedings (Davenport, Ed. 1979; Kot and Morris. Eds. 1979: Kot and Bramfitt, Eds. 1981).

Steel Compositions

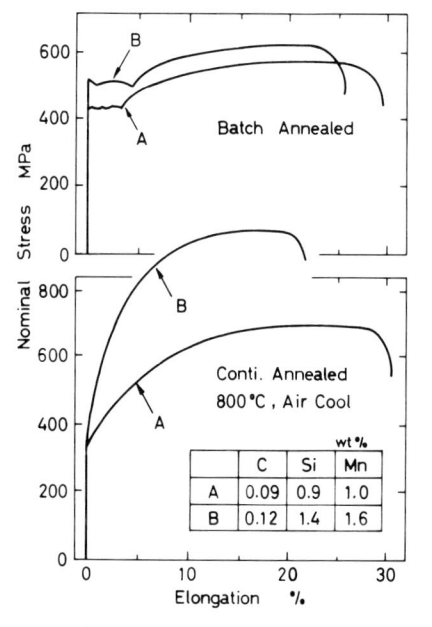

Fig. 6. Stress-elongation curves (Imamura and Furukawa, 1976).

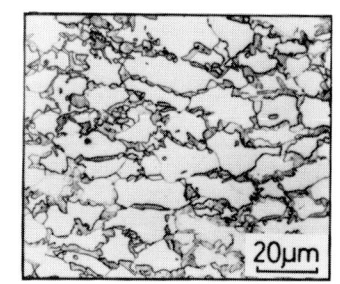

Fig. 7. Optical micrograph of a dual-phase steel.

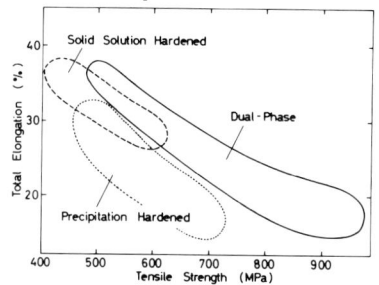

Fig. 8. Relation between tensile strength and elongation (Hayami and Furukawa, 1975).

Typically, a dual-phase structure can be produced in hot or cold rolled, carbon (0.1 percent or less)-manganese (0.4 to 2 percent) steels by continuous annealing in the alpha-gamma temperature range followed by cooling at a rate of 10^0 to 10^3 degC/s. Usually, the steel composition can not afford sufficient hardenability to produce the martensite phase if cooled at a mild cooling rate from the gamma single phase region. However, if heated into the intercritical temperature range, enough carbon is partitioned into the austenite pools to produce the required level of hardenability. Naturally, a leaner alloy requires faster cooling to result in a dual-phase structure. On the other hand, the intercritical heating in box anneal followed by furnace cooling can afford a dual-phase structure if the steel contains 2 percent or more of manganese (Matsuoka and Yamamori, 1975; Okamoto and Takahashi, 1981). Some dual-phase steels contain a small amount (0.1 percent or less) of vanadium (Bucher and Hamburg, 1977; Rashid, 1976) or molybdenum (Morrow and Tither, 1978), to improve the austenite hardenability (Repas, 1979), rather than to produce carbonitride precipitates.

Structure Formation and Structure-Property Relationships

Fig. 9 shows a schematic illustration of structure formation, with reference to schematic phase diagram. Martensite volume fraction is controlled by composition, continuous annealing temperature and time, and cooling rate. It has been shown that a linear relationship exists between tensile strength and martensite volume fraction (Bucher and Hamburg, 1977; Davies, 1978; Imamura and Furukawa, 1976; Matsuoka and Yamamori, 1975). This indicates that tensile

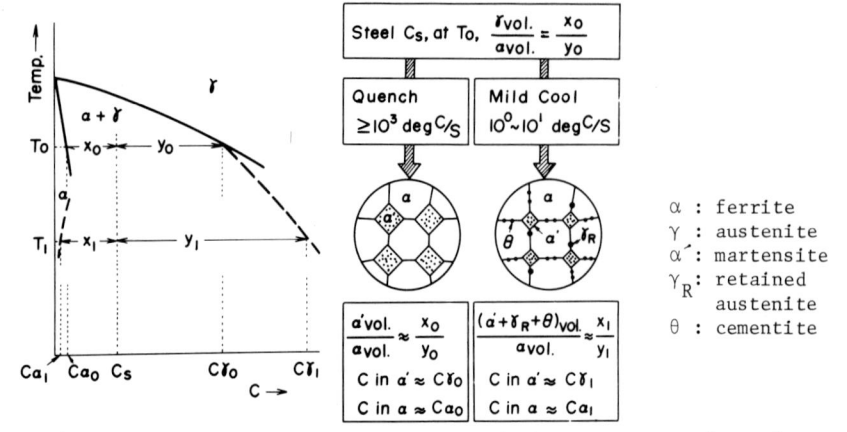

Fig. 9. Schematic illustration of dual-phase structure formation.

strength roughly obeys a rule of mixtures between a hard martensite and a soft ferrite. The low yield strength and smooth yielding characteristics are thought to result from the elastic strain and mobile dislocations introduced into the ferrite matrix by the strain due to martensitic transformation of the austenite (Imamura and Furukawa, 1976; Rigsbee and VanderArend, 1977), as indicated by transmission electron microscopy shown in Fig. 10.

Retained austenite is generally present either as interlath films in the martensite particles or as isolated particles. An example is shown also in Fig. 10. The amount of retained austenite varies from 1 to 10 percent in volume, depending on the steel composition and cooling rate. It tends to increase in steels of high austenite hardenability after rather slow cooling. Some investigators (Furukawa and others, 1979; Marder, 1977; Rigsbee and Vander-Arend, 1977) believe that the transformation of retained austenite during plastic straining contributes to the increase in work-hardening rate resulting in an improved uniform elongation. However, this is still controversial (Speich, 1981; Eldis, 1979) because the retained austenite transforms at low strains so that the work hardening at only low strains may be improved, as indicated in Fig. 11.

As previously noted, the as-annealed, undeformed dual-phase steels are generally non-ageing at room temperature but exhibit notable strain ageing behaviour after deformation. According to the study by Tanaka and others (1979), non-ageing property of undeformed dual-phase steels stems from nonuniform distribution of dislocations which are mostly confined to the ferrite regions adjacent to the martensite particles (Fig. 10). As a result, carbon diffusion needs to occur over long distances from the grain interior to the highly dislocated regions in order to attain sufficient dislocation pinning. This requires a long time at room temperature so that practically non-ageing behaviour is observed. After deformation, carbon atoms are readily available to pin dislocations sufficiently because deformation makes the dislocation distribution much more uniform.

New Production Techniques for As-Hot-Rolled Dual-Phase Steels

Although dual-phase steels of hot-rolled gauges can be produced by using

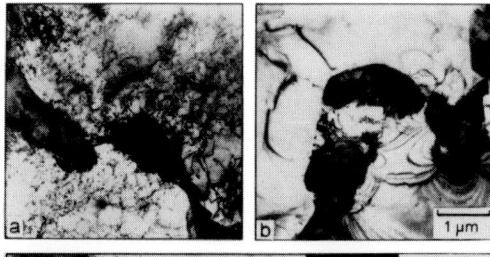

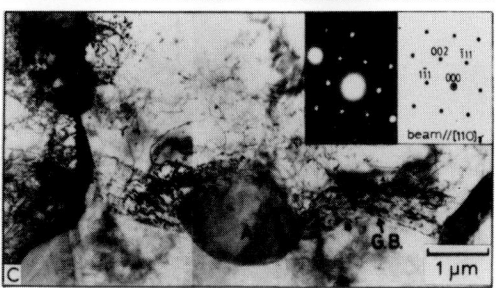

Fig. 10. Transmission electron micrographs of: (a) martensite islands surrounded by dislocations, and (b) bend-extinction contours in the ferrite matrix, and (c) retained austenite, in a 0.1 % C, 2 % Mn steel air cooled from 750°C (Furukawa and others, 1979).

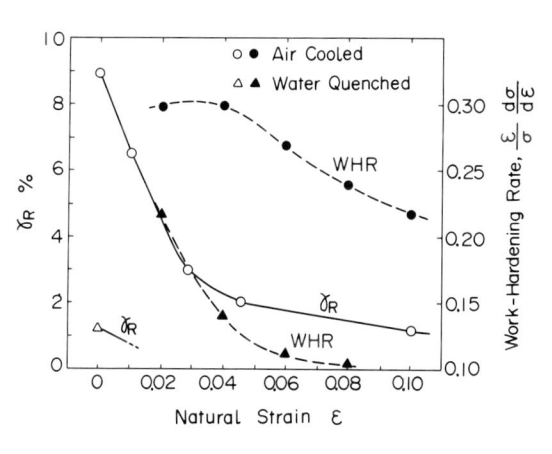

Fig. 11. Volume fraction of retained austenite and work-hardening rate as a function of strain, in a 0.1 % C, 2 % Mn steel held at 750°C for 2 min and air cooled or water quenched (γ_R: retained austenite, WHR: work-hardening rate) (Furukawa and others, 1979).

continuous heat treatment, an as-hot-rolled method is desirable from the viewpoint of production economy. Coldren and Tither (1978) proposed a low carbon, manganese-silicon-chromium-molybdenum steel as an as-hot-rolled dual-phase material whose hot rolling practice consists of a finishing at about 850°C and a coiling at about 550 to 600°C where ferrite coexists with austenite. In this steel austenite hardenability is strongly modified by the alloy additions.

Furukawa and others (1979, 1981) employed a different approach consisting of a low finishing temperature and a very low coiling temperature, using a steel of plain chemistry. The idea underlying this approach was that the steel could be made into a two-phase (ferrite and austenite) structure at the finishing temperature, then cooled down to a low temperature where the martensitic transformation of the austenite, at least in part, has already taken place, as indicated schematically in Fig. 12. In this technique, a silicon

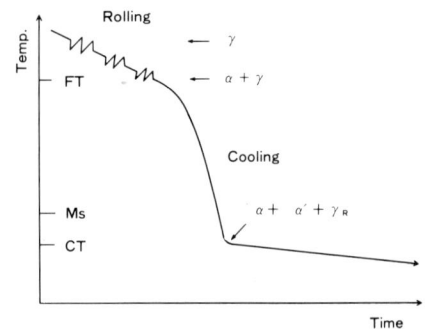

Fig. 12. Description of an as-hot-rolled dual-phase steel production method (FT: finishing temperature, CT: coiling temperature) (Furukawa and others, 1979).

addition was found to be beneficial since it greatly enlarges the optimum finishing temperature range and enhances the hardenability of the partitioned austenite, as shown in Fig. 13 and Fig. 14.

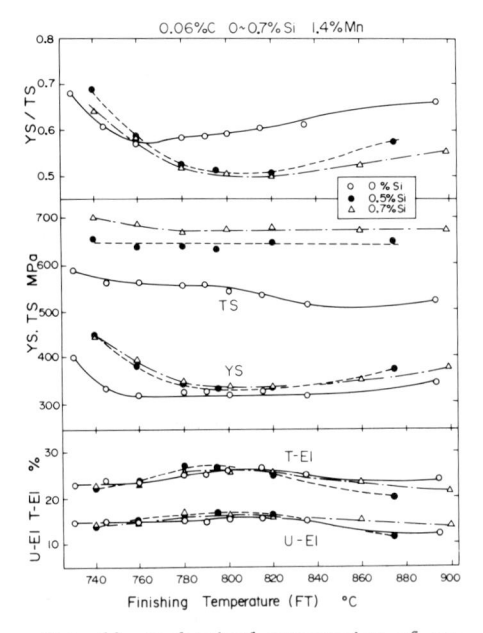

Fig. 13. Mechanical properties of as-hot-rolled dual-phase steels as a function of finishing temperature (Furukawa and others, 1981).

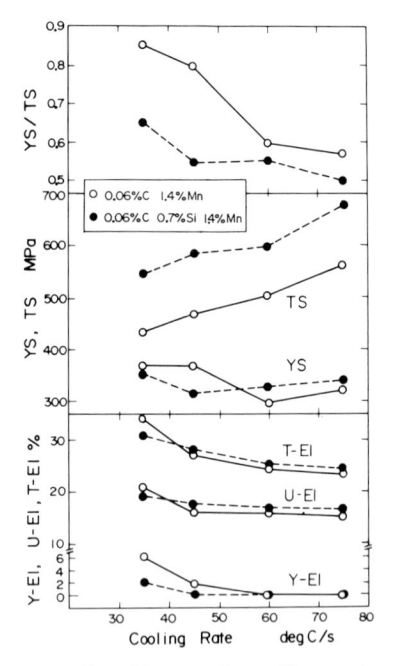

Fig. 14. Effects of cooling rate after finishing (Furukawa and others, 1981).

SOME CURRENT PROBLEMS

There has been an increasing necessity of applying high-strength formable sheet steels to cold pressings, due to the trend toward improvements in

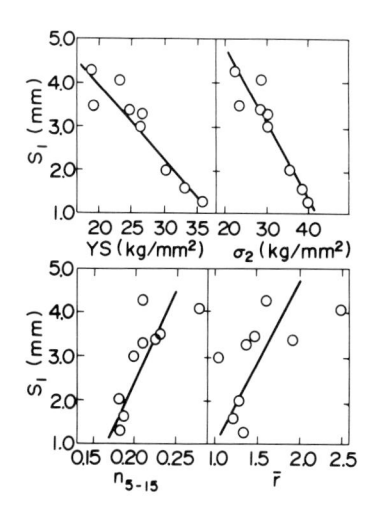

Fig. 15. Relation between mechanical properties and surface wrinkling tendency (Sl: punch stroke to the occurrence of wrinkles) (Oki and others, 1981).

safety and weight reduction of motor vehicles. In pressing cold-rolled high-strength steels into shallow panels such as fenders and door panels, it is reported that occurrence of surface wrinkles or deflections is a serious problem, which has proved to be closely related to yield strength, r value and n value (Takechi and others, 1981). A parameter, r/yield strength, is considered to be a suitable estimation value for press performance (Yoshida and others, 1981). Fig. 15 shows an example of experimental results indicating how mechanical properties affect the occurrence of wrinkles (Oki and others, 1981). It can be seen that the smaller strengths at small strains, the larger n values at medium strains and the larger r values are in favour of press performance. Therefore, low yield strengths and high n and r values are required not only for deep drawing steels but also for high-strength cold-rolled steels to be pressed into shallow panels.

On the other hand, denting resistance of formed products requires a high yield strength after forming. Inasmuch as a high degree of work hardening is not expected in shallow panels, strain-age hardenability at temperatures of paint baking becomes important, with the concomitant fulfillment of non-ageing requirement before forming.

To date, phosphorus added aluminium-killed steels (rephosphorized steels) are becoming popular because of their low cost. Phosphorus added, interstitial-free titanium steels have an excellent formability, however, they have no strain-ageing response after forming. The amount of phosphorus added is usually limited to be less than 0.1 percent so as not to aggravate weldability. Therefore tensile strengths of these classes of steels hardly exceed 400 MPa. Although some of the rephosphorized steels are still produced by using batch annealing, the continuous annealing is employed as a major production technique, based on the idea of scavengeing as described earlier. Dual-phase steels are characteristic of a high n value and a high response of strain-age hardening. However, the shortcoming in them is poor r value, being about 0.8 to 1.0.

Typical examples of mechanical properties of the steels mentioned above are listed in Table 3.

TABLE 3 Typical Mechanical Properties of Cold-Rolled High-Strength Steels
(Takahashi and others, 1982a; Takechi and others, 1981)

Steel	Yield Strength (MPa)	Tensile Strength (MPa)	Elongation (%)	Bake Hardening* (MPa)	$\bar{r}$
Rephosphorized Steel	230	390	37	20	1.7
Phosphorus added Titanium Steel	201	393	41	none	2.1
Dual-Phase Steel	234	466	38	55	0.9

* Yield strength increase resulting from ageing at 170°C for 20 min, in 2 % prestrained steel (excluding strain hardening due to prestrain).

Okamoto and Takahashi (1981) obtained a cold-rolled dual-phase steel with an r value of about 1.4 by employing a slow-heating batch anneal, using an aluminium-killed steel containing 2 to 2.4 percent manganese. However in this case the strain-ageing response at about 170 to 200°C must have been largely lost. According to Nishimoto and others (1981a), the r value decreases with increasing hardness and volume fraction of the martensite phase. Sudo and others (1982) have also confirmed this, and have proposed a way to obtain a dual-phase steel having a high r value. Their method consists of a batch annealing (hence the ferrite matrix texture is optimized) followed by the intercritical continuous annealing. They claim that an improved level of r values is obtained by introducing bainite plus martensite, instead of mere martensite, as the second phase. Some of their results are shown in Fig. 16.

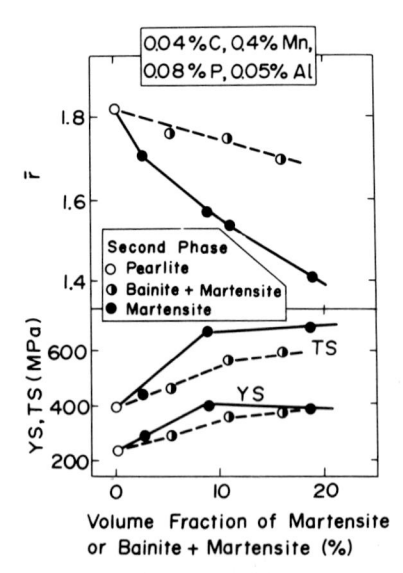

Fig. 16. Mechanical properties as a function of volume fraction of the second phase consisting of bainite plus martensite or martensite (Sudo and others, 1982).

Stretch flangeability of a dual-phase steel is reported to be improved by lowering of the second phase hardness by means of tempering (Davies, 1981; Nishimoto and others, 1981b) or controlled cooling after the intercritical annealing (Sudo and others, 1982). Probably this is related to the improvement in post-uniform elongation due to tempering of dual-phase steels (Speich and Miller, 1981). This kind of property change is quite possible since the ductile fracture behaviour is closely related to the nature of the second phase particles. Of course there are some trade-offs for this improvement, such as a decrease in tensile strength (because the martensite strength is decreased by tempering) and an increase in the yield strength (because the strain introduced into ferrite due to the martensitic transformation is largely lost by tempering of martensite) leading to a decreased n value, so that a compromised treatment may be required depending on the type of forming.

CONCLUDING REMARKS

As described above, various classes of sheet steels are now available for press-forming. However, especially in the area of high-strength steels, further studies regarding such items as follows would be required: (a) natural ageing and strain ageing behaviour in dual-phase steels, (b) effects of phosphorus and silicon additions on ductility and texture formation, because of the frequent employment of these elements as solid-solution strengtheners, (c) fracture behaviour of dual-phase steels with various second-phase hardness and morphology, and (d) texture control and plastic anisotropy behaviour in dual-phase steels.

Although the topics have been related with only the sheet steel materials, developments in forming techniques, by means of studies on the assessment of sheet metal forming severity (Hobbs, 1981), are undoubtedly important as well.

ACKNOWLEDGEMENT

The author is grateful to Dr. M. Abe, Fundamental Research Laboratories, Nippon Steel Corporation, for helpful discussions.

REFERENCES

Abe, M., M. Okamoto, N. Arai, and S. Hayami (1979). Tetsu-to-Hagane, 65, 418-424.
Abe, M., Y. Kokabu, Y. Hayashi, and S. Hayami (1980). J. Jap. Inst. Met., 44, 84-94.
Abe, M., Y. Kokabu, N. Arai, and S. Hayami (1981). J. Jap. Inst. Met., 45, 942-947.
Adams, M.A., and J.R. Bevan (1966). JISI, 204, 586-593.
Araki, K., S. Fukunaka, and K. Uchida (1977). Trans. ISIJ, 17, 701-709.
Bucher, J.H., and E.G. Hamburg (1977). SAE paper 770164.
Coldren, A.P., and G. Tither (1978). J. Metals, 30 (April), 6-9.
Davenport, A.T. (Ed.) (1979). Formable HSLA and Dual-Phase Steels, TMS-AIME, New York.
Davies, R.G. (1978). Met. Trans., 9A, 671-679.
Davies, R.G. (1981). In R.A. Kot and B.L. Bramfitt (Eds.), Fundamentals of Dual-Phase Steels, TMS-AIME, New York. pp. 265-277.
Eldid, G.T. (1979). In R.A. Kot and J.W. Morris (Eds.), Structure and Properties of Dual-Phase Steels, TMS-AIME, New York. pp. 202-220.
Furukawa, T., H. Morikawa, H. Takechi, and K. Koyama (1979). idem, pp.281-303.
Furukawa, T., and M. Tanino (1981). In R.A. Kot and B.L. Bramfitt (Eds.), Fundamentals of Dual-Phase Steels, TMS-AIME, New York. pp. 221-248.
Hayami, S., and T. Furukawa (1975). In M. Korchynsky (Ed.), Microalloying 75, 2A, pp. 78-87.

Hobbs, R.M. (1981). In Proceedings of International Symposium on New Aspects on Sheet Metal Forming, The Iron and Steel Inst. Japan, Tokyo, pp. 102-124.
Imamura, J., and T. Furukawa (1976), Seitetsu Kenkyu, No.289, pp.11952-11961.
Inagaki, H. (1976). Tetsu-to-Hagané, 62, 1000-1008.
Keh, A.S., and S. Weissmann (1963). In G. Thomas and J. Washburn (Eds.), Electron Microscopy and the Strength of Crystals, Interscience, New York. pp. 231-300.
Kot. R.A., and B.L. Bramfitt (Eds.) (1981). Fundamentals of Dual-Phase Steels, TMS-AIME, New York.
Kot, R.A., and J.W. Morris (Eds.) (1979). Structure and Properties of Dual-Phase Steels, TMS-AIME, New York.
Marder, A.R. (1977). In A.T. Davenport (Ed.), Formable HSLA and Dual-Phase Steels, TMS-AIME, New York. pp.87-98.
Matsuo, M., S. Hayami, and S. Nagashima (1971). In Proceedings of International Conference on the Science and Technology of Iron and Steel, Part II, The Iron and Steel Inst. Japan, Tokyo. pp.867-871.
Matsuoka, T., and K. Yamamori (1975). Met. Trans., 6A, 1613-1622.
McMahon, C.J. (1966). Acta Met., 14, pp. 839-845.
Mizunuma, S., S. Yamaguchi, M. Abe, and S. Hayami (1980). Tetsu-to-Hagané, 66, 221-230.
Morrison, W.B. (1966). Trans. ASM, 59, 824-846.
Morrow, J., and G. Tither (1978). J. Metals, 30 (March), 16-19.
Nagumo, M., Y. Abe, S. Yamaguchi, K. Ohoka, M. Akazawa, and H. Nakajima (1972). Seitetsu Kenkyu, No.274, pp. 9997-10009.
Nishida, M., K. Hashiguchi, I. Takahashi, T. Kato, and T. Tanaka (1978). In Proceedings of IDDRG 10th Biennial Congress, pp.211-213.
Nishimoto, A., Y. Hosoya, and K. Nakaoka (1981a). Trans. ISIJ, 21, 778-782.
Nishimoto, A., Y. Hosoya, and K. Nakaoka (1981b). In R.A. Kot and B.L. Bramfitt (Eds.), Fundamentals of Dual-Phase Steels, TMS-AIME, New York. pp.447-463.
Ohashi, N., T. Irie, S. Satoh, O. Hashimoto, and I. Takahashi (1981). SAE paper 819927.
Okamoto, A., and M. Takahashi (1981). In R.A. Kot and B.L. Bramfitt (Eds.), Fundamentals of Dual-Phase Steels, TMS-AIME, New York. pp.427-445.
Oki, T., Z. Shibata, and M. Sudo (1981). SAE paper 810030.
Rashid, M.S. (1976). SAE paper 760206.
Repas, P.E. (1979). SAE paper 790008.
Rigsbee, J.M., and P.J. VanderArend (1977). In A.T. Davenport (Ed.), Formable HSLA and Dual-Phase Steels, TMS-AIME, New York. pp.56-86.
Speich, G.R., and R.L. Miller (1979). In R.A. Kot and J.W. Morris (Eds.), Structure and Properties of Dual-Phase Steels, TMS-AIME, New York. pp.145-182.
Speich, G.R. (1981). In R.A. Kot and B.L. Bramfitt (Eds.), Fundamentals of Dual-Phase Steels, TMS-AIME, New York. pp.3-45.
Speich, G.R., and R.L. Miller (1981). idem, pp.279-304.
Stevenson, R. (1980). Met. Trans., 11A, 1909-1913.
Sudo, M., I. Tsukatani, and Z. Shibata (1982). To be published in Proceedings of Conference on Metallurgy of Continuous-Annealed Sheet Steel, TMS-AIME, New York.
Takahashi, N., M. Shibata, Y. Furuno, H. Hayakawa, K. Kakuta, and K. Yamamoto (1982a). idem.
Takahashi, N., M. Abe, O. Akisue, and H. Katoh (1982b). idem.
Takechi, H., M. Usuda, N. Iwasaki and Y. Hayashi (1981). In Proceedings of International Symposium on New Aspects on Sheet Metal Forming, The Iron and Steel Inst. Japan, Tokyo. pp. 66-81.
Tanaka, T., M. Nishida, K. Hashiguchi, and T. Kato (1979). In R.A. Kot and

J.W. Morris (Eds.), Structure and Properties of Dual-Phase Steels, TMS-AIME, New York. pp. 221-241.

Yoshida, K., H. Hayashi, K. Miyauchi, M. Hirata, T. Hira, and S. Ujihara (1981). In Proceedings of International Symposium on New Aspects on Sheet Metal Forming, The Iron and Steel Inst. Japan, Tokyo. pp.125-150.

Deformation at High Strains

M. Hatherly

School of Metallurgy, University of New South Wales, Kensington, NSW, Australia

ABSTRACT

Metals deform by two basic processes, slip and twinning, but at high strain levels plastic instability develops and shear bands form. The development of the microstructure and the morphology of the shear bands are different in each case and depend on factors such as crystal structure, stacking fault energy, crystal orientation, deformation temperature and strain rate. Some effects of these variables on the microstructure are described for the deformation range $\varepsilon \sim 0.02$-5.0. The nature of the cells and microbands formed in metals that deform by slip is discussed and it is suggested that attempts to relate these features to the flow stress by a Hall-Petch type expression are of doubtful value. The relationship between the deformed microstructure and the deformation texture is described and the significance of that microstructure to the nucleation of recrystallized grains discussed.

KEYWORDS

Microstructure; shear bands, microbands; deformation twinning; deformation texture; recrystallization texture; nucleation of recrystallized grains.

INTRODUCTION

Investigations of the nature of deformed metals tend to follow well defined historical paths that are dictated by the nature of the techniques available for examination or currently in vogue. In the period prior to 1950 most investigations were based on optical microscopy and etched specimens. In order to see the effects of the imposed stresses relatively high strains were generally required. During the same period the growing use of x-ray diffraction techniques stimulated an interest in the preferred orientations developed during deformation and annealing. The pattern began to change in the period 1945-1955 when the search for experimental evidence of dislocations prompted the examination of very lightly deformed metals and the change was accelerated in 1956 by the appearance of thin foil electron microscope techniques. The situation was compounded by the difficulty of preparing suitable specimens of heavily deformed metals. With the exception of a few notable papers (e.g. Embury and others, 1966a, b; Langford and Cohen, 1969 and Hu, 1969) electron microscope examination of heavily deformed metals virtually ceased. The exceptions were successful because the specimens were examined in geometries that required special preparation techniques. Rolled specimens for example, were examined by means of thin sections normal to the rolling direction (RD) or transverse direction (TD). These techniques were quickly taken up and over the last 10 years a number of significant observations have emerged. Some of these were old matters that have been rediscovered, e.g. the magnitude of the inhomogeneity of deformation and the contribution of shear bands to deformation at high strain levels. Others were already suspected and have now been confirmed, e.g. the role of twinning as a common deformation mode in fcc metals and alloys of low stacking fault energy (SFE). Still others appear to contradict currently accepted theories and await confirmation and general acceptance, e.g. some aspects of the microbands found in metals that deform primarily by slip. As understanding of the microstructure of deformed metals increased so too has understanding of the origin of deformation and recrystallization textures. In many metals and alloys, deformation to high, or even to moderate, strain levels is accompanied by annealing affects that may be dynamic or may occur subsequently to deformation. Recovery effects are commonly observed even in metals with moderately high melting points, e.g. copper, iron and in the more extreme cases, (e.g. zinc) dynamic recrystallization occurs. The present paper is concerned primarily with the development of microstructure and texture in heavily deformed metals, the relationship of that microstructure to the flow stress and to a lesser extent with the origin of recrystallized grains.

THE MICROSTRUCTURE OF DEFORMED METALS AND ALLOYS

For many years, descriptions of the development of microstructure in highly strained metals were bedevilled by the absence of an agreed terminology but it is now possible to use terms which are generally accepted. These will be defined as they emerge in the discussion that follows.

Within the context of this paper there are two basic deformation processes. Metals deform either by slip or by twinning. Because of the crystallographic nature of these processes it might be assumed that the deformation occurring in the individual crystallographic units or grains of a polycrystalline metal should be homogeneous. This, of course, is not so. The constraints exercised on any grain by the need to conform to a prescribed macroscopic shape change and the presence of neighbouring grains ensure that the strain state varies from region to region within individual grains and this results in local misorientation. The regions of different orientation in a single grain are known as <u>deformation bands</u>. The boundary of a deformation band is a region of continuous orientation change that is called a <u>transition band</u>. In certain cases, the orientation is identical on either side of a deformation band and the band is then known as a <u>kink band</u>. These three features are characteristic of the microstructure of all deformed metals and occur independently of crystal geometry, crystal structure or whether deformation occurs by slip or twinning. Bands of each type develop profusely at low strains but their incidence decreases as the texture develops with strain and the various volumes of material rotate towards the limited number of stable end orientations. The incidence is determined also by such factors as the initial grain size, deformation temperature, strain rate, etc. It will be clear that the presence of these various inhomogeneities makes the significance of an individual grain rather meaningless in even moderately deformed metals.

Deformation by Slip

This category includes all the fcc metals and alloys of medium and high stacking fault energy (SFE) such as copper and aluminium with values of

SFE ~60 and ~200mJm^{-2} respectively. Included also are iron and its bcc alloys and probably all other metals with this crystal structure. Specifically excluded are such fcc materials as silver, 70:30 brass and austenitic stainless steels which have SFE <20mJm^{-2} and probably all of the cph metals.

For all metals and alloys within this group deformation begins by dislocation glide on the most favourably oriented slip system and the results of this can be seen as slip lines and slip bands on previously polished surfaces. The first of these terms is used here to denote the individual steps of the elementary structure described by Wilsdorf and Kuhlman-Wilsdorf (1953) and the second to describe the clusters of larger steps that are visible in the optical microscope at $\varepsilon > 0.01$. One of the more significant areas of recent progress has been the growing ability to relate the various manifestations of deformation revealed by quite different experimental techniques. In this respect the slip bands that are so readily seen on polished surfaces have proved difficult. There are no simple etching techniques that produce equivalent markings on metallographic specimens of iron or aluminium and although reagents exist that reveal evidence of prior deformation in copper no one-to-one relationship has been established with slip bands. Luckily, transmission electron microscopy has provided a partial answer. In specimens strained by moderate amounts (e.g. $0.1 < \varepsilon < 0.3$) the dislocations are observed to lie in tangled networks that define an equiaxed cell structure ~0.5-2μm in diameter. Adjacent cells are misoriented by up to 2° and the misorientation is non-cumulative. Superimposed on this structure is a series of long straight bands that lie on the slip planes of the grains involved and often extend across the full width of the grain (Fig. 1). These bands, which will be discussed below, correspond to the surface slip bands. They are always ~0.1-0.2μm wide and the term <u>microband</u> is now generally used to describe them. The origin of the equiaxed cell structure is not clear. It was accepted for some time that this structure was a genuine feature of the deformed structure because the cells developed gradually as the dislocation content increased with strain and because the cell walls at low levels of strain are often parallel to low index planes. Unfortunately, there appears to be no correlation between the long slip lines of the elementary surface structure and the cells which are an order of magnitude larger than the spacing of the slip lines (~0.03μm). Many slip planes must have operated within the volume of every cell before the cell structure was established. Moreover, the elementary structure is already fully developed before the cell structure has emerged. For these and other reasons it seems likely that the cell structure is merely a relaxation configuration that develops in the bulk material when dislocation movement ceases.

Microbands have been observed in rolled single crystals of copper at strain levels <0.02, Fig. 1 (Malin and others, 1981). In general, however, strains >0.1 are required for foils examined in conventional microscopes. At higher levels of reduction ($\varepsilon \sim 0.2$) microbands are often observed on more than one slip plane in a grain and the intersections are then marked by prominent shear displacements. At still higher levels of strain ($\varepsilon \sim 0.8$) the microbands are observed to cluster and to be rotated towards alignment with the rolling plane. This alignment continues with further rolling and is virtually complete at $2.5 < \varepsilon < 3.0$ (Fig. 2). Areas of equiaxed cells can still be observed between the clusters of microbands even after 99% reduction ($\varepsilon = 4.6$). STEM analysis has shown that the orientation within a microband does not vary by more than 1° or 2° and that the mean orientation is usually within 2° of the adjacent equiaxed cell regions. The interiors of the microbands have a high dislocation density and the dislocations in the boundaries form a clear but diffuse boundary at low strains. At $\varepsilon \sim 1.4$ recovery effects lead to a sharpening of the boundaries of both cells and microbands in copper and the

Fig. 1 Microbands in copper crystal (ε=.02)

Fig. 2 Microbands in rolled copper (ε=3.5)

Fig. 3 Slip bands formed on copper surface at ε=0.51.

(Longitudinal sections; 1µm marker parallel to RD)

dislocations within both features adopt a regular array. Similar effects occur in iron at a strain level of ~2 during rolling.

The above account of microband development in the microstructure of deformed metals leaves several unanswered questions viz.(i) how do microbands form and do they operate more than once; (ii) is microband formation the only deformation mode and (iii) how do microbands that have been sheared by other microbands rearrange their boundaries to form the long aligned, straight bands that are so prominent a feature of highly strained metals like copper and iron? There is no clear answer to any of these questions. With respect to the first an experiment by Malin and Hatherly (1979) raises some major problems. A polished surface perpendicular to TD was prepared on a specimen that had previously been rolled to ε=0.5 and the specimen was then given a further light reduction. High resolution TEM replicas were taken from the previously polished surface; the resolution in the replicas was thought to be better than 0.01µm. Figure 3 shows an area in which slip bands have formed on two intersecting systems. It is apparent first of all that elementary slip is not a contributing factor to the deformation process at this strain level. Much more surprising is the fact that within the limits of experimental resolution the shear in the bands appears to be homogeneous. The generally accepted view of the structure of a slip band is that it consists of a number of discrete slip lines of more than average height and spaced at ~0.03µm. Figure 3 is incompatible with this view. It is also incompatible with the notion of a Frank-Read source generating glide dislocations on discrete slip planes. There was at one time considerable discussion as to the nature of a dislocation source that could provide a uniform shear in a relatively wide band. One such source, the so-called "cone source" of Bilby (1955) is distinguished from the Frank-Read source by virtue of the fact that the Burgers vector of the pole dislocation is not restricted to the operative slip plane. Any configuration involving a component perpendicular to that plane produces a homogeneous deformation as the sweeping dislocation climbs around the pole. Perfect dislocations produce a homogeneous shear, imperfect dislocations lead to twinning. Figure 3 is consistent with such a model. The second question is somewhat easier. As far as can be seen from relatively simple calculations all of the deformation during rolling in the range $0.01<\varepsilon<1$ occurs by slip, i.e. with the formation of microbands. This may seem surprising in view of the relative infrequency of microbands seen in most investigations but in this respect, it should be noted that Willis (1982) has shown that microbands are much more profuse in the thicker areas of specimens examined in high voltage microscopes than in thinner regions or in foils prepared for conventional 100Kv microscopes (Fig. 4). It seems probable that dislocation rearrangements occur in the thinner areas of some foils with a resultant loss of the microband structure. The third question poses a difficult problem. The long, aligned, sharply-defined microbands seen at high strain levels show no evidence of the numerous intersections and displacements that occur at lower strains, but still have the same width as the microbands first formed. It is possible that the serrated array of small segments coalesces and recovers so that the internal boundaries are lost but there is no experimental evidence for this view. Dillamore (1979) has suggested that a microband may be merely a micro-kink band formed from dislocation debris on the active slip planes but if this view is correct there should be orientation difference between the matrix and the bands. Such a change has not been seen in any of the STEM experiments so far reported. All of these show inconsistent small orientation differences and this interpretation is also inconsistent with the homogeneous shear shown in Fig. 3.

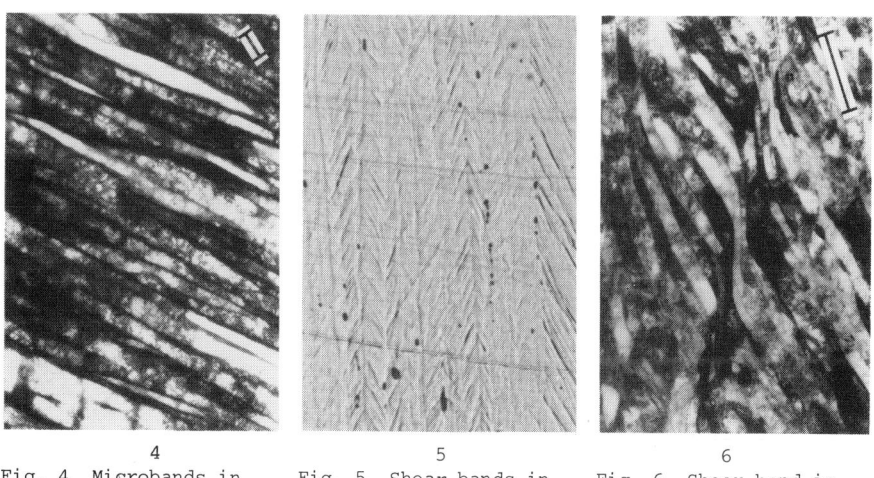

4 5 6
Fig. 4 Microbands in rolled iron ($\varepsilon=0.2$) (Willis, 1982) Fig. 5 Shear bands in rolled copper ($\varepsilon=2$) x250. Fig. 6 Shear band in rolled iron ($\varepsilon=2$)(Willis, 1982)

(Longitudinal sections; 1μm marker parallel to RD - Figs. 4, 6).

There seems to be little doubt that a microband is the electron microscope manifestation of a surface slip band but the extent of the association is debatable. The failure to observe microbands in very thin foils and the increased population seen in HVEM studies point to the occurrence of recovery effects capable of eliminating microbands from the thin foil specimens used. In this regard Willis (1982) found that microbands were absent from {100}<011> grains of iron and similar results have been reported many times for the complementary orientation {110}<001> in rolled copper. It is inconceivable that slip does not occur in such grains. If recovery can eliminate microbands completely from some grains the observed structure of others must be suspect. Bulk specimens can now be examined in scanning electron microscopes in the back scattered electron mode and features of similar shape and size to microbands have been observed (Ridha and Hutchinson, 1981). Unfortunately, no further evidence has been provided about the nature of the actual deformation process.

At strain levels $>\sim 1.2$ a new deformation mode, shear band formation, begins to contribute to the total strain and thereafter plays an increasingly significant role. These bands are easily revealed by etching in copper (Fig.5), aluminium and iron and are also easily detected in TEM specimens (Fig. 6). In rolled materials they form independently of the local orientation on planes that are $\sim \pm 35°$ to the rolling plane and parallel to TD. In the metals under consideration shear bands form in colonies in each of which only one set of parallel bands develops. The bands in alternate colonies are in opposite senses so that an overall herringbone pattern is developed (Fig. 5) The colonies often extend across the thickness of many grains and the shear bands cross grain boundaries without deviation. The shear strain in the bands is large and values as high as 6 have been reported in iron (Willis, 1982); values of 2-3 are more common. The structure of a shear band is very clearly resolved in Fig. 6; the nearly aligned microbands typical of heavily rolled iron have been swept into the shear band and local elongation and thinning have accompanied the shear strain. At still higher levels of strain ($\varepsilon \sim 3.5$) larger shear bands develop which cross the rolled sheet from top to bottom and when a uniform relatively widely spaced population of these exists, failure occurs along the bands.

Deformation by Twinning

There are two broad groups of materials in which slip is limited to quite small strain levels and in which twinning becomes a major deformation mode. These are the fcc metals of low SFE and the cph metals. Twinning may also occur in bcc metals and in fcc metals of high SFE if deformation occurs at high strain rates or low deformation temperatures but these special cases will not be considered. Brief mention will be made of alloys such as the low zinc brasses and austenitic stainless steels which have intermediate values of SFE and of the high silicon, copper-silicon alloys which have very low values of SFE.

(i) FCC metals with low SFE. Most of the recent experimental work in this category has been done with 70:30 brass and interpretation has been made easier by the ease with which several modes of deformation can be revealed by metallographic etching. Features such as those shown in Fig. 7 develop on {111} planes at low-medium strain levels and are called strain markings. Although their nature was a subject of controversy for many years it is now known that they consist of clusters of fine ($\sim 0.02 \mu m$ thick) deformation twins spaced at $\sim 0.1 \mu m$ (Fig. 8). TEM examination shows that from the very beginning of deformation the dislocations are dissociated into partials and that planar arrays of stacking faults develop. There are no microbands and no equiaxed cell structure. The first twins appear in rolled brass at $\varepsilon \sim 0.05$ and clustering into bands follows almost immediately. As rolling proceeds the twins are rotated into alignment with the rolling plane and in many grains and deformation bands this process is complete at $\varepsilon \sim 0.8$-1.0. At $\varepsilon \sim 0.8$ shear bands begin to form. These bands are readily etched and develop initially in those regions in which twin alignment is most perfect. In contrast with the behaviour of metals deforming basically by slip two sets of intersecting shear bands are usually observed in a particular volume (Fig.9). The shear on the bands is high and values of shear strain up to 10 have been

reported (Duggan and others, 1978). It was shown that the bands operate only
once and that after formation they too are rotated towards alignment with the
rolling plane. Shear bands in these materials are easily seen in longitudinal TEM specimens as bands of very fine elongated units derived from the
aligned twin-parent matrix structure (Fig.10). These units or crystallites
are ~0.02-.05µm thick and have aspect ratios of 2:1 or more. STEM analysis
has shown that the orientation varies widely from one crystallite to another
but there is a marked preference for {110}<001> which corresponds to a major
component of the fcc shear texture after the necessary change of reference
axes. Shear band formation appears to be the major mode of deformation in
low SFE materials in the strain interval 0.8-2.0 but texture studies indicate
that thereafter slip begins in the recovered shear band crystallites and
becomes dominant (Hutchinson and others, 1978). At a later stage, and in
common with all other materials, a new set of shear bands emerges which
traverse the sheet from top to bottom and eventually the sheet fractures
along these bands.

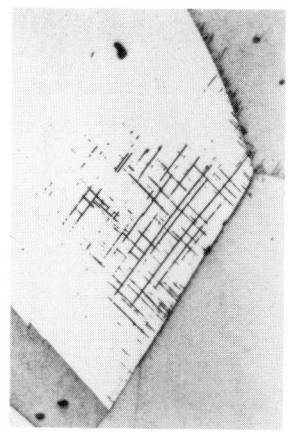

Fig. 7 Strain markings in rolled 70:30 brass (ε=0.06) x500.

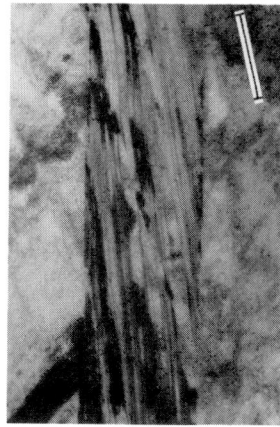

Fig. 8 Clustered twins in rolled 70:30 brass (ε=0.35).

Fig. 9 Shear bands in rolled 70:30 brass (ε=1.4) x500. (Longitudinal sections; 1µm marker parallel to RD - Fig. 8).

(ii) <u>FCC metals with intermediate SFE.</u> The marked differences in the microstructure of deformed copper and 70:30 brass suggest that mixed deformation
modes should occur in alloys of intermediate SFE. This expectation is borne
out by the work of Blicharski and Gorczyca (1978) with stainless steel and
more particularly that of Wakefield and Hatherly (1981) with 90:10 brass. In
the latter case it was shown that while some grains or parts of grains deform
by slip in a manner typical of copper others deform by twinning in a fashion
characteristic of 70:30 brass. The shear bands formed subsequently are also
characteristic of the regions in which they develop. The complexity of the
situation is well illustrated by Fig. 11 in which the first-formed lower
twin has been sheared by the slip associated with the near-vertical microbands. The subsequent re-establishment of twinning is indicated by the displacement of the microband by the upper twin. The somewhat meagre results
available can be summed up as follows: twinning is not a preferred
deformation mode during rolling in regions oriented at {110}<001> (Blicharski

Fig. 10 Shear band in rolled 70:30 brass (ε=1.4).

Fig. 11 Microband and twins in rolled 90:10 brass (ε=0.2).

Fig. 12 Faults and twins in rolled Cu-8.8Si (ε=0.2)

(Longitudinal sections; 1μm marker parallel to RD)

and Gorczyca (1978) and {110}<112> (Malin and others (1981). In rolled 90:10 brass twinning is not observed in {112}<111> grains at low strains but at ε=0.8 this orientation is copiously twinned (Wakefield and Hatherly, 1981).

(iii) Copper-silicon alloys. Saturated or near-saturated Cu-Si alloys have very low values of SFE (~3mJm^{-2}) and it is to be expected that twinning and shear band development will be significant processes during deformation. Malin and others (1982a) have shown that in the near saturated 8.8at.%Si alloy faulting and twinning occurred on all four {111} planes during rolling to ε=0.4 (Fig. 12). In such circumstances the gradual alignment of a single set of twins with the deformation axes is clearly impossible and in the absence of any alternative deformation mode shear band formation began at strain levels of 0.4-0.5. The alloy cracked along macroscopic shear bands when rolled to ε~1.2.

(iv) CPH metals. The deformation behaviour of cph metals is controlled by two important factors. Both the number of available slip systems and the magnitude of the twinning shears are considerably less than for metals with cubic structures. Except for the special case of beryllium the actual slip system favoured is determined by the c/a ratio so that basal slip is predominant for metals having c/a >1.633 (ideal) and prismatic slip for metals having c/a <ideal. In either case the magnitude of the deformation occurring by slip is limited and twinning is common at strains as low as 0.05. The twins are large and easily observed. The subsequent behaviour is determined primarily by the melting point of the metal. Blicharski and others (1979) have shown that in rolled titanium (M.P. 1945K) deformation occurs by twinning and slip until at ε~0.5 the size of the twinned units becomes too small for further twinning. Subsequent deformation occurs by slip but at ε~2.5 shear band formation becomes the major deformation mode. In rolled zinc (M.P. 692K) Malin and others (1982b) found that whilst twinning was favoured at ε<0.1 it was replaced by slip in the twinned regions at ε=0.17. Shear bands were common at ε=0.5 but in all specimens examined the bands were recrystallized. With further rolling to ε=1.6 the volume of recrystallized material increased to almost 100% and the structure stabilised as an array of equiaxed recrystallized grains of 30-50μm diam. Twinning was prominent as a deformation mode in many of these grains.

DISCUSSION

Two matters that are of considerable importance to the theme of this conference emerge from the above review. One of these is the much discussed possibility of a relationship between the size of the structural units in deformed metals and the flow stress, and the second, the significance of shear band formation as a deformation mode at medium-high levels of strain.

The Hall-Petch Relationship in Heavily Deformed Metals

The microstructure of metals that deform by slip consists of microbands and equiaxed cells and there have been several attempts to relate the size of these features to the flow stress by an expression of the Hall-Petch type. Following the earlier work of Embury and Fisher (1966), Embury, Keh and Fisher (1966) found that such a relationship existed for a wide range of ferrous materials and for copper. However, the subsequent results of Langford and Cohen (1969) with drawn iron did not correlate with the reciprocal square root of the "cell" size ($\bar{d}$) and instead a linear dependence on $(\bar{d})^{-1}$ was found. Nuttall and Nutting (1978) who also examined a wide range of materials recognized that the dislocation morphology depended on such factors as SFE and homologous temperature and suggested that in the presence of a relatively uniform high dislocation density the flow stress would be given by an expression of the type

$$\sigma_f = \sigma_o + K\rho^{1/2}$$

where ρ is the dislocation density. If, on the other hand, the structure consisted of an array of cells a normal Hall-Petch relation was suggested

$$\sigma_y = \sigma_o + K^1 (\frac{1}{\bar{d}})^{1/2}$$

All of these results must be regarded with caution. Nuttall and Nutting (1978) examined rolling plane foils in which identification of the substructure is virtually impossible. Each of the other investigations was based on the examination of longitudinal specimens but the "cell" size was determined by counting the number of "cell" wall intercepts along a fixed traverse. So far as can be determined by comparison with the published photographs every intercept was counted. It was implied moreover that the "long elongated features" present at high strain levels developed by elongation of the equiaxed cells. In the terminology of the present paper, microbands are already present at very low levels of strain and they appear to remain relatively constant in thickness at ~0.2μm. Equiaxed cells are also relatively constant in size with diam. 0.5-1.0μm. The long elongated features seen at heavy reductions cannot have developed from such cells. In simple terms, the microstructure of rolled copper consists of microbands in a matrix of equiaxed cells. The number of microbands increases with strain and it follows naturally that values of "cell" size based on linear intercepts must decrease with increasing strain. Unfortunately, the equiaxed cells appear to be formed by a relaxation process of some sort and may have no significance during the cold working of bulk specimens. Furthermore, as Willis (1982) has shown, the number of microbands seen depends on the specimen thickness and the operating voltage of the microscope used. It is highly probable that the flow stress of heavily deformed metals will prove to be related to the dimensions of the substructure but until the nature of that substructure is better understood, measurements of the dimensions would seem to be of little value. In the meanwhile it should be borne in mind that analyses of the type just discussed become meaningless in low SFE metals where cells do not form and the structure consists of very thin deformation twins.

Shear Bands

Although shear bands were clearly described as long ago as 1922 (Adcock) and subsequently reported from time to time in non-ferrous rolled products their significance largely escaped recognition until the work of Brown (1972) on aluminium and Mathur and Backofen (1973) on steel. Since then, there have been many reports of experimental observations of shear bands and several attempts to elucidate the conditions under which they develop. In this regard it is important to note 5 points: (1) shear bands are a form of plastic instability; (2) the geometry of shear bands is not dictated by considerations of crystallography; (3) in the case of heavily rolled metals shear bands develop at ~35° to the rolling plane and not at the expected angle of 45°; (4) shear bands may be short and confined to a single grain or group of grains or may traverse the full thickness of rolled sheet; (5) a particular volume may contain complementary sets of positive and negative shear bands or a single set only. In the latter case alternate volumes have bands of opposite sense.

In plane strain deformation (or rolling) the condition for instability is that $\frac{1}{\sigma} \cdot \frac{d\sigma}{d\varepsilon} \leq 0$. Following Dillamore and co-workers (1978, 1979) we write this condition as

$$\frac{1}{\sigma} \cdot \frac{d\sigma}{d\varepsilon} = \frac{n}{\varepsilon} + \frac{m}{\dot{\varepsilon}} \cdot \frac{d\dot{\varepsilon}}{d\varepsilon} + \frac{1+n+m}{M} \cdot \frac{dM}{d\varepsilon} - \frac{m}{N} \cdot \frac{dN}{d\varepsilon} \leq 0$$

where n and m are the strain hardening and strain rate exponents, M is the Taylor factor and N the density of mobile dislocations. Of the terms in the equation that in ε is always positive, that in $\dot{\varepsilon}$ is usually positive and that in N is negative. It is the last of these that accounts for the formation of Luders bands (also a form of plastic instability) during the yielding of mild steel. It follows that if an instability is to occur at medium-high strain levels, and in the absence of strain rate effects, it must be because the term in M becomes negative through a negative value of $dM/d\varepsilon$. Such a condition corresponds to geometric softening. $dM/d\varepsilon$ can be calculated from $dM/d\theta$ (the rate of change of hardness with orientation) and $d\theta/d\varepsilon$ (the rate of change of orientation with strain) and Dillamore and others (1979) show that this factor is both negative and a minimum at ~35° for typical, rolled metals with well developed textures.

Analyses of this type assume that the whole volume is involved with the formation of full-thickness shear bands but there is no reason why individual grains or groups of grains should not deform by the operation of shorter bands. Such bands would represent the response of a limited volume to the combination of the imposed macroscopic strain and the displacements occurring in neighbouring volumes. Both Dillamore and others (1979) and Van Houtte and others (1979) have considered this problem. The former point out that the short shear bands developed in any grain may be positive or negative in inclination or may consist of an array of both sets and argue that the choice is determined by the orientation of the volume concerned. In {110}<001> and {110}<112> grains both sets of bands are predicted. This prediction is apparently confirmed by the observation that sets of intersecting shear bands are a feature of the microstructure of rolled 70:30 brass; the rolling texture of this material is usually described as a mixture of {110}<112> and {110}<001> components. However, this texture emerges only at high strains

(ε>2) whereas the shear bands develop first at ε~0.8. At this stage the rolling texture of 70:30 brass is better described as diffuse with a major component {112}<111> (Duggan and others, 1978) and the prediction remains unconfirmed. It was pointed out earlier that the development of deformation bands and other inhomogeneities within the grains of a deformed metal makes the concept of grain size rather meaningless. This situation is compounded when shear bands begin to form. As Solomon and others (1982) have shown, the short shear bands that develop in heavily rolled copper appear to cross several grains and similar results are common with other metals.

Macroscopic shear bands are a common feature of the microstructure of industrially formed products, e.g. the chips removed during machining. Many of these are distinguished by high rates of strain and in the more extreme cases such as high speed punching, ballistic impact and explosive fragmentation there appears to be no doubt that adiabatic effects are significant. Any discussion of "adiabatic" shearing is open to argument and two points need to be emphasized. (1) Completely adiabatic deformation is a thermodynamic ideal. The term is used here in the more general sense first used by Zener and Holloman (1944) i.e. if insufficient time is available for dissipation some of the heat generated in a deforming volume must be contained therein. (2) Adiabatic shearing occurs when an already localised strain is concentrated further by material or geometric considerations; it does not result, per se, from a state of instability that causes a uniform strain to localize. Adiabatic shear bands have been reported in many metals and alloys after ballistic deformation but the best known example is the so-called "white-etching" bands found in some steels. These bands which have high hardness and a very fine structure are generally believed to occur as a result of adiabatic heating to above the critical temperature and very rapid self quenching. The shear strain in the bands may be as high as 100. Some heat must be developed in any shear band and Backman and Finegan (1973) suggest that adiabatic shear bands should be classified as either deformed or transformed. Examples of the former include the shear bands found in rolled copper or brass whilst the shear bands reported during machining of a Fe-18.5Ni steel by Lemaire and Backofen (1972) are clearly of the latter class. The case of rolled zinc then becomes interesting. As reported by Malin and others (1982) the shear bands developed in this low melting point (692K) metal are always recrystallized and this occurs, presumably, because the heat developed in the band is sufficient to cause dynamic recrystallization without major contribution from strain rate effects. It could be that many other cases of so-called dynamic recrystallization or hot working have a similar origin.

DEFORMATION TEXTURES

Understanding of the development of microstructure during deformation has provided a new opportunity to examine the origin of deformation textures. Although several attempts to correlate these factors have been made in the past (e.g. Mathur and Backofen, 1973, Hu, 1969 and the more theoretical analysis of Dillamore and Katoh, 1974) the discussion of the rolling texture of 70:30 brass by Hutchinson and others (1979) was probably the first to relate the development of microstructures to the development of rolling texture over a wide range of strain. At strain levels up to ~0.4 deformation occurs mainly by slip and the rolling texture of brass is very similar to that of copper. Subsequently, twinning becomes the dominant deformation mode and at ε~0.8 shear bands begin to form. The competition between twin alignment and twin destruction by shear bands leads to a maximum in the value of I_{111} at ε~1.5 and subsequently I_{110} rises sharply as slip occurs again in the recovered crystallites of the shear bands. It follows that the well-known texture transition in fcc metals must be due to the effects of SFE and orientation on the deformation mode of individual volumes in the deforming metal. In 90:10 brass, for example, some regions deform basically by slip and others by twinning (Wakefield and Hatherly, 1981) and it was proposed that the relevant copper- and brass-type textures develop therein. The close association between microstructure and rolling texture has been shown even more clearly at this conference by the work of Carmichael and others who determined the textures developed in the individual grains of polycrystalline 95:5 brass. Regions in which slip predominated had $\{112\}<111>$ and $\{123\}<634>$ textures whilst regions associated with twinning had $\{110\}<112>$ and $\{110\}<001>$ textures. The nature of the texture transition has been systematically examined by Lücke and co-workers (1978,1981) who analysed the changes that occur as a function of SFE in terms of 4 major texture components. These were identified as the copper component $\{112\}<111>$; the S component, $\{123\}<634>$; the Goss component, $\{110\}<001>$ and the brass component $\{110\}<112>$. The first two of these adequately describe the rolling texture of copper whilst the last two are relevant to the texture of 70:30 brass. Almost every account of the origin of the texture transition has adopted the view that the brass component, $\{110\}<112>$ is the terminal texture at the low SFE, alloy end of the transition. The recent work of Malin and others (1982) with a rolled Cu-8.8Si alloy has shown that this is not so. In this very low SFE alloy ($3mJm^{-2}$) the major rolling texture component was $\{110\}<001>$ at 4.61xR whereas the strength of the $\{110\}<112>$ component was only 2.8xR. It will be necessary to include these marked effects in the data of Lücke and co-workers but their significance to discussions of the origin of the texture transition requires a clearer understanding of the very confused microstructure of this alloy.

There has been less recent interest in the textures of high SFE metals that deform by slip. As Dillamore and Katoh (1974) have shown such textures can be predicted with considerably accuracy; those deviations from experimental results that do exist are almost certainly attributable to the operation of shear bands in the later stages of deformation.

RECRYSTALLIZATION

Each of the strain free grains that comprise the microstructure of a recrystallized metal has grown from a nucleus that must have been identifiable in the deformed microstructure. The potential sites for nucleation include each of the inhomogeneities of deformation discussed above, the complex boundary regions between inhomogeneities of various types, the grain boundaries of the deformed metal and any interphase interfaces present. Of the available inhomogeneities shear bands have attracted considerable recent attention and this has taken two forms. There has been an interest in the mechanism of nucleation and an interest in the role of shear band nucleation in the development of recrystallization textures.

Nucleation

In heavily rolled low SFE materials recrystallization begins in the shear bands (Duggan and others, 1979b) and this is followed by a rapid sequence of growth events that involves multiple twinning stages. The nucleation stage is complex and involves the prior formation of recovery twins. The morphology and development of these twins has been described by Huber and Hatherly, 1979; 1980. In cold rolled ($\varepsilon=2.3$) 70:30 brass they appeared first after 3 minutes at 200°C as colonies of very thin (0.002-0.01µm) twins within the shear bands (Fig.13). The dislocation density in the volumes associated with the recovery twins gradually decreased so that the volumes became strain free but during this change there was no clear boundary between those volumes and the cold worked matrix. At the same time, the twins thickened slightly. Nuclei, i.e structural units with clearly defined boundaries developed only after further heating. In rolled brass the twins developed preferentially in {110}<112> regions and nuclei of that orientation eventually emerged therefrom. Figure 14 shows all three twin types, viz., deformation, recovery and annealing twins. It was proposed originally that recovery twins formed

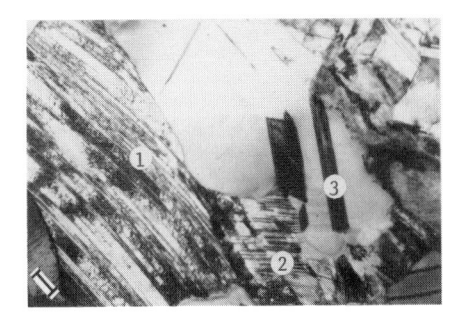

Fig. 13 Recovery twins in rolled 70:30 brass (ε=2.3). Heated 3 min. at 260°C.

Fig. 14 Twins in rolled 70:30 brass. (ε=1.4). Heated at 2°C min^{-1} to 300°C, showing 1.deformation twins, 2.recovery twins and 3.annealing twins.

(Longitudinal sections; 0.2μm marker parallel to RD)

because of a need to provide a boundary at which the excess dislocations generated during working could be annihilated and because the crystallites in the shear bands were too small to permit normal growth by bulging. Recently Verbraak, (1981) has suggested that the formation of recovery twins is the low SFE equivalent of polygonisation in high SFE materials. It was proposed also that partial dislocations piled up between deformation twins could cross the twin boundaries during heating, annihilate the excess dislocations within those twins and thereby create the recovery twins. This model successfully predicts some of the observed experimental features but ignores the observation that recovery twins form in the shear bands of low SFE metals and not in the clusters of deformation twins. Recovery twins are occasionally seen in as-rolled material at very high levels of strain, e.g. Hu and others, (1966) in Cu-4Al, Slakhorst and Verbraak,(1976) in silver, Nuttall and Nutting, (1978) in silver, and Köhloff and Malin (1981) in a copper single crystal rolled at 77K. They have been observed also in the brass-type shear bands of a heated 90:10 brass by Wakefield and Hatherly (1982) and more interestingly in the copper-type microbands of the same alloy. It seems that they are associated with a recovery mechanism that has general application to nucleation in low SFE materials.

Recrystallization Textures

The observation that recrystallized grains appear first in the shear bands of heavily deformed metals has prompted suggestions that shear band nucleation must be significant to the development of recrystallization textures. Two examples will be discussed to show that such views are oversimplified. Carmichael and others (1982) have found that recrystallization in 95:5 brass rolled to ε=1.4 began in the shear bands of grains that had deformed by twinning etc. in a brass-like fashion. The first nuclei appeared within 2 minutes at 350°C and recrystallization was complete in these grains after 6 min.; the recrystallized grains were very small. In other grains in which slip had been the major deformation mode recrystallization began only after 6 hours or more and the new grains were much larger. Growth of either set of new grains into each other or into neighbouring undeformed grains was minimal and neither set of grains had orientations compatible with the recrystallization texture in this alloy. Grains corresponding to the major component of this texture, {3,7,11}<1$\bar{2}$1> (Virnich and Lücke, 1978) developed in a previously unrecognized type of structure in the deformed material. Once again, nucleation began in the shear bands but only after 3 hours at temperature. These grains had the capacity to grow into neighbouring volumes of recrystallized grains but even so the small grains nucleated within the first 2 min. were still present after 24 hours. Although formed first these grains did not grow and were eventually consumed by other grains. Shear band nucleation is clearly significant to the final texture in this case; the difficulty lies in recognizing the appropriate type of shear band, or rather, the base structure in which such shear bands develop. A current examination of 70:30 brass is producing similar results. Experiments like these are not easily reconciled with that of Noda and others (1978). These workers determined the orientations of 300 freshly nucleated grains (<2μm diam.) in a partly annealed 70:30 brass and reported that the composite pole figure was similar to that of a fully recrystallized specimen.

Following earlier work by Noda and Huber (1978) with an Fe-Ni alloy Ridha and Hutchinson (1981) have shown that the cube texture grains in annealed copper grow by a boundary-bulging process from long thin bands in the deformed structure. The bands are oriented near {100}<001> and growth occurs only if a high misorientation exists relative to the neighbouring substructure. These bands have been identified as transition bands and the observations lend considerable support to the model proposed by Dillamore and Katoh (1974) for the origin of the cube texture. When rolling conditions were altered so as to produce an abundance of shear bands the long thin bands were absent from the deformed structure and a strong cube texture did not develop. The necessary bands had been destroyed, presumably by the shear bands. Nucleation occurred preferentially in the shear bands with a wide orientation spread and the final texture was diffuse. These results support the contention that shear band nucleation controls the final texture and also provide a better understanding of the origin of the cube texture.

REFERENCES

Adcock, F. (1922). J.Inst.Met., 27, 73-101.
Bilby, B. (1955). In "Defects in Crystalline Solids", Phys.Soc., London, pp.124-133.
Backman, M.E. and S.A. Finnegan (1973). In "Metallurgical Effects at High Strain Rates", Plenum Press, New York, pp.531-543.
Blicharski, M. and S. Gorczyca (1978). Met.Sci., 12, 303-312.
Blicharski, M., S. Nourbakhsh and J. Nutting (1979). Met.Sci., 13, 516-522.
Brown, K. (1972). J.Inst.Met., 100, 341-345.
Carmichael, C.M., A.S. Malin and M. Hatherly (1982). This conference.
Dillamore, I.L. and H. Katoh (1974). Met.Sci., 8, 21-27.
Dillamore, I.L. (1978). In Fifth International Conference on Textures of Materials, Aachen, Springer Verlag, 1, pp.67-80.
Dillamore, I.L., J.G. Roberts and A.C. Bush (1979). Met.Sci., 13, 73-77.
Duggan, B.J., M. Hatherly, W.B. Hutchinson and P.T. Wakefield (1978a). Met.Sci., 12, 343-351.

Duggan, B.J., W.B. Hutchinson and M. Hatherly (1978b). Scrip.Met., 12, 293-295.
Embury, J.D. and R.M. Fisher (1966a). Acta Met., 14, 147-159.
Embury, J.D., A.S. Keh and R.M. Fisher (1966b). Trans.AIME, 236, 1252-1260.
Hu, H. (1969). In J. Grewen and G. Wassermann (Eds.), "Textures in Research and Practice", Springer, Berlin, pp.200-226.
Hutchinson, W.B., B.J. Duggan and M. Hatherly (1979). Met.Tech. 6, 398-403.
Huber, J. and M. Hatherly (1979). Met.Sci., 13, 665-669.
Huber, J. and M. Hatherly (1980). Z.Metallk., 71, 15-20.
Köhloff, G. and A.S. Malin (1981). Private Communication.
Langford, G. and M. Cohen (1969). Trans. ASM, 62, 623-638.
Lemaire, J.C. and W.A. Backofen (1972). Met.Trans. 3, 477-481.
Lücke, K. and K.H. Virnich (1978). In Fifth Conf. on Textures of Materials, Aachen, Springer-Verlag, 1, pp.397-404.
Lücke, K. (1981). Private communication.
Malin, A.S. and M. Hatherly (1979). Met.Sci. 13, 463-472.
Malin, A.S., J. Huber and M. Hatherly (1981). Z.Metallk., 72, 310-317.
Malin, A.S., M. Hatherly, P. Welch and J. Huber (1982a). Z.Metallk. In press.
Malin, A.S., M. Hatherly and V. Piegerova (1982b). This conference.
Mathur, P.S. and W.A. Backofen (1973). Met.Trans., 4, 643-651.
Nuttall, J. and J. Nutting (1978). Met.Sci., 12, 430-437.
Noda, T. and J. Huber (1978). Z.Metallk. 69, 570-574.
Noda, T., B. Plege and J. Grewen (1978). In Fifth Int. Conf. on Textures of Materials, Aachen, Springer-Verlag, 1, pp.443-452.
Ridha, A.A. and W.B. Hutchinson (1981). In Sixth Int. Conf. on Textures of Materials, Tokyo. In press.
Slakhorst, J.W. and C.A. Verbraak (1976). In G.J. Davies (Ed.) "Texture and the Properties of Materials", Met.Soc. London, pp.160-167.
Solomon, R., A.S. Malin and M. Hatherly (1982). This conference.
Van Houtte, P., J.G. Sevillano and E. Aernoudt (1979). Z.Metallk., 70. 426-432.
Verbraak, C.A. (1981). In Sixth Int. Conf. on Textures of Materials, Tokyo. In press.
Wakefield, P.T. and M. Hatherly (1981). Met.Sci., 15, 109-115.
Wakefield, P.T. and M. Hatherly (1982). Met.Sci. In press.
Willis, D.J. (1982). Submitted to Met.Sci.
Wilsdorf, H. and D. Kuhlman-Wilsdorf (1953). Acta Met., 1, 394-413.
Zener, C. and J.H. Holloman (1944). J.Appl.Phys., 15, 22-31.

Doing more with less Materials: Perspectives for Research and Development in the Eighties

Dieter G. Altenpohl

Swiss Aluminium Ltd., Feldeggstrasse 4, PO Box 495, CH-8034 Zurich, Switzerland

ABSTRACT

The industries producing and using engineering materials are in a period of significant changes. A main trend is for instance more efficient use of primary and composite materials as well as systems which are less energy intensive. Conservation of primary materials by substituting them with knowhow, will be one of the important tasks of the applied research in the near future.

KEYWORDS

Efficient use of materials and energy; consumption per capita of various metals; composite materials; the trend to more sophisticated materials and systems.

1. Introduction

To understand the future goals and targets of materials research, let us look first at a few facts regarding the demand pattern of the most commonly used metals and materials in the last decades and in the foreseeable future. We recognize the S-curve for the per capita consumption in figure 1. Developing countries are still far below in this S-curve and newly industrializing countries enter now stage II, accelerated growth.

Many highly industrialized countries, however, are in the upper part of the S-curve and are even witnessing a decline of consumption per capita of common materials during this decade.

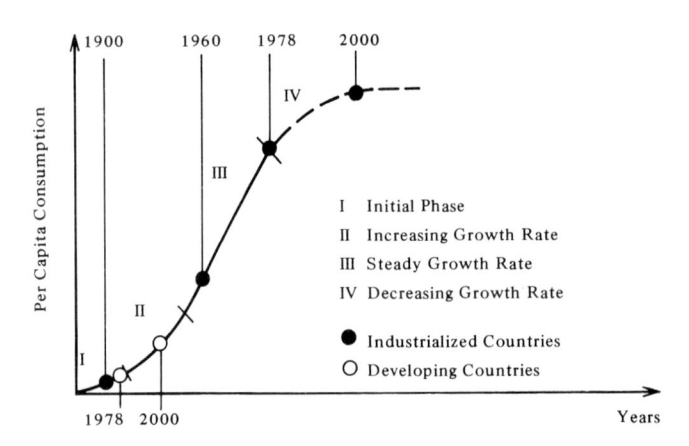

Figure 1: Development of Per Capita Raw Material Consumption (Schematic)

Table 1 explains that growth rates of consumption of 5 major materials are declining drastically in the last quarter of this century.

	1951/75 %	1976/2000 %
Crude Steel	5.4	2.5 **
Aluminium	8.6	4.5 **
Copper	4.4	2.8 *
Zinc	4.6	2.9 *
Plastics	13.0	4.5 **

Sources: * W. Malenbaum
** Various

Table 1: World Demand for Metals:
Weighted Average Annual Growth Rates (based on 5 year periods)

To make an assessment why this has happened, we must look for the underlying reasons for this slow-down in the use of materials in industrialized countries, which still consume three-quarters of all engineering materials. There is a number of different reasons why materials conservation, which means less use of primary materials, takes place in the OECD countries. I will just name the more important ones:

- Primary materials are energy intensive. The costs for a kilo of steel comprises close to 40 % energy costs, for a kilo of aluminium about 50 %, and similar figures can be found for most common materials, plastics included.

Discounting inflation, prices for all primary energies have doubled or quadrupled within one decade. This serves as a powerful stimulus to use less primary materials.

- Ecological movements all over the world have increased awareness to avoid pollution. Pollution is nothing else than misplacement of materials, solid, liquid or in gas form. The decades of 'planned obsolescence', which were specifically manifest in the car industry, but also in the packaging industry, are over. Now a variety of ways for conservation of materials are being actively pursued, gaining more momentum every year.

- Miniaturization
  Caused by more ingenuity in application of materials, miniaturization takes place in many different products. The semi-conductor applications of today weigh only small fractions of their predecessors a decade ago. Less material is used for vehicles of all kinds, as well as for electrical and mechanical systems.

These are some of the reasons and consequences related to the slow-down in consumption of materials. But, it is very important to understand, why this happened.

It is a logical sequence to look first for the 'why', the underlying causes, and then for what ought to be done, and only then, how to achieve implementation.

To better understand, why today's technology is demanding materials science to provide new and improved systems and materials for the technologies of the next generation, let us now look at the changes in the total environment of the metal industry.

This environment has changed considerably in the last two decades. The most important factors are:

- raw materials
- energy
- environmental protection
- social arena

Since the influence of the social arena is not easy to define - although it might be of utmost importance - we will concentrate on the other three quantifiable factors. These are raw materials, energy and environmental protection, shown in the problem triangle in figure 2. There is a strong interdependence between these three factors.

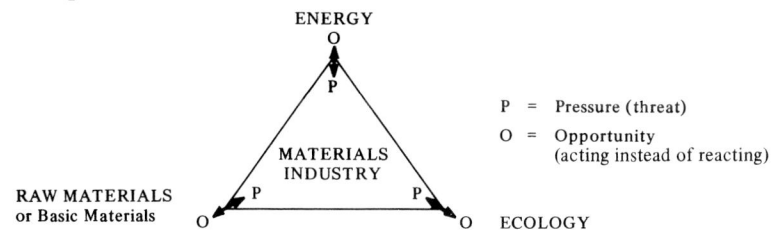

```
                    ENERGY
                       O
                       ↕
                       P
                      / \
                     /   \
                    /     \
                   / MATERIALS \
                  /  INDUSTRY   \
RAW MATERIALS    P               P
or Basic Materials O ——————————— O  ECOLOGY
```

P = Pressure (threat)
O = Opportunity
    (acting instead of reacting)

Figure 2: Problem Triangle of the Materials Industry

2. Energy and Materials

Actually, materials and energy are inextricably bound together. This holds true in two directions:

- all new or 'alternative' energy, regardless its production, storage or transmission, is more or less under material constraint (e.g. unsufficient material properties and/or high material cost).
- the production of many materials involves considerable amounts of energy. This is illustrated in table 2, indicating that 20 % of the total US energy use are required for the acquisition and processing of metals and chemicals.

	% of total national energy use	
Metals	8	}
Chemicals	6	} 14 %
Energy-end-use materials	4	
Non-materials, etc.	2	
	20 %	

Source: A. Bueche, 1980

Table 2: Energy used for Materials - US

As the late A.Bueche pointed out, the opportunity for energy conserving in the provision of steel, aluminium and plastics is so great that it deserves special attention.

The close connection between materials and energy also becomes evident from a look at the share of energy cost in total production cost of some materials (table 3).

	Electric Power	Fuel	Total
Nickel	12 %	35 %	47 %
Iron ore pellets	13	25	38
Aluminium	30	7	37
Zinc	18	3	21
Copper	9	6	15

Source: A. Bueche, 1980

Table 3: Energy Costs as a % of Total Producing Costs

The energy cost increment to produce primary metals is going up further.
For aluminium, it is today approximately 50 %. But the energy cost for the
production of materials is only one aspect.

The total life cycle of a material-intensive product is shown schematically
in figure 3. Energy is involved in two ways: First, as a resource necessary for production. Second, as a factor involved with the application and
final disposing of the product.

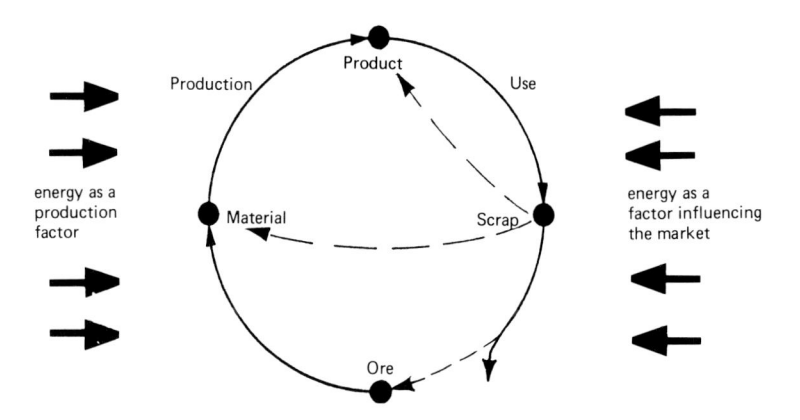

Figure 3: Life Cycle of a Product

This life cycle can be closed in different ways. Accordingly, the metal
industry is in a transition stage today, whereby the dissipative use of
primary materials and energy will be decreased.

Two important opportunities result from this, which are closely wedded to
energy technology. They are materials conservation and materials substitution.

Figure 4 outlines the possibilities for conservation of materials. This
figure gives numerous starting points for industrial research. The most
important trend is, to avoid unnecessary waste of materials. This means
less dissipative use, which can be achieved by either redesigning or through
different ways of recycling.

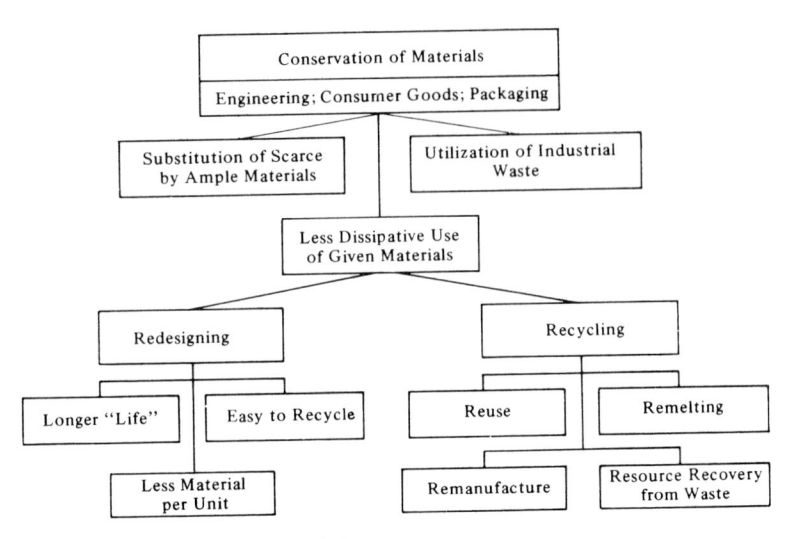

```
                    ┌─────────────────────────────┐
                    │  Conservation of Materials  │
                    ├─────────────────────────────┤
                    │Engineering; Consumer Goods; Packaging│
                    └─────────────────────────────┘
         ┌────────────────┬────────────────────────┐
┌──────────────────┐  ┌──────────────────┐  ┌──────────────────────┐
│Substitution of Scarce│ │Less Dissipative Use│ │Utilization of Industrial│
│ by Ample Materials   │ │ of Given Materials │ │        Waste         │
└──────────────────┘  └──────────────────┘  └──────────────────────┘
```

Figure 4: Conservation of Materials

(Redesigning: Longer "Life", Easy to Recycle, Less Material per Unit; Recycling: Reuse, Remelting, Remanufacture, Resource Recovery from Waste)

Let us look at automobiles as an example. If we compare the materials' mix of today's cars with those of the future, we can realize that different fields of action come into play, for example:

- substitution of materials
- redesigning cars for less weight, longer life, or with materials that are more suitable for recycling.

One can predict that there will be very interesting new tasks for applied research in the field of redesigning. The same can be said about recycling, whereby 're-use and/or remanufacture' will probably grow at a high rate. For example, aluminium engine blocks will not be remelted but re-equipped with new precision parts, like cylinder liners. It must be emphasized that there are many interactions correlated to figure 4, that cannot be described in detail here.

3. Resulting Trends for the Eighties

Practically for all materials, the main trends are:

o Recycling and re-use
o Conservation and substitution
o Miniaturization
o Using less materials and/or energy intensive systems

The main driving forces for doing more with less are:

o Basic materials will be more expensive
o Energy will be scarce and more costly
o The developing world will need hardware
o Man's inborn desire to innovate and improve his skills

Intensity of use of common engineering materials

Generally, we could say that material consumption per capita slows down with increasing GNP per capita in developed countries. This is schematically illustrated in figure 5, for four different materials. It is easy to recognize that the intensity of use for steel and copper are already declining. Which means that the yearly growth rate for these metals is less than that of GNP or GDP in industrialized countries. Aluminium and plastic are believed to be exceptions with a yearly growth rate of around 3 - 4,5 %. This demonstrates the development towards materials with less specific weight. They push aside materials like cast iron and steel, particularly in the automobile sector.

Remark: 1980/81 demand for all primary metals dropped sharply by the recession. Primary aluminium use dropped around 8,5 % over these two years. But this is mainly temporary and not due to structural change in the world's need for primary aluminium.

Intensity of Use (I-U) Versus GDP per Capita
(Schematic)

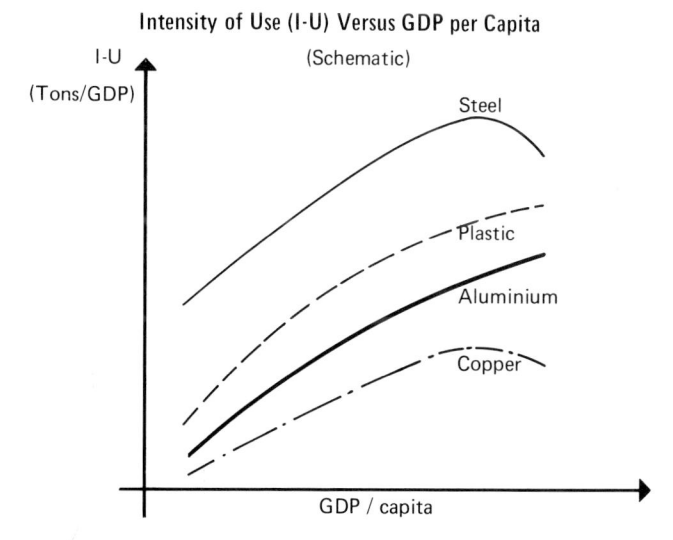

Figure 5: Intensity Use (I-U) versus GDP per Capita
GDP = Gross Domestic Product;   GNP = Gross National Product

Doing more with less materials

Figure 6 shows a sophisticated use of aluminium extrusions with an intricate shape. Such extrusions are recently used in large number of rail vehicles like passenger cars and are now in trial runs of coal cars for bulk transport. This is a typical example, how by a proper design, higher stiffness can be combined with less weight.

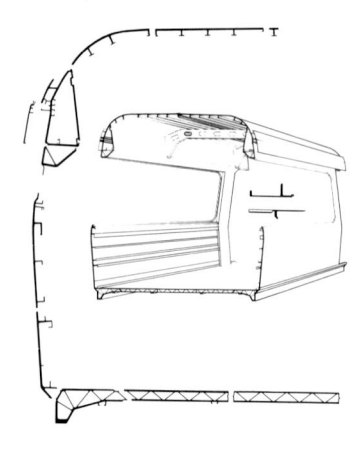

Figure 6: Extruded Profiles for the Construction of welded Railroad Cars. AlMgSi-Alloy

In an article which appeared in 'The Economist', 17/5/80, the materials of the future were described as

- End of the blacksmith's era
- Stretching materials to the limit.

Let me explain with a few words what the blacksmith era means. Let us take for example a drive shaft of a truck which has weight of well over 20 kg. It is made from a heavy steel tube and even needs, in the center of the drive shaft, an additional bearing to take care of the vibration of this tube. It is a real 'blacksmith type' technology. Recently, the same drive shafts have been made from fibre-reinforced plastics with metallic end-pieces. These new drive shafts have only 1/3 of the weight of steel shafts and do not need the center bearing because they have vibration dampening characteristics.

Increased use of composite structures

The previously mentioned example of a drive shaft, which is now in testing with several truck and car companies, is only one example from a very long list of new composite structures.

Composites have been first used on large scale in aero-space structures, where mainly high-quality glass fibres and carbon fibres are in use, often embedded in epoxy or other engineering plastics.

Since well over a decade, a large segment of high-quality composite structures for terrestrial applications has emerged. Besides sporting goods, carbon fibres embedded in a resin matrix are also used for military bridges in combination with aluminium alloy profiles to achieve a high stiffness combined with low weight.

In advanced skies, up to five different material components are combined, including aluminium alloy sheet with the highest possible yield strength, carbon fibre and plastic foam.

We will limit ourselves to three examples where composite structures are now rapidly increasing in use by achieving the desired stiffness of the structure and other desirable properties with less metal than in the past.

I. Composite sandwich structure

In the lower part of figure 7, a sandwich structure is shown, where thin aluminium sheets are on the outside and the core material is either a thermoplast or duroplast. The plastic can be applied compact or as a foam and/or fibre-reinforced.

Schematic representation of the stress distribution under elastic bending between solid Al-sheet and a composite sheet (under equal load P).
Typical thickness for Al-PE-Al composite
d_T = about 3 to 8 mm
d_A = about 0.2 to 0.6 mm
d_K = 2–7 mm
Z = Tensile stress
D = Compression stress
GFUP = glass fiber reinforced polyester
PE = Polyethylene

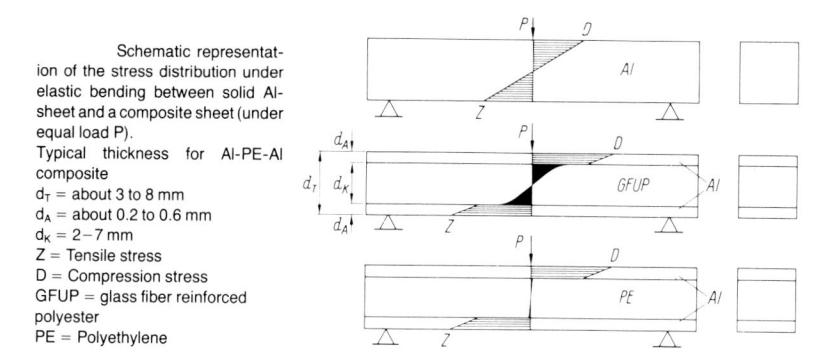

Figure 7: Stress distribution under elastic bending between solid Al-sheet and a composite sheet

Composite sheets of this kind can be deformed in different ways and even deep-drawn. In Europe, USA and Japan there exists today a production of such composite sheets which is well over 2 million m^2 /year.

The application is mostly in buildings, display, but also in vehicles.

II. Composite tubes

Aluminium tubes, reinforced on the outside with glass fibre-reinforced plastics are today used on an increasing scale for pressurized gases. The aluminium seamless tubes are closed on both sides. Half of the stresses in the hoop's direction are taken by the outer layer of fibre-reinforced plastics. There is a rather sophisticated technology involved. For instance the wrapped tube is expanded in its diameter by plastic deformation and thus the glass fiber is under high elastic stress, like steel in prestressed concrete structures.

III. Composite foams made from duroplasts

In figure 8, the relative weight of sheets from different materials, but with identical flexural strength is plotted.

As can be seen, unsaturated polyester foam, glass fibre-reinforced, has the lowest relative weight. Further, it has to be kept in mind that this material is not costly.

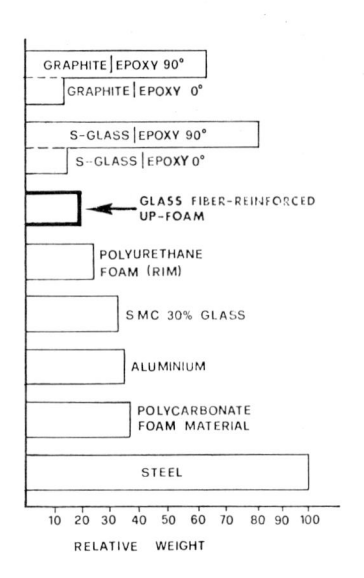

Figure 8: Relative weight of sheets
UP = Unsaturated Polyester; S-Glass = High strength glassfiber

Therefore, fibre-reinforced polyester foams find rapidly increased uses in the transportation sector.

At the same time, conventional fibre-reinforced plastics (unfoamed) is making inroads in the transportation sector. But in every case where foamed plastic is acceptable, it will be preferred, because of less weight and less cost per unit of application.

Increasing importance of software for the use of materials

In order to be compatible with the needs of the future market, those products and systems will be in demand which require less energy and less primary material than their predecessors.

This trend is schematically shown in figure 9a.

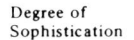

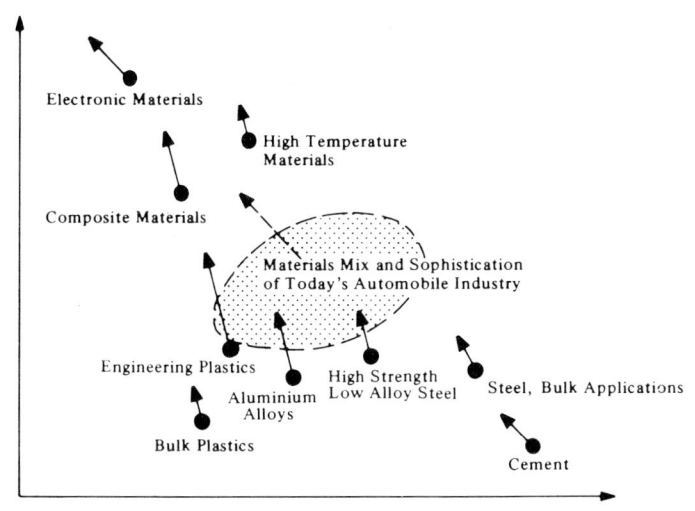

Figure 9a: Heuristic Model providing Correlation between the Quantity of Material in a given Product and 'Information' attached to the Material and its Use

Degree of sophistication means know-how ('software'). In addition to the relative position of the material, this plot also shows the position of a material-intensive industry, e.g. automotive. It is easy to recognize that practically all the vectors in this 'heuristic model' point in the same general direction, i.e. towards less material consumption and more sophistication per unit (nearly all of the vectors point to the upper left corner).

Figure 9b explains desirable trends for one specific company A of the materials industry. It is often irrelevant, whether the amount of primary energy or primary material is used as yardstick for the resource input (one kilo raw material contains a fixed high energy cost increment, as explained above). Trend A ⟶ a is today a mass movement in all materials industries to reduce costs by more rational use of energy and materials (best example: the car industry). Even more important is trend A ⟶ b, where additional 'matière grise' comes into play (best example for this trend are materials for electronic applications like the 'chips'). Trend A ⟶ c is an old one, to achieve for instance better properties of a given material. Trend A ⟶ d will be rather rare (example: much higher added value by additional conversion steps).

Conclusions drawn from figure 9b are of vital importance for many industries and firms and their need to innovate into the right direction. A company under free market forces remaining static at the previous position (A) is going 'to die slowly' - examples can be found in shocking numbers in the steel and nonferrous industry. And a car producer, remaining static would even die very rapidly. To move into directions a and/or b, c, is therefore of utmost importance for applied materials research and - engineering for practically every materials intensive industry.

SOPHISTICATION

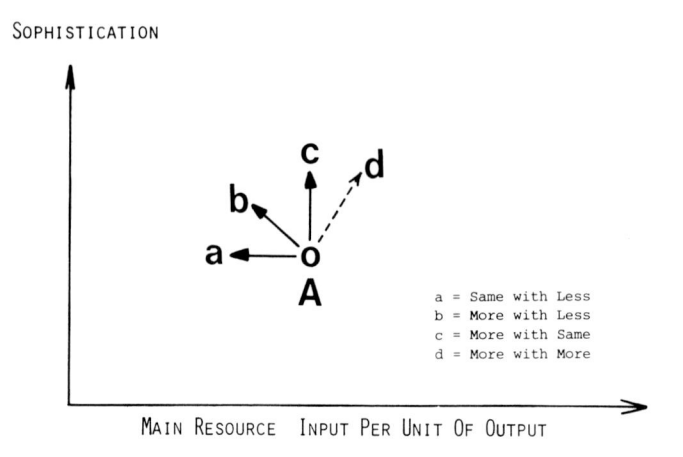

a = Same with Less
b = More with Less
c = More with Same
d = More with More

MAIN RESOURCE INPUT PER UNIT OF OUTPUT

Figure 9b: Correlation between Sophistication and Input of Resources like primary Energy or primary Metal for the Product Mix of a given Firm A

4. Less Material, less Energy

The goal defined previously, is to develop new products which, during manufacture and use, need less material and less energy. Furthermore, they should be compatible with the changing preferences of the market. According to study of the US National Academy of Sciences there will be important developments in the next five years for the following materials:

- Semi-conductors
- Materials used for high temperature (turbine blades)
- New light metal alloys
- Fibre-reinforced plastics
- Composite materials
- New ceramic materials
- Metallic glasses
- Biomedical materials

In the field of information, materials development will lead to a third industrial revolution.

'Doing more with less' calls for innovative spirit and effort.
For industrial implementation, many criteria have to be considered. The four key issues for future innovations are, according to IRI (May 1980) the following:

a) Energy saving —— trend conforming
b) Compatible with existing expertise and/or equipment
c) Know-how intensive
d) Technology: innovative and available

These 'golden rules' are valid for quite different materials industries.

5. Outlook

Let us come back to figure 9a again. The quintessence of our heuristic model is: Materials, including energy, will be substituted by 'matière grise'. But what are driving forces ?

1. Changes in the environment relating to each industry have to be recognized and the selection for new activities has to be made.

2. To save energy and material, by substituting them with knowledge, will be the task of many projects in applied research.

3. In order to accelerate this process, we have to include 'trained intelligence' to enhance the possibilities of the human brain through the use of electronic data processors. A good example of this is CAD, 'Computer Aided Design'.

4. Of course, the researcher as a human being must remain in the centre. His motivation will be an important task and a main concern.

A process of changing attitude is under way. For a company within the materials industries, it makes a big difference whether one reacts too late – with his back to the wall – or anticipates and acts ahead of time.

The interaction between research and industrial innovation has always been a two way street: R and D results created new technology, applied in production and/or the market place. This new technology initiates a new round of R and D effort. This process can be described as the 'research and technology spiral', which is self-propelling.

We stand now at the end of the blacksmith age. A hint for the future trend in applying materials can be found in Homer's Odyssey:

„ὀλίγοσ τε φίλη τε"

in a free translation: LESS, BUT WITH CARE

This will be the key issue of many materials industries – and not only in the eighties.

Literature

D. Altenpohl   Materials Research and Engineering
Volume 1 : Materials in World Perspective
(200 pages in English)
Springer Verlag, 1981

M. Bueche   Materials and Energy
Lecture presented at ASME sixth Inter-American
Conference on Materials Technology,
San Francisco, California, 12 August, 1980

Post-Conference Papers

Cyclic Response of the Directionally Solidified Superalloy 73C

M. A. Abdellatif* and A. Lawley**

*Associate Professor of Production Engineering, Ain Shams University, Cairo, Egypt
**Professor of Materials Engineering, Drexel University, Philadelphia, Pa. 19016, USA

ABSTRACT

The deformation behavior under cyclic loading of the directionally solidified superalloy 73C has been characterized at room temperature in the as-grown condition and following post solidification isothermal exposure or thermal cycling. Fatigue tests were performed in tension-compression loading at 0.167 Hz.

A unique hysteresis behavior for the as DS material; flow stress in the compression half-cycle was much higher than the tensile reflecting considerable strength differential. This is caused by thermal residual stress resulting from thermal expansion mismatch between matrix and fibers. However, it generally disappeared after post solidification heat treatment due to relaxation of thermal stresses by recovery processes and/or matrix creep. Cyclic hardening generally occurred for the as DS and post solidification heat treated materials due to their high initial strain hardening exponents. Isothermal annealing at 1186K and thermal cycling between 352 - 1186K enhanced the fatigue resistance due to the development of a fine scale precipitate of $(Co,Cr)_{23}C_6$. On the other hand, annealing at 1394K and thermal cycling between 352 and 1394K decreased the fatigue resistance due to fiber coarsening and/or degradation.

KEYWORDS

Directionally solidified superalloy; hysteresis behavior; cyclic hardening or sofening; strength differential; thermal cycling.

INTRODUCTION

The directionally solidified superalloy 73C (DSS 73C) is a potential material for turbine blade production due to its superior static properties (Lemkey, 1978). Apart from static stress response, cyclic response is very important. In addition to isothermal high-temperature exposure, aircraft gas-turbine

components are subjected to severe forms of thermal cycling. Thus the
integrity and mechanical behavior of DSS 73C under these conditions should
be confirmed.

In this work, the cyclic response of DSS 73C has been characterized at room
temperature, both in the as DS condition and following post solidification
isothermal exposure or thermal cycling. Cyclic loading response has been
rationalized in terms of changes in microstructure and strength accompanying
thermal treatment. The cyclic stress-strain curves were obtained under all
conditions and compared with the monotonic.

EXPERIMENTAL PROCEDURE

Material of overall composition Co-41 w/o Cr-2.4 w/o C was directionally
solidified at a G/R = 35 x 10^8 °C s/m^2 to yield a rod-like reinforcement of
(Cr, Co)$_7$C$_3$ in a cobalt-rich Co, Cr matrix of a volume fraction V_f = 0.3
(Thompson and Lemkey, 1970). Four thermal treatment regimes were imposed on
the DS material (1) Isothermal exposure at 1186K (T/Tm = 0.75) (2) at 1394K
(T/Tm = 0.87) (3) Thermal cycling 352 - 1186K and (4) 352 - 1394K. These
treatments were designed to delineate the possible effects of cyclic thermal
fatigue, mismatch stresses arising from differential thermal expansion/
contraction, precipitation and physiochemical instability.

Cylinders 12.5 mm diam. and 108 mm long were EDM from the ingots, then
carefully ground and polished to hour glass shaped specimens with 5 mm gage
diameter.

Fatigue tests were performed on an MTS electro-hydraulic closed loop machine
in tension-compression loading with a fully reversed sinusoidal wave form at
a frequency of 0.167 Hz. All tests were carried out under a total diametral
strain control of ± 0.35% at room temperature using specially constructed
grips. The hysteresis loop was continuously monitored on an x-y recorder.
Load-time charts were also obtained for these tests. In tests designed to
determine the cyclic stress strain curves by the incremental step method
(Landgraf, Morrow and Endo, 1969), a data track programmer was coupled to
the MTS machine.

RESULTS AND DISCUSSION

A unique hysteresis behavior was observed for the as DS material. Flow
stress in the compression half-cycle was much higher than the tensile
reflecting considerable strength differential, Fig. 1. This is caused by
thermal residual stresses resulting from thermal expansion mismatch between
matrix and fibers (Thompson and others, 1970; Koss and Copley, 1971). Fig. 1
also shows that the as DS material cyclically hardens. This is expected in
view of its high initial strain hardening exponent (Landgraf, 1970). The
tensile and compressive cyclic stress-strain curves for the as DSS 73C are
compared with the monotonic in Fig. 2. The figure clearly illustrates the
two phenomena just discussed above, namely: strength differential and cyclic
hardening. Fig. 3 shows a comparison between the monotonic and cyclic tensile
stress-strain curves for the isothermally annealed and thermally cycled DSS
73C. The cyclic curves were obtained by subjecting the specimens to blocks
of gradually increasing and then decreasing strain amplitudes. The
hysteresis loops throughout a block are shown in Fig. 4 for the isothermally annealed DSS 73C at 1186K for 25 days. It is depicted from these

figures that not all post solidification treatments resulted in cyclic hardening. Material thermally cycled between 352 - 1186K up to 2500 cycles and isothermally annealed at 1186K up to 25 days exhibited cyclic hardening due to its high initial strain hardening exponent. However, material thermally cycled between 352 - 1394K up to 2500 cycles and isothermally annealed at 1394K up to 26 days showed stable behavior due to its low exponent. The tensile and compressive (whenever strength differential exists) cyclic stress-strain curves for the DSS 73C subjected to the four regimes of thermal treatments are shown in Figs. 5 - 8. The tensile and compressive cyclic stress-strain curves are included in every figure for the sake of comparison.

The cyclic stress-strain curves of the isothermally annealed material at 1186K, Fig. 5, indicate that strength differential decreases after 5 days annealing (compare curves 4 and 6 with 1 and 5). This may be attributed to the relief of the thermal residual stresses in the as DS material by recovery processes and/or matrix creep. It decreases with increasing annealing time and totally disappears for 15 and 25 days materials (curves 2 and 3) which represent the average stress (tension or compression).

It is also observed that the tensile cyclic stress-strain curve increases in level with annealing time. This is due to the strengthening effect associated with the microstructural changes in the material as a result of the annealing treatment, represented by a degeneration reaction of the $(Cr, Co)_7 C_3$ fibers (Lin, Abdellatif and Lawley, 1978: Abdellatif and Lawley, 1982) and the precipitation of fine particles of $(Cr, Co)_{23} C_6$, Fig. 9.

The cyclic stress-strain curves for the thermally cycled DSS 73C between 352 - 1186K, Fig. 6 exhibited strength differential for all cycles. This means that the time "spent" by the material above the stress relaxation temperature (Koss and Copley, 1971) was insufficient for thermal residual stress relief. However, the strength differential decreased with increased number of cycles indicating more available time for stress relief. The tensile and compressive cyclic stress-strain curves for the thermally cycled material had a higher level than the as DS due to the precipitation of fine particles of $(Cr, Co)_{23} C_6$ as in the isothermal annealing.

The cyclic stress-strain curves of the isothermally annealed DSS 73C at 1394K, Fig. 7, and the thermally cycled between 352 - 1394, Fig. 8, did not show any strength differential since these thermal treatment regimes involved temperature considerably higher than the stress relaxation temperature (1033K). The curves exhibited a lower level than the as DS material. The deterioration in the fatigue resistance could be explained by the carbide fiber coarsening in the annealing treatment, Fig. 9, and fiber degradation in the thermal cycling treatment, Fig. 10.

Fig. 11 shows the tensile, compressive and total stress range variation with number of fatigue cycles for the as DS material at a total diametral strain range of 0.7%. Apart from the strength differential and cyclic hardening phenomena previously stated, the stress was found to stabilize very early in life, only after a few cycles consistent with the observations of other workers (Henry, 1974). The stabilized tensile and compressive stress is plotted in Fig. 12 as a function of annealing time and in Fig. 13 as a function of thermal cycles. These plots indicate that the fatigue resistance is degraded by isothermal annealing at 1394K and thermal cycling between 352 - 1394K but is enhanced by isothermal annealing at 1186K and thermal cycling between 352 - 1186K. Strength differential exists only for the lower temperature treatment regimes where the temperature and/or the treatment time are not sufficient for stress relaxation to occur.

CONCLUSION

Post solidification thermal treatment may degrade or enhance the fatigue resistance of the DSS 73C. Isothermal annealing at 1394K and thermal cycling between 352 - 1394K degrades the fatigue resistance due to fiber coarsening associated with annealing and fiber degradation associated with thermal cycling. Both isothermal annealing at 1186K and thermal cycling between 352 - 1186K enhance the fatigue resistance. This enhancement effect is related to the microstructural changes associated with both treatment regimes and represented by a degeneration reaction of the $(Cr, Co)_7 C_3$ carbide fibers and the precipitation of fine particles of $(Cr, Co)_{23} C_6$.

The as DS, isothermally annealed at 1186K and thermally cycled between 352 - 1186K materials exhibited cyclic hardening. This may be attributed to their high initial strain hardening exponents. Materials, isothermally annealed at 1394K and thermally cycled between 352 - 1394K exhibited a stable behavior due to their low initial strain hardening exponents.

Materials exhibiting strength differential were the as DS, the thermally cycled between 352 - 1186K and isothermally annealed at 1186K for 5 days, since the treatment temperature and/or time was insufficient for stress relaxation by recovery processes and/or matrix creep. For materials isothermally annealed at 1394K or thermally cycled between 352-1394K, stress relaxation occurred and strength differential disappeared.

REFERENCES

Lemkey, F.D. (1978). Private Communication
Thompson, E.R., and F.D. Lemkey (1970). Met. Trans., 1, 2799 - 2806
Landgraf, R.W., J.D. Morrow, and T. Endo (1969). JMLSA, 4, 176 - 188
Koss, D.A., and S.M. Copley (1971). Met. Trans, 2, 1557 - 1560
Thompson, E.R., D.A. Koss, and J.C. Chesnutt (1970). Met Trans., 1, 2807-2813
Landgraf, R.W. (1970). ASTM STP, 467, 3 - 36
Lin, L.Y., M.H. Abdellatif, and A. Lawley (1978). Proc. 2nd. Inst. Conf. on Comp. Mat., Toronto, Canada.
Abdellatif, M.H. and A. Lawley (1982), Unpublished work.
Henry, M.F. (1974). GE Tech. Inf. Series, 74CRD235.

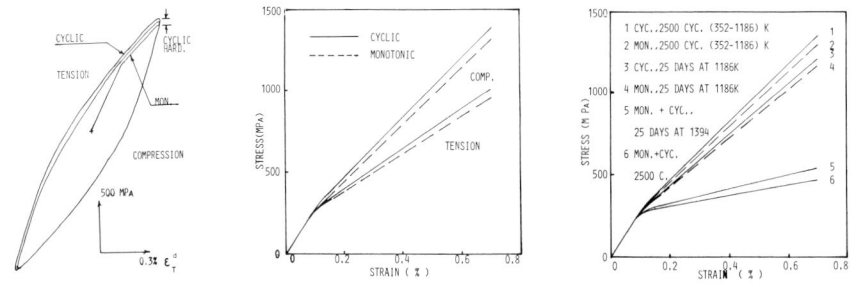

Fig. 1 *(left)* Hysteresis loop showing cyclic and monotonic behavior for the DSS 73C; Fig. 2 *(centre)* Monotonic and cyclic stress-strain curves in tension and compression for as DSS 73C; Fig. 3 *(right)* Monotonic and cyclic tensile stress-strain curves for isothermally annealed and thermally cycled DSS 73C.

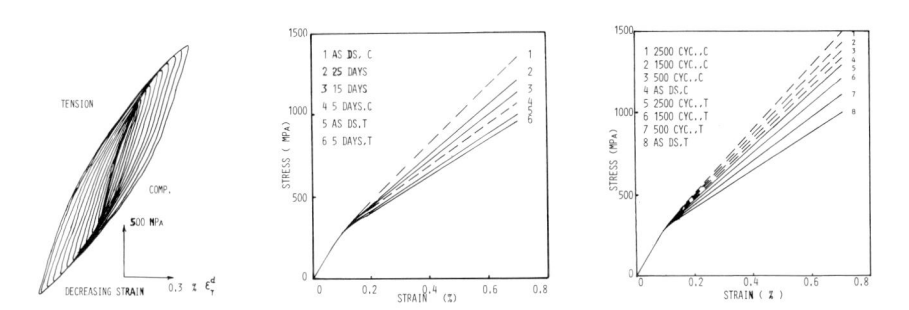

Fig. 4 *(left)* Hysteresis behavior of incremental step cycling of isothermally annealed DSS 73C at 1186K and 25 days; Fig. 5 *(centre)* Cyclic stress-strain for isothermally annealed DSS 73C at 1186K; Fig. 6 *(right)* Cyclic stress-strain curves for thermally cycled DSS 73C between 352 - 1186K.

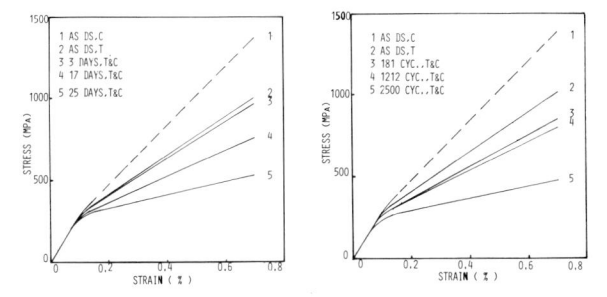

Fig 7 *(left)* Cyclic stress-strain curves for isothermally annealed DSS 73C at 1186K; Fig. 8 *(right)* Cyclic stress-strain curves for thermally cycled DSS 73C between 352 - 1394K.

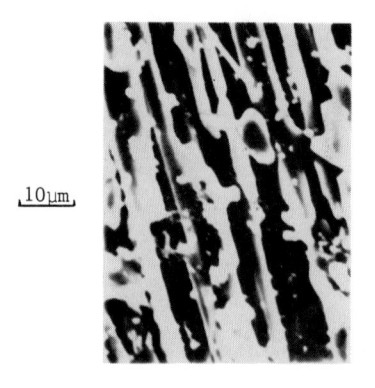

Fig. 9. SEM micrograph for DSS 73C after isothermal annealing at 1186K for 25 days.

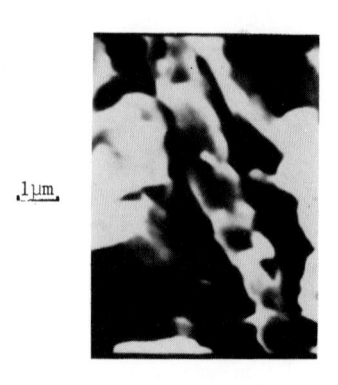

Fig. 10. SEM micrograph for DSS 73C after thermal cycling between 352 - 1394K for 2500 cycles.

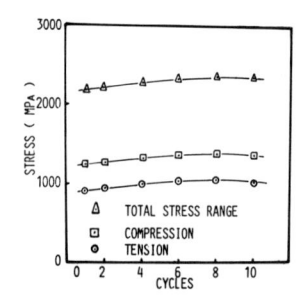

Fig. 11. Stress variation with fatigue cycling for as DSS 73C at $\Delta\varepsilon_T^d = \pm\ 0.35\ \%$

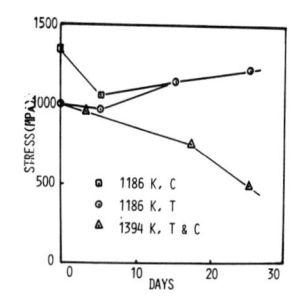

Fig. 12. Stabilized stress at $\Delta\varepsilon_T^d = \pm\ 0.35\%$ for isothermally annealed DSS 73C.

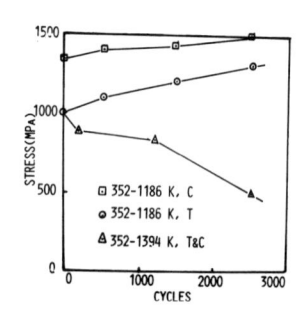

Fig. 13. Stabilized stress at = $\pm$ 0.35% for thermally cycled DSS 73C.

The Role of Deformation Character on Fatigue Crack Growth in Titanium Alloys

John E. Allison and James C. Williams

Carnegie-Mellon University, Pittsburgh, PA 15213, USA

ABSTRACT

Deformation character has been shown to exhibit a strong effect on the fracture-related properties of a wide range of alloys. In this study, the role of slip planarity on fatigue crack growth in titanium alloys has been investigated. The principal variables of interest were aluminum content (Ti-4Al and Ti-8Al) and aging treatment, with an emphasis on fatigue crack growth in the low growth rate, near-threshold region. Fatigue crack growth rates were significantly decreased with increasing aluminum content however this benefit was substantially reduced upon aging the Ti-8Al. The results are discussed in terms of the role of deformation mode on fatigue crack growth and the influence of such factors as slip reversibility, environmental susceptibility and roughness-induced fatigue crack closure.

KEYWORDS

Fatigue crack growth; titanium-aluminum alloys; slip character; planar slip; fatigue crack closure; fracture surface roughness; environmental susceptibility; residual stresses; fractography.

INTRODUCTION

A wide range of microstructures can be produced in titanium alloys through thermomechanical processing. It is known that this wide variation in microstructure can have a profound influence on fatigue crack growth (FCG) in these alloys (Thompson and co-workers, 1982; Peters and co-workers, 1981), however the mechanism for this effect is unclear. One proposed mechanism (Williams and Lütjering, 1981) is that the coarse planar slip present in many titanium alloys and large slip lengths (represented by large grain or colony sizes) combine to reduce fatigue crack growth rates (FCGR) because of improved slip reversibility. A number of investigators have considered the influence of deformation character on FCG in age-hardenable nickel (Hornbogen and Zum Gahr, 1976), aluminum (Gysler and co-workers, 1979; Coyne and Starke, 1979) and titanium alloys (Gysler and co-workers, 1979; Chakrabortty and Starke, 1979). In all cases, when alloys were underaged to produce coherent, shearable particles which led to planar slip, FCG rates were reduced relative to overaged conditions. The purpose of this paper is to present recent FCG results for binary Ti-Al alloys and to analyze these data in terms of the role of deformation character on FCG. Ti-Al alloys were selected for this study because the deformation mode can be varied from homogeneous, wavy slip in the solid solution condition (represented by Ti-4Al) to coarse, planar slip in the unaged Ti-8Al. A higher degree of intense planar slip was also achieved by aging the Ti-8Al to precipitate fine, coherent α_2 particles.

EXPERIMENTAL PROCEDURE

Two Ti-Al alloys containing 4 and 8 weight percent aluminum were prepared by Reactive Metals Inc., Niles, Ohio. The chemical compositions of these alloys are shown in Table 1. Both alloys were hot-rolled in the α-phase field to produce a fine grain size after recrystallization and a predominately basal texture. Compact specimens (H/W = 0.6) were machined so that the loading axis was either perpendicular (TL) or parallel (LT) to the rolling direction. The specimen dimensions were 63.5mm wide(W), 76.2mm tall(2H) and 9.5mm thick (B). Machined specimens were given a solution treatment to recrystallize the deformed grains and water quenched to give an equiaxed grain size of 40μm. Selected specimens were aged 48 hours at 550°C. Solution treatment conditions and tensile data from the work of Kim (1981) on the same plate of material are summarized in Table 2.

TABLE 1. Alloy Compositions of Ti-Al Alloys (in weight percent)

Alloy	Al	Fe	O	N
Ti-4Al	3.91	0.053	0.060	0.011
Ti-8Al	8.00	0.050	0.080	0.001

TABLE 2. Tensile Properties of Ti-Al Alloys (Kim, 1981)

Alloy	Solution Treatment	σ_{ys}(MPa)	σ_{uts}(MPa)	R.A.(%)
Ti-4Al (ST+Q)	900°C, 2.5hr/WQ	524	629	39
Ti-8Al (ST+Q)	950°C, 3.0hr/WQ	721	793	36
Ti-8Al (Aged)	950°C, 3.0hr/WQ (550°C, 48hr/WQ)	779	810	10

FCG measurements were performed on a servohydraulic, closed-loop testing machine using tension-tension sinusoidal loading with a minimum/maximum load ratio (R) of 0.1 at a frequency of 30 hz. All tests were conducted at room temperature in laboratory air. Crack lengths were measured using both a DC electric potential technique and an optical traveling microscope with stroboscopic illumination. An emphasis was placed on the low growth rate, near threshold region of FCG and thus load shedding was used to achieve growth rates of about 1x10^{-10}m/cycle prior to commencing constant load testing. In accordance with ASTM specification E647-78T (ASTM, 1980) the cyclic stress intensity, ΔK, to produce a growth rate of 1x10^{-10}m/cycle was used as an operational definition of the threshold cyclic stress intensity, ΔK_{TH}.

RESULTS

FCG data are presented in Fig. 1 for Ti-4Al and Ti-8Al in the solution treated and quenched (ST+Q) condition. The experiments revealed a significant effect of aluminum content on FCG which extends over four decades of da/dN. As can be seen, increasing the aluminum content from 4Al to 8Al led to an increase in the average ΔK_{TH}, from 5.5MPa$\sqrt{m}$ to 12.5MPa$\sqrt{m}$. At low ΔK(13MPa$\sqrt{m}$ the

Ti-8Al exhibited growth rates a factor of 300-400 times lower than the Ti-4Al and at higher $\Delta K(25MPa\sqrt{m})$ the growth rates for Ti-8Al were an order of magnitude lower than the Ti-4Al. At high $\Delta K(50MPa\sqrt{m})$ the growth rates for the two alloys converged. Data is shown for specimens with both TL and LT orientations; a slight increase in FCGR was observed for the LT orientation.

FCG results for aged Ti-8Al are compared with the ST+Q conditions of Ti-4Al and Ti-8Al in Fig. 2. The aging treatment led to substantially higher FCG rates than the unaged Ti-8Al and lower growth rates than Ti-4Al. The aged Ti-8Al exhibited a ΔK_{TH} of $7.5MPa\sqrt{m}$, and near this threshold the aged Ti-8Al had growth rates which were an order of magnitude lower than those of Ti-4Al at the same ΔK. At a ΔK of $13MPa\sqrt{m}$, growth rates in the aged Ti-8Al were only a factor of 2 lower than those in the Ti-4Al and a factor of 100 times faster than growth rates in the unaged Ti-8Al. The FCG response of the aged Ti-8Al crossed over that of the Ti-4Al at ΔK of $25MPa\sqrt{m}$. At this ΔK, FCG rates for the aged Ti-8Al were a factor of 10 faster than the unaged Ti-8Al.

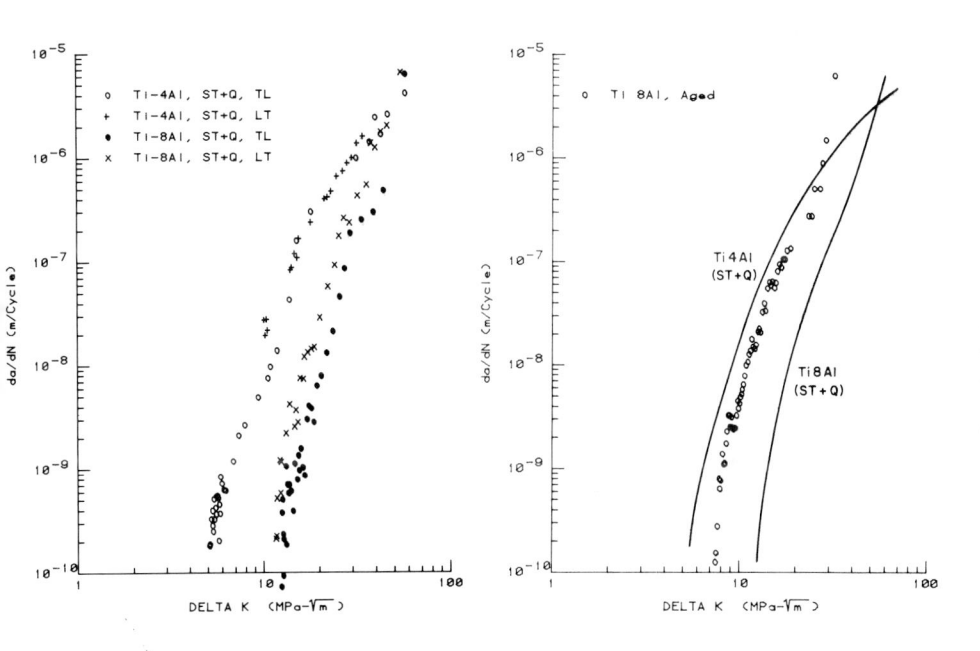

Fig. 1. FCG behavior for Ti-4Al and Ti-8Al in ST+Q condition.

Fig 2. FCG behavior of aged Ti-8Al compared to ST+Q conditions.

Post-fracture investigations revealed that both Ti-8Al conditions exhibited macroscopic fracture surfaces which were rough in nature as compared to the smooth macroscopic surfaces of the Ti-4Al specimens. At higher magnification, the low growth rate fracture surfaces of the ST+Q conditions of both Ti-4Al and Ti-8Al had similar features as shown in Fig. 3a and 3b. They consisted predominately of transgranular, faceted regions interspersed with areas of ductile tearing. The facets appeared to be crystallographic in nature and some contained well developed river patterns (Fig. 3a), while others were smoother and featureless. In addition to these faceted regions and regions of ductile tearing, the low growth rate fracture surfaces of the aged Ti-8Al (Fig. 3c) exhibited faceted regions intersected by perpendicular steps, indicating that the crack path changed orientation frequently. In this condition, large, featureless areas were observed which gave rise to the rougher macroscopic appearance of the fracture surface. At higher growth rates, both ST+Q conditions of Ti-4Al and Ti-8Al produced fracture surfaces comprised of transgranular striations and regions of ductile tearing (Fig. 3d and 3e). In addition, the Ti-8Al(ST+Q) exhibited scattered regions of intergranular fracture (Fig. 3e) and some grain boundary decohesion. The high growth rate fracture surfaces of the aged Ti-8Al were primarily composed of intergranular facets and ductile tearing (Fig. 3f). Significant amounts of secondary cracking due to grain boundary decohesion were also observed, as well as occasional regions of transgranular striations.

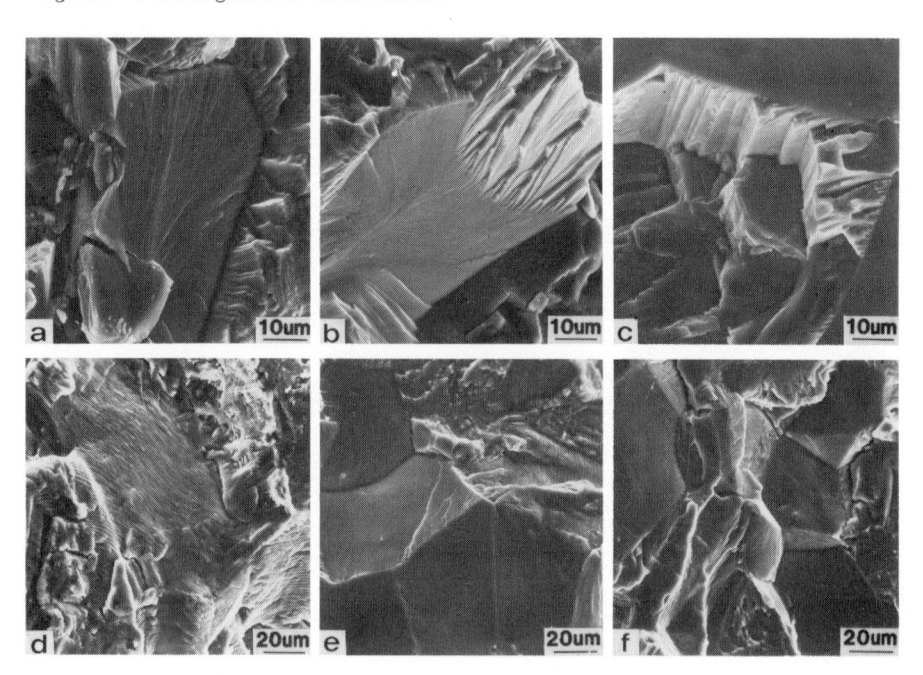

Fig. 3. SEM fractography at:
1×10^{-10} m/cycle in a) Ti-4Al(ST+Q), b) Ti-8Al(ST+Q), c) Ti-8Al (aged); and 1×10^{-6} m/cycle in d) Ti-4Al(ST+Q), e) Ti-8Al(ST+Q), f) Ti-8Al (aged. (Direction of crack propagation is from bottom to top)

DISCUSSION

For the Ti-Al alloys investigated, the results show clearly that increasing aluminum content has a profound effect on FCG, that is, increasing aluminum content leads to significantly reduced growth rates. It is appealing to attribute this reduction strictly to the decreased plastic strain accumulation due to the increased slip reversibility of the Ti-8Al. However, this does not allow a satisfactory explanation for the FCGR results of the aged

Ti-8Al, nor does it account for important factors such as environmental susceptibility, crack closure effects and residual stresses.

The specimens in this study were solution treated and quenched after they had been machined and thus residual compressive stresses due to differential thermal contraction during the water quench would be expected. It has been shown (Gysler and co-workers, 1981; Vosikovsky and co-workers, 1980) that such residual stresses can reduce growth rates especially in the near threshold region. Gysler and co-workers(1981) have also shown that these residual stresses account for a decrease in ΔK_{TH} of 2-3MPa$\sqrt{m}$ in both Ti-4Al and Ti-8Al. Thus, for the data in Fig. 1 the difference between the ST+Q conditions is not thought to be affected by residual stresses even though the absolute values are shifted to higher values of ΔK. The 48 hour aging treatment at 550°C is, however, more than adequate to relieve these residual stresses (Gysler and co-workers, 1981) and thus in comparing these data with those of the ST+Q conditions this must be born in mind. The ΔK_{TH} for the aged Ti-8Al would then be expected to be about 3MPa$\sqrt{m}$ lower than that of the unaged Ti-8Al. In fact, the difference is closer to 5MPa$\sqrt{m}$ and, moreover, the increased FCGR of the unaged Ti-8Al extends over the entire growth rate regime.

The differences between the aged and unaged Ti-8Al must also be considered in light of the influence of environment. It has been shown that FCG in alloys which exhibit planar deformation is more susceptible to environmental influences (Coyne and Starke, 1979; Chakrabortty and Starke, 1979; Gysler and co-workers, 1981). Gysler and co-workers(1981) observed a significant increase in FCGR attributable to environmental influences in Ti-8.6Al which had been subjected to long term aging to produce a high degree of planar slip. Much less environmental influence was found in a shorter aging time condition with less pronounced planar slip. Since laboratory air can be viewed as an aggressive environment for titanium alloys (Peters and co-workers, 1981; Gysler and co-workers, 1979), the results shown in Fig. 2 are consistent with this notion, that is, the aged Ti-8Al has substantially higher growth rates than the unaged Ti-8Al because it is more susceptible to environmental influences.

Recently, attention has been given to fatigue crack closure which arises due to irregular or rough fracture surface morphologies (Halliday and Beevers, 1981; Asaro and co-workers, 1981; Mayes and Baker, 1981; Minikawa and McEvily, 1981). Because the fracture surfaces of the Ti-8Al specimens were macroscopically rough in comparison to the Ti-4Al specimens, the contribution of this factor must also be considered. The previously cited investigations have reported that rough fracture surfaces lead to incomplete closure of the fatigue crack and that this leads to a lower effective ΔK and ultimately to reduced FCGR. This is especially true in titanium alloys where extremely rough fracture surfaces are possible depending on the underlying microstructure or deformation mode. Recent unpublished results of the authors indicate that a substantial portion of the difference between FCG rates in Ti-4Al and Ti-8Al(ST+Q) is attributable to this roughness-induced closure. Increased slip planarity therefore appears to have an indirect effect on FCG in producing a rough fracture surface which mechanically retards crack growth. This is in addition to its contribution to improved slip reversibility and reduced crack tip plastic strain accumulation.

At high growth rates both Ti-8Al conditions have FCG rates which cross over that of the Ti-4Al condition. At a growth rate of about 6×10^{-10}m/cycle the three rank, in increasing ΔK, as follows: Ti-8Al(aged), Ti-8Al(ST+Q) and Ti-4Al(ST+Q). This is the same as the ranking of the ductilities for these three conditions (Table 2) and is consistent with correlations between ductility and fracture toughness.

CONCLUSIONS

1. Fatigue crack growth rates of solution treated and quenched Ti-4Al and Ti-8Al decreased dramatically with increasing aluminum content. This is attributed to the increasing tendency for planar slip with increasing aluminum content.

2. Aged Ti-8Al exhibited increased fatigue crack growth rates compared to Ti-8Al in the solution treated and quenched condition, thus substantially reducing the benefits derived from the increased aluminum content. This increase in growth rates is attributed to the enhanced environmental susceptibility of the aged condition. Residual compressive stresses may also play a role at low ΔK.

3. Deformation character plays an important role in determining FCG properties. However, in addition to its role in controlling crack tip plastic strain accumulation, deformation character also appears to have an indirect effect on FCG related to the macroscopic fracture surface morphology it produces. This factor and others such as environment must be taken into account to completely understand the role of deformation character on fatigue crack growth.

ACKNOWLEDGMENT

The authors would like to acknowledge helpful discussions with Dr. G. Lütjering. This research was supported by the U.S. Air Force Office of Scientific Research.

REFERENCES

ASTM Specification E647-78T (1980) Annual Book of ASTM Standards, Part 10, American Society for Testing of Materials, 749-769.
Asaro, R. J., L. Hermann, and J. M. Baik (1981) Met. Trans., A, 12A,1133-1135.
Chakrabortty, S. B., and E. A. Starke, Jr. (1979) Met.Trans., A, 10A,1901-1911.
Coyne, E. J.,Jr., and E. A. Starke,Jr. (1979) Int. J. Fracture, 15, 405-417.
Gysler, A., J. Lindigkeit, and G. Lütjering (1979) Strength of Metals and Alloys, Vol. 2, P. Haasen, V. Gerold, G. Kostorz, eds., Pergamon Press, 1113-1118.
Gysler, A., J. E. Allison, J. C. Williams, and G. Lütjering (1981) Fatigue Thresholds, J. Bäcklund, A. Brom, C. J. Beevers, eds., EMAS Publ. Ltd.
Halliday, M. D., and C. J. Beevers (1981) J. of Testing and Eval., 9, 195-201.
Hornbogen, E., and K. H. Zum Gahr (1976) Acta Met., 24, 581-592.
Kim, H. M. (1981) The Effect of Slip Character on Low Cycle Fatigue Behavior of Ti-Al Alloys, Ph.D. Thesis, Carnegie-Mellon University.
Mayes, I. C., and T. J. Baker (1981) Fatigue of Eng. Mat. and Struc., 4, 79-96
Minikawa, K., and A. J. McEvily (1981) Scripta Met., 15, 633-636.
Peters, M., A. Gysler, G. Lütjering (1981) Titanium '80, Science and Technology, Vol. 3, H. Kimura and O. Izumi, eds., TMS-AIME, 1777-1786.
Thompson, A. W., J. C. Williams, J. D. Frandsen, and J. C. Chesnutt (1982) Titanium and Titanium Alloys, Vol. 1, J. C. Williams and A. F. Belov, eds. Plenum Press, 691-704.
Vosikovsky, O., L. D. Trudeau, and A. Rivard (1980) Int. J. of Fracture, 16, R-187.
Williams, J. C., and G. Lütjering (1981) Titanium '80, Science and Technology, Vol. 1, H. Kimura and O. Izumi, eds., TMS-AIME, 671-681.

The Development of a High Strength Manganese Steel

R. D. Jones[*], G. R. Palmer[*], V. Jerath[**], S. Kapoor[***] and R. J. Yeldham[****]

[*]Department of Metallurgy and Materials Science,
University College Cardiff, Wales, UK
[**]British Steel Corporation, Teesside Laboratory,
Ladgate Lane, Middlesbrough, UK
[***]British Steel Corporation, Swinden Laboratory,
Moorgate, Rotherham, UK
[****]Texaco Oil Company, Pembroke, Dyfed, UK

ABSTRACT

The aim of this study has been the development of a high strength Fe-Mn steel based on the principles used in Fe-Ni maraging steels. It is shown that appreciable precipitation hardening in a fully martensitic matrix can be achieved through additions of Mo to an Fe-Mn base but that strengthening is accompanied by a large drop in impact toughness. Improved impact toughnesses after maraging result when the martensitic matrix contains a dispersion of reverted or retained austenite. An optimum composition range for this type of steel is indicated.

KEYWORDS

Iron-manganese steels, maraging steels, substitute steels, reverted austenite, retained austenite, precipitation hardening, martensite.

INTRODUCTION

About twenty years ago, a family of high strength steels known as 'maraging steels' was introduced by International Nickel (Decker and others, 1962). The steels are based on an air hardenable Fe-Ni-Co martensitic (α') matrix which can be further strengthened by an age hardening process carried out at 450-500°C. During age hardening, intermetallic compounds of Fe and Ni with Mo and Ti are precipitated. The maraging temperatures specified are too low for reverse shear transformation of α' to austenite (γ). Instead, a slow, diffusion-controlled γ reversion process occurs.

Subsequently, many attempts have been made to produce lower cost maraging steels by replacing with cheaper alternatives some of the alloying additions in the conventional maraging grades (e.g. Patterson and Richardson, 1966, Manenc, 1970, Squires and others, 1974). This approach has only achieved limited success. The main difficulty encountered has been the retention of toughness at high strength levels. The present study, which began in 1970, represents an attempt to develop a maraging-type steel in which Mn adopts the role taken by Ni.

Mn and Ni are both γ stabilising elements when added to Fe and both reduce the critical cooling speed to form α' following an austenitising treatment. Thus, with sufficient alloying addition, the steel can become air hardening. In Fe-Ni alloys containing 5-25% Ni and in Fe-Mn alloys containing about 2-10% Mn, the martensite formed is bcc α'. However, in Fe-Mn alloys with more than 10% Mn, some hcp ε- martensite is formed (Holden, Bolton and Petty, 1971). Further increases in Mn content lead to the retention of some γ at room temperature following solution treatment. At Mn concentrations above 15% the structure comprises only γ and ε- martensite.

MATERIALS AND METHODS OF INVESTIGATION

During the course of the development programme a large number of experimental steels have been produced using both air and reduced pressure inert gas melting of electrolytic and mild steel bases to which were added electrolytic manganese flake and various ternary elements of commercial purity. In this paper we shall limit our attention to binary Fe-Mn and ternary Fe-Mn-Mo alloys. Chemical analyses of the representative vacuum melted alloys reported on are given in Table 1.

Table 1   Composition of Alloys (Weight %)

Alloy	Mn	Mo	C	Si	S	P
Fe - 10Mn	9.70		0.005	-	0.01	0.002
Fe - 10Mn-6Mo	9.75	6.40	0.006	-	0.02	0.004
Fe - 11.5Mn-6Mo	11.46	6.02	0.05	0.009	0.018	0.006
Fe - 12Mn-4Mo	12.00	4.20	0.004	-	0.01	0.003
Fe - 12.5Mn-6Mo	12.35	5.94	0.10	0.013	0.026	0.006
Fe - 13.5Mn-6Mo	13.56	5.96	0.07	0.012	0.024	0.006
Fe - 14Mn-6Mo	13.92	6.01	0.10	0.011	0.025	0.007
MM125/H	12.60	4.48	0.10	0.22	0.018	0.017
MM125/L	12.30	5.00	0.01	0.05	0.007	0.004

The alloy ingots were homogenised at 1150°C for 24 hours and then hot rolled to various bar and slab sizes. Some slabs were further cold rolled to 0.5mm sheet. This served as starting material for the production of specimens for transmission electron microscopy (TEM) and transmission Mössbauer spectroscopy (TMS). Tensile properties were determined with Hounsfield No. 12 specimens and impact tests were carried out using standard Charpy V-notch specimens. Quantitative measurements of the amounts and nature of the phases present were obtained using TMS and X-ray diffractometry. Conventional TEM was used for the examination of microstructures while fracture surfaces were studied using scanning electron microscopy (SEM).

RESULTS AND DISCUSSION

Maraging Reactions

Initially the maraging influence of a number of ternary additions to martensitic Fe-Mn alloys was studied. The most pronounced hardening response was achieved with Mo additions. It was also found possible to augment this response with additions of Co (Jones and Kapoor, 1973) in an analogous way to the synergistic Co-Mo reactions occurring in Ni maraging steels (Decker and others, 1962). However, the Co alloyed steels had an extremely low impact toughness, even in the solution treated condition, while the Co free alloys tended to experience a large drop in impact toughness following a maraging treatment. An example of this is shown for an Fe-10Mn-6Mo alloy in Fig. 1. There is some evidence that the toughness of this alloy increases again after long ageing times, particularly at higher temperatures. This effect has also been noted in Fe-12Ni-6Mn steels by Squires and Wilson (1972) and attributed to the formation of reverted γ on grain boundaries.

Toughness Improvement

The influence of γ in promoting improved toughness was examined in more detail using an Fe-10Mn alloy. This alloy is almost completely α' in the solution treated condition. Reheating to temperatures in the range 500-600°C brings about the formation of stable reverted γ. This effect in binary Fe-Mn alloys has been reported previously (Stannard and Baker, 1979). The amount of reverted γ formed is both time and temperature dependent and has a strong effect in improving toughness (Fig. 2).

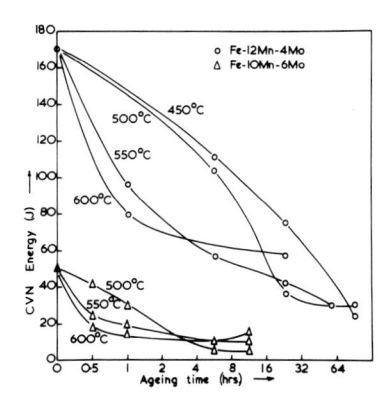

Fig. 1. Effect of ageing on impact toughness of Fe-10Mn-6Mo (fully γ') and Fe-12Mn-4Mo (5% retained γ). Prior solution treatment 1 h at 950°C.

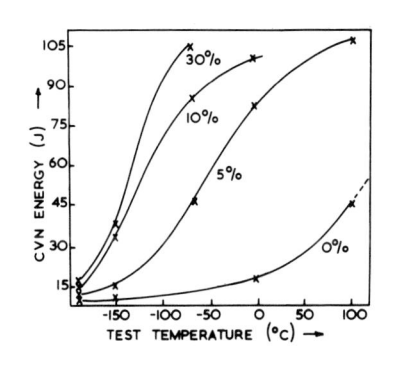

Fig. 2. Variation of impact toughness with testing temperature for Fe-10Mn.
0%γ-solution treated 2h at 900°C (ST).
5%γ-ST and 10h at 500°C
10%γ-ST and 10 hr at 525°C.
30%γ-ST and 10 h at 575°C.

In a conventional maraging reaction, overageing with a loss of strength occurs before γ forms in sufficient amounts to improve toughness. It was thus decided to examine the influence on toughness and strength of γ retained in higher Mn content alloys at room temperature after solution treatment.

Retained Austenite Effects

To retain γ after a conventional solution treatment it is necessary to ensure that the M_f temperature is below ambient. This was achieved by increasing the Mn content of the steels to about 12%. In such alloys although toughness dropped during ageing, the levels remaining after significant strengthening were acceptable. This effect is shown for an Fe-12Mn-4Mo alloy in Fig. 1. The corresponding age hardening response is given in Fig. 3. As ageing proceeds the amount of γ present in the structure increases presumably from a diffusion-controlled reversion process (Fig. 4). In Fig. 4 any ϵ-martensite present has been combined with the γ component to give an overall content for the close-packed phases. Note that the larger initial γ contents are associated generally with higher Mn levels although the Fe-11.5Mn-6Mo alloy has more retained γ than the Fe-12Mn-4Mo alloy by virtue of its higher Mo content.

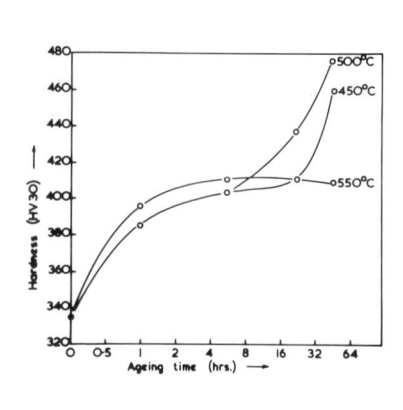

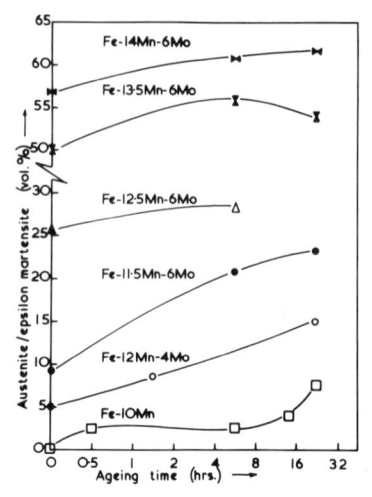

Fig. 3. Ageing response of Fe-12Mn-4Mo. Prior solution treatment 1h at 950°C.

Fig. 4. Influence of ageing time at 500°C on the γ/ε-martensite content of Fe-Mn steels.

As precipitation hardening will only occur in the α' component of the solution treated structure, strengthening effects on ageing are less in the higher γ content alloys. To ensure that adequate strengths are attained after ageing, the retained γ content needs to be regulated through composition control. For example, higher C content melts need to be compensated for by lowering the Mn contents so that excess γ is not present in the solution treated state.

Structural Changes

As solution treated the structure of an Fe-12Mn-4Mo alloy comprises a heavily dislocated lath α' in a 'brickwork' arrangement (Bogachev, Khadayev and Nemirovskii, 1975) (Fig. 5a). The individual α' crystals are arranged into quite distinct bands whose length is considerably greater than their width.

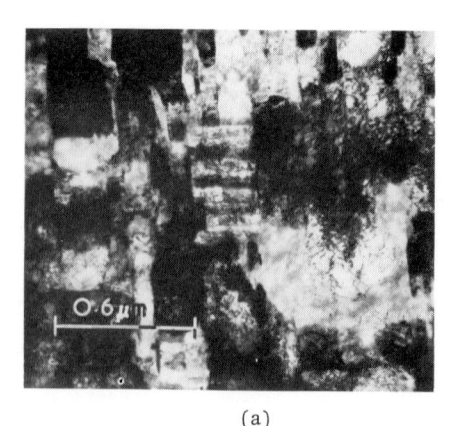

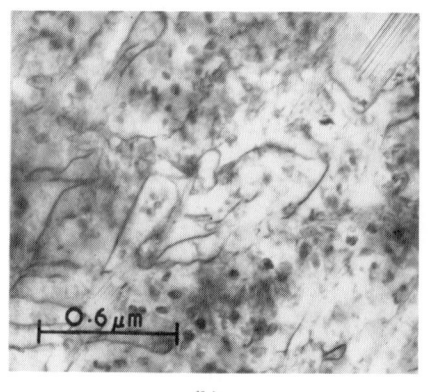

Fig. 5. Fe-12Mn-4Mo (a) solution treated 1 h at 950°C (b) solution treated and aged 100 h at 500°C. (TEM).

Precipitation was barely detectable in material aged to give optimum mechanical properties. However an overageing treatment produces the dense array of precipitates shown in Fig. 5b. Spheroidal intergranular precipitates and ribbon-like precipitates at lath boundaries are both evident. These have been identified using Mössbauer spectroscopy as Fe_2Mo (Jones and Kapoor, 1972). Some faulted γ and/or ε-martensite is also present. Since ε-martensite can be considered as γ with a stacking fault on every alternate plane there is some difficulty in differentiating between these phases using TEM.

Fracture appearance
The fracture surface of Charpy specimens which had shown high toughness contain an array of large and small dimples with some associated MnS inclusions (Fig. 6a). In steels with insufficient retained γ after solution treatment, intergranular failure occurred in specimens aged to peak hardness (Fig. 6b). Binary alloys that could be tempered to produce γ by reversion displayed a gradation of fracture appearances with increase in tempering time from an initially intergranular surface through a microdimpled intergranular fracture (Fig. 6c) to a fully dimpled morphology. A microdimpled intergranular fracture appearance has also been reported by Stannard and Baker (1979) in high Mn steels and associated with the presence of grain boundary γ films formed during tempering.

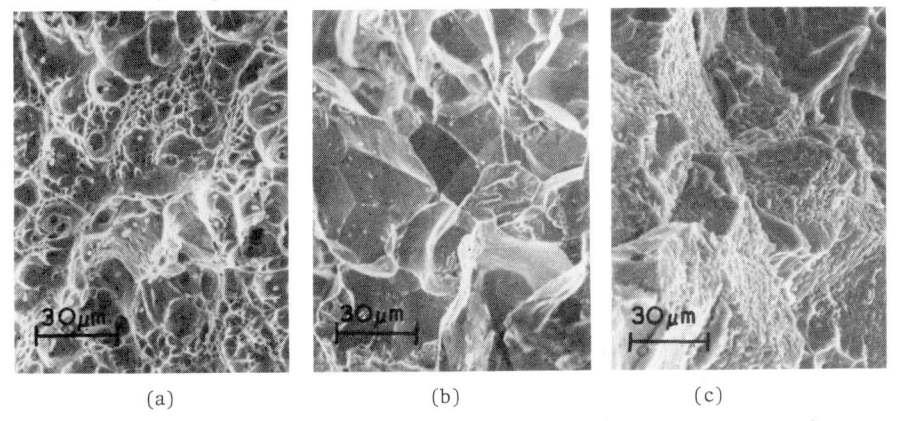

(a) (b) (c)

Fig. 6. (a) Fe-12Mn-4Mo, solution treated 1h at 950°C, aged 5h at 450°C.
 (b) Fe-10Mn-6Mo, solution treated 1h at 950°C, aged 6h at 500°C.
 (c) Fe-10Mn, solution treated 2h at 900°C, tempered 10h at 525°C impact tested at -70°C. (SEM).

Mechanical Properties
Table 2 summarises the mechanical properties obtained on two larger scale melts based on an Fe-12.5Mn-5Mo target composition. The analysis of these steels has been given in Table 1.

The higher C and Mn contents of alloy MM125/H have generated a higher retained γ content, 64%, compared with 37% in MM125/L. This is reflected in the higher proof stress and hardness values of MM125/L both in the solution treated and aged conditions.

TABLE 2 Mechanical properties of MM125/H and MM125/L

Steel	Condition	0.2%P.S. N mm⁻²	T.S. N mm⁻²	Elong. %	Red.in area %	CVN J	HV30
MM125/H	S.T.	333	1350	26	49	138	305
	Aged	765	1350	25	49	116	364
MM125/L	S.T.	642	1148	19	70	156	366
	Aged	1123	1286	21	65	101	434

S.T. : Solution treated 1 hour at 950°C, air cooled.
Aged : S.T. and aged for 5 hours at 450°C.

Optimum Composition Range

It has been shown that these steels require Mo for age hardening and retained γ for toughness retention, the amount of retained γ following solution treatment being determined by composition, particularly Mn and C contents. We have established through our development programme the following optimum composition range for this type of steel (weight %):

11.5 - 13Mn, 4-6Mo, 0.2max.C, 0.4max.Si, 0.02max.S, 0.03max P, bal. Fe.

This range covers air and vacuum melts with both high and low purity stock. It also permits realisation of an appropriate balance of toughness and strength to meet specific needs.

ACKNOWLEDGEMENTS

Financial support for this work has been provided by University College Cardiff (S.K. and R.Y), Ministry of Defence (V.J.), Science Research Council (G.R.P.) and British Technology Group (G.R.P.). We thank B.T.G. for permission to publish this paper.

REFERENCES

Bogachev, I.N., M.S. Khadayev and Yu. R. Nemirovskii (1975). Martensitic transformations in Fe-Mn alloys. Phys. Met. Metallogr., 40, (4), 69-77.
Decker, R.F., J.T. Eash and A.J. Goldman (1962). 18% nickel maraging steel. Trans. A.S.M., 55, 58-76
Holden, A., J.D. Bolton and E.R. Petty (1971). Structure and properties of Fe-Mn alloys. J.I.S.I., 209, 721-728.
Jones, R.D., and S. Kapoor. (1973). A synergistic Co-Mo age-hardening interaction in Fe-10Mn martensite. J.I.S.I., 211, 226-228.
Manenc, J. (1970). Influence du manganese sur le durcissement d'aciers martensitiques du type maraging. Rev. de Met., 67, 443-450.
Patterson, W.R., and L.S. Richardson (1966). The partial substitution of Mn for Ni in maraging steel. Trans. A.S.M., 59, 71-84.
Squires, D.R., and E.A. Wilson (1972). Ageing and brittleness in an Fe-Ni-Mn alloy. Met. Trans., 3, 575-581.
Squires, D.R., F.G. Wilson, and E.A. Wilson (1974). The influence of Mo and Co on the embrittlement of an Fe-Ni-Mn alloy. Met. Trans., 5, 2569-2578.
Stannard, D.M., and A.J. Baker (1979). The inter-relationship of microstructural and mechanical properties in high manganese steels. I.C.S.M.A.5, 2, 1371-1376.

The Effects of Composition and Temperature on the Dislocation Structure of Cyclically Deformed Ti-Al Alloys

H. M. Kim* and J. C. Williams**

*National Research Council Fellow, Air Force Materials Laboratory,
Wright-Patterson, OH 45433, USA
**Carnegie-Mellon University, Pittsburgh, PA 15213, USA

ABSTRACT

The role of alloy composition and test temperature on the cyclic deformation behavior of hcp Ti-Al alloys has been studied in Ti-4 and 8wt%Al alloys. At room temperature it is seen that the Ti-4Al alloy forms cells, whereas the Ti-8Al alloy forms localized shear bands. At 300°C this tendency persists, whereas at 550°C both alloys tend to form tangles. Helices and multi-poles are especially prevalent in the Ti-8Al. The origins of these variations in structure are qualitatively accounted for by the presence of Short Range Order in the Ti-8Al alloy.

KEYWORDS

Fatigue; low cycle fatigue; dislocation structure; transmission electron microscopy; cyclic loading.

INTRODUCTION

For more than two decades, basic mechanisms of cyclic deformation have been studied by numerous investigators. Because cyclic deformation is more complex than monotonic deformation, many of these studies have been conducted on single phase fcc metals and alloys which are well characterized. Moreover, most of these studies have been confined to room temperature deformation behavior since variations in deformation temperature also add to the complexity of the deformation structures.

Recently, though, the monotonic and cyclic deformation behavior of hcp alloys such as Ti-Al has been studied. Room temperature results have been reported (Kim, Paris, Williams, 1980) recently and it was shown that both Al content and loading history affect the dislocation arrangements in these alloys. Very recently the effect of temperature on dislocation arrangements after both cyclic and monotonic loading have been studied. The results of this work will be summarized here.

EXPERIMENTAL

The alloys and experimental procedures used here were identical to those reported elsewhere (Kim, Paris, Williams, 1980). The alloys studied contained 4 and 8wt%Al and both had ∿ 600wt ppm oxygen. In addition, a special furnace and modified extensometers were used for the elevated temperature (300 and 550°C) cyclic tests. A resistance heated furnace which has less than 5°C gradient along the gage length of specimen was used and quartz extension rods were attached to a conventional extensometer for these tests.

RESULTS AND DISCUSSION

Since cyclic deformation structures are often compared to monotonic loading behavior, the monotonically deformed structures will be briefly mentioned first. These structures have been described elsewhere (Kim, Paris, Williams, 1980) to consist of homogeneously distributed long, straight screw dislocations in Ti-4Al after deformation at 25°C. In Ti-8Al, these long, straight screw dislocations are inhomogeneously grouped in localized bands. In contrast, deformation at 550°C results in fairly homogeneously distributed dislocations of mixed character for both alloys. These changes indicate a reduced tendency for inhomogeneous slip and an increased tendency for cross-slip at the elevated temperatures. It is, also, found that both pyramidal and basal slip are more frequently observed at 550°C than at 25°C where prismatic slip is dominant. This agrees well with earlier single crystal studies by (Paton, Baggerly, Williams, 1976). For purposes of comparison, typical dislocation structures of Ti-4 and 8Al alloys after cyclic deformation ($\frac{\Delta\varepsilon_p}{2} \cong 0.4\%$) at 25°C are shown in Figs. 1a and b. Poorly developed cell structures are seen in Ti-4Al, whereas inhomogeneously distributed bands of dislocations are seen in Ti-8Al. These structures also have been described in greater detail elsewhere (Kim, Paris, Williams, 1980). After cyclic deformation ($\frac{\Delta\varepsilon_p}{2} \cong 0.3\%$) at 300°C, dislocation structures of Ti-4 and 8Al alloys were qualitatively similar to those formed at 25°C (Figs. 2a and b). In Ti-4Al (Fig. 2a), well-developed, elongated cell structures are more frequently observed than for 25°C. Similar to the cell structures formed at 25°C, cell walls consist mainly of dislocation debris such as dislocation loops or dipoles. As in the room temperature deformation, $\bar{c}+\bar{a}$ dislocations were seldom observed at this strain amplitude, especially in grains where dislocation cells were formed. In contrast with the Ti-4Al, the cyclically deformed ($\frac{\Delta\varepsilon_p}{2} \cong 0.3\%$) Ti-8Al contained inhomogeneously distributed slip bands (Fig.4b). These bands appear to be more diffuse than those seen after room temperature deformation and consist mainly of dipoles and prismatic loops with a few dislocation tangles. Substantially fewer straight screw dislocations are observed in the 300°C specimens than for 25°C. In Ti-8Al, which was solution treated in α-phase region and aged at 550°C for 48 hours to form ordered α_2 precipitates (ST&A), slip bands appear to be sharper than those of solution-treated Ti-8Al (ST) and these slip bands mainly consist of straight screw dislocations.

After cyclic deformation ($\frac{\Delta\varepsilon_p}{2} \cong 0.3\%$) at 550°C, homogeneously distributed tangled dislocations are present in both Ti-4 and 8Al alloys (Figs. 5a and b). In Ti-4Al, early stages of elongated cell formation is sometimes observed in addition to the tangles. In comparison with lower deformation temperatures, the less dislocation debris is observed. As in the case of both 25°C and 300°C deformation, dislocation density is much higher after cyclic deformation (e.g., $\frac{\Delta\varepsilon_p}{2} \cong 0.3\%$) than after even much larger tensile deformation (e.g. $\varepsilon_p \cong 0.3\%$). Hexagonal networks consisting of three distinct $\bar{a}$-type dislocations are occasionally observed. With increasing plastic strain amplitude up to 1%, slip character is generally unchanged, except for the anticipated increase in dislocation density. Better developed cells, which often occur in fcc alloys with increasing strain amplitude, were not observed here. In Ti-8Al, the tangled dislocations are somewhat less homogeneously distributed than in Ti-4Al. That is, diffuse slip traces were often observed although these were much less pronounced and not as planar as those seen after deformation at lower temperatures. In these regions, fringe contrast is frequently observed in conjunction with basal slip traces. It is not clear whether this contrast is due to stacking faults or residual multipole contrast. Such contrast features are observed after cyclic deformation at all temperatures but they are almost never seen after monotonic deformation at any temperature. Since multi-poles also only form during cyclic deformation, the occurrence of this fringe contrast only after cyclic deformation does not help resolve its origin. Nevertheless, it does show a basic difference in the deformation character during monotonic and cyclic deformation.

Another interesting feature observed in cyclically deformed Ti-8Al is the high density of helical dislocations, as shown in Figs. 4a and b. These are observed after cyclic deformation at all temperatures but are more prevalent for 300°C. Since non-conservative dislocation motion creates excess vacancies which aid in the formation of helices, it is suggested that cyclic deformation causes more extensive non-conservative dislocation motion than monotonic deformation. Further, these helical dislocations are rarely seen in Ti-4Al. This suggests that deformation by motion of long, straight screw dislocations, which is more pronounced in Ti-8Al, favors the formation of helical dislocations. It is also seen (e.g. Fig. 4b) that trails of dislocation loops are formed near the straight dislocations and the direction of these trails is parallel to the axis of the helical dislocations.

The differences in overall slip character after cyclic deformation at 25, 300 and 550°C are summarized in Table 1 for Ti-4 and 8Al alloys. From this it can be seen that the slip character of these alloys, especially the higher Al content alloys such as Ti-8Al, does not change continuously with increasing temperature from 25°C to 550°C. Instead, it changes rather abruptly between 300°C and 550°C. This also has been confirmed by observing surface deformation structures of Ti-8Al which was cyclically deformed at 25°C, 300°C and 550°C (Figs. 5a and c). Here, inhomogeneously distributed planar slip bands are seen at 300°C and at 25°C, although the slip bands are a little less intense at 300°C than at 25°C. In contrast with this, fine, homogeneously distributed wavy slip bands are seen at 550°C. This, also, demonstrates the abrupt change of slip character between at 300°C and at 550°C. It is well-known that increasing the deformation temperature promotes the dispersal of planar slip by increasing the propensity for cross-slip and dislocation climb. Since 300°C and 550°C correspond to about $0.29T_m$ and $0.42T_m$ of these alloys, respectively, only a gradual change in planar slip dispersal with increasing temperature might be expected. However, present observation of an abrupt change in this alloy system suggests that some mechanism other than this general effect of temperature must be operative also. In this regard, it is suggested that the temperature dependence of the intensity of Short Range Order (SRO), which is generally agreed to exist in this alloy system (Blackburn, 1967) at low temperatures, is responsible for the change. Since the effect of SRO on the deformation is to cause inhomogeneous, planar slip (Cohen, Fine, 1962), disappearance of SRO over a small temperature range could lead to an abrupt change in slip character. If the temperature at which SRO disappears in Ti-8Al is above 300°C, then the slip inhomogeneity at 25°C and 300°C would be similar, although the combined effects of slightly less intense SRO at 300°C and a greater propensity for cross-slip due to thermal activation can account for the reduced sharpness of the planar bands.

At 550°C ($\simeq$0.4 Tm) the SRO disappears nearly completely permitting profuse cross-slip which leads to the homogeneously distributed dislocation tangles observed at this temperature.

As in the earlier 25°C monotonic and cyclic deformation studies (Kim,et al), the present work has shown that cyclic deformation modes of Ti-Al alloys have been found to be qualitatively similar to the monotonic deformation modes even at 550°C. In the high Al content alloys such as Ti-8Al, slip character during cyclic deformation is virtually the same as that of monotonic deformation,although more extensive dislocation debris and a higher dislocation density is observed after cyclic deformation. In Ti-4Al,cyclic straining results in the formation of dislocation cell structures. These are best developed at 300°C, which indicates that increasing temperature may promote the cell formation by more frequent occurrence of cross-slip. However, more poorly developed cell structures at 550°C suggest that cell formation again becomes difficult at higher temperatures. This may be due to the propensity for hexagonal net formation because of attractive junctions at these temperatures. This may compete with cell formation. Figs. 6a and b were taken from the same area of cyclically deformed Ti-4Al at 550°C but with different $\bar{g}$ vectors operating. This shows that, when cell structures form, nearly all the dislocations and dislocation debris have one $\bar{a}$ type Burgers vector. When the density of dislocations with different $\bar{a}$ type Burgers vectors such as those few shown in Fig. 6b becomes more equal, it is suggested that networks form instead of cells. The formation of networks during monotonic loading at 600°C has been reported previously (Paton, Rauscher and Williams, 1973).

Although little effect of plastic strain amplitude on the formation of cell structures has been found, there seems to be a certain strain amplitude below which cell structures are not developed. That is, cell structures would not be expected in the high cycle (long life) fatigue region.

SUMMARY

1) Cyclic deformation modes of Ti-Al alloys, expecially Ti-8Al have been found to be qualitatively similar to the monotonic deformation modes both at room temperature and at elevated temperatures. However, closer scrutiny showed that more dislocation debris and helical dislocations formed during cyclic deformation. 2) Cyclic deformation structures at 550°C showed very homogeneously distributed dislocation structures were observed at 300°C. 3) Rather abrupt change of cyclic slip characters above 300°C is attributed to the disordering temperature of Short Range Order (SRO) at this temperature range.

REFERENCES

Blackburn, M. J. (1967), Trans. TSM-AIME, vol. 239, p 1200.
Cohen, J. B. and M. E. Fine (1962), J. Phys. Radium, vol. 23, p 749.
Kim, H. M., H. G. Paris, and J. C. Williams (1980), Proc. of Fourth International Conference on Ti-Alloys, Kyoto, Japan, TMS-AIME.
Paton, N. E., R. G. Baggerly, and J. C. Williams (1976). Rockwell Int. Report.
Paton, N. E., J. C. Williams, and G.P. Rauscher, "The Science,Technology and Applications of Ti", R.I. Jaffee and H.M. Burte, Eds., Plenum Press,1973, 1049

Table 1. Summary of Cyclic Slip Character of Ti-4 and 8Al Alloys

Alloys	Ti-4Al	Ti-8Al
25°C	Poorly developed cells	Inhomogeneous slip bands
300°C	Fairly well-developed cells	Inhomogeneous slip bands (a little broader slip bands than at 25°C)
550°C	Homogeneous tangled dislocations with very early stage of cell formation	Homogeneous tangled dislocations with slip trace

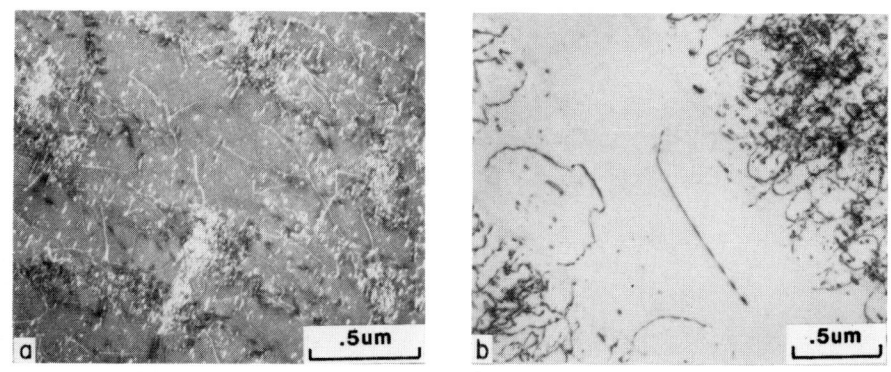

Fig. 1. Cyclic deformation structures; $\frac{\Delta\varepsilon_p}{2} \cong 0.4\%$ at 25°C. (a) Ti-4Al weak beam dark field, (b) Ti-8Al bright field.

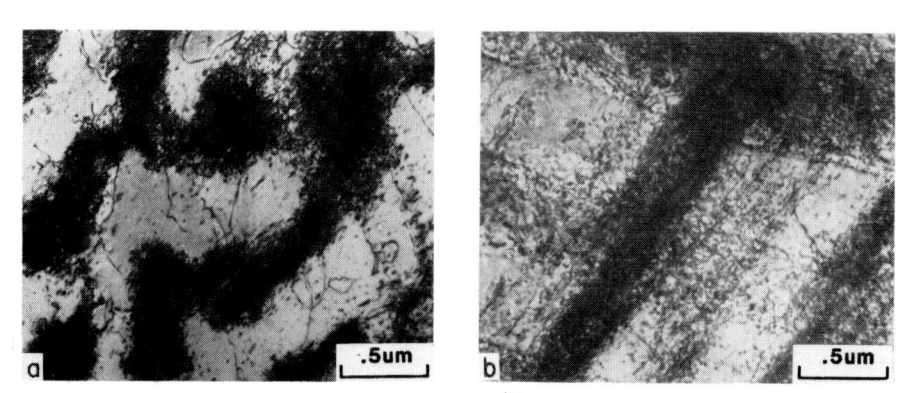

Fig. 2. Cyclic deformation structures; $\frac{\Delta\varepsilon_p}{2} \cong 0.3\%$ at 300°C. (a) Ti-4Al, (b) Ti-8Al.

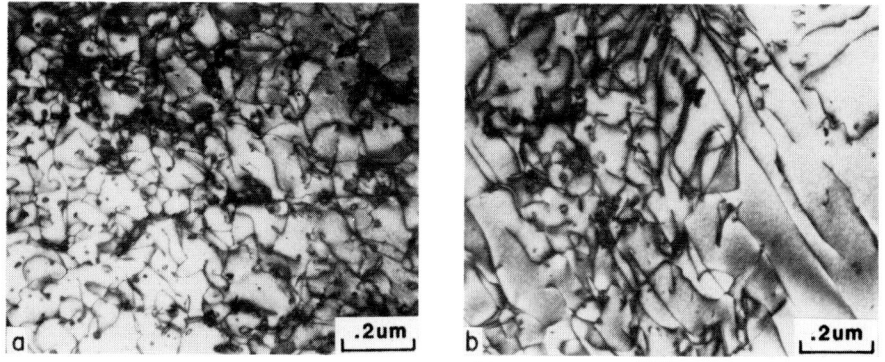

Fig. 3. Cyclic deformation structures; $\frac{\Delta\varepsilon_p}{2} \cong 0.3\%$ at 550°C. (a) Ti-4Al, (b) Ti-8Al.

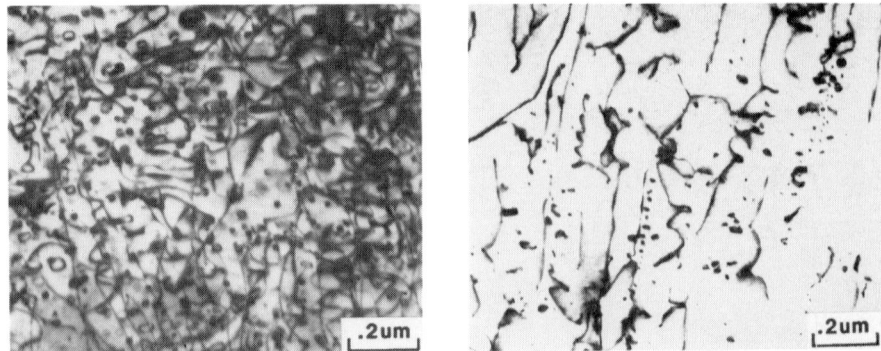

Fig. 4. Cyclic deformation structures in Ti-8Al tested at 550°C, showing dislocation loops and helices.

Fig. 5. Surface slip structure (SEM) in Ti-8Al after $\frac{\Delta\varepsilon_p}{2} \cong 0.4\%$ at (a) 25°C, (b) 300°C and (c) 550°C.

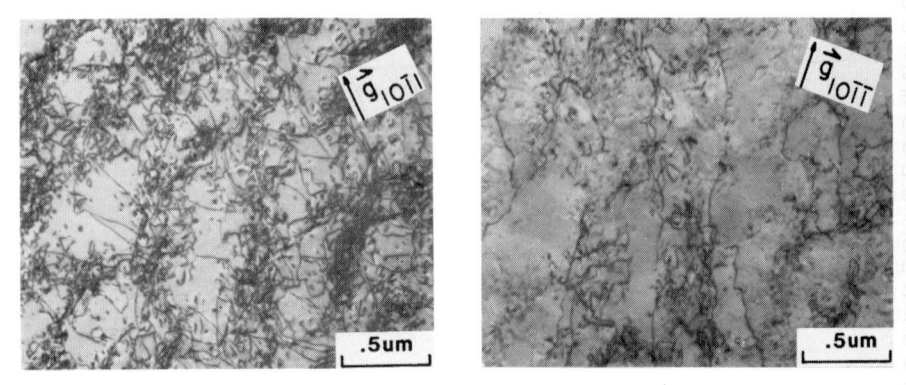

Fig. 6. Cyclic deformation structures in Ti-4Al after $\frac{\Delta\varepsilon_p}{2} \cong 0.3\%$ at 550°C, showing the beginning of cell formation.

Relationship between Microstructure and Strength Properties of Copper-Tin-Nickel Alloys Prepared by Powder Metallurgy

N. C. Kothari

Reader in Materials Science, James Cook University of North Queensland, Townsville, Qld. 4811, Australia

ABSTRACT

Studies have been carried out to determine the effect of powder metallurgy process variables on the strength and microstructure of Copper-8Sn-8Ni alloys (composition / wt. percent). Alloys were pressed and sintered to a density of 7.50 g/cc. The sintered specimens were re-pressed and re-sintered to a density of 8.20 g/cc, and then given various treatments. Alloys sintered and cooled slowly had a tensile strength of 289 MPa with 4 percent elongation, while alloys sintered, re-pressed, re-sintered, quenched and re-pressed showed a tensile strength of 351 MPa with 2 percent elongation.

KEY WORDS

Powder metallurgy, density, pressing, sintering, spinodal decomposition.

INTRODUCTION

Powder metallurgy has been widely used in the manufacture of bearing and structural components from prealloyed and preblended copper-tin powder material. In spite of numerous investigations on the sintering characteristics of copper-10wt. percent tin bronze (Bell et al, 1958; Berry, 1972; Pound et al, 1960; Sheppard and Greasley, 1978), this material still poses considerable problems especially in the manufacture of high density structural parts with good machinability and mechanical properties. In the case of copper-10Sn P/M bronze alloys, during sintering at 800°C or above, the tin, due to its low melting point, melts first and alloys with copper forming Cu_3Sn (a delta phase) an intermetallic compound which tends to make the material very brittle and also causes excessive growth in the specimen decreasing the finished density by 6 to 8 percent. To avoid this irregular growth a low-temperature short-time sintering cycle coupled with other secondary operations is used. This increases the operational cost and the resultant products have low mechanical strength with very poor ductility. To circumvent this, various additions such as zinc, phosphorus, iron, nickel etc. are added to improve machinability, increase mechanical properties and lower cost. However, none of these have been successful except nickel (Fetz et al, 1974; Kothari, 1974 and 1980). It has been shown that in cast copper-tin alloys, nickel addition up to 15 weight percent improves casting quality and increases mechanical properties (Eash and Upthegrave, 1933; Fox, 1948; Price et al, 1928; Wise, 1928; Wise and Eash, 1934). Furthermore the precipitation

hardening kinetics of cast Cu-Sn-Ni bronzes have been examined in detail by inve
tigators (Badia, 1962; Bastow and Kirkwood, 1971; Ditchek, 1978; Kato and Schwar
1979; Kato et al, 1980; Koumani, 1972; Lefevre et al, 1978; Leo, 1967; Plewes, 1
Schwartz and Plewes, 1974; Schwartz et al, 1974). Plewes (1974, 1975) has shown
that cast Cu-Sn-Ni alloys can be successfully age-hardened to obtain yield stren
greater than 1000 MPa by modification of the Ni-Sn ratio and the thermomechanica
processing technique. The age-hardening in Cu-6Sn-9Ni alloys is accompanied by
spinodal decomposition of the α-phase, and furthermore, aging at higher temperat
results in discontinuous precipitation of the θ phase, (CuNi)₃Sn, a needle-like
phase similar to the one observed by Leo (1967). The mechanical properties of c
and age-hardened Cu-Sn-Ni alloys seem to depend upon the balance between spinoda
decomposition and discontinuous precipitation.

The work done by Kothari (1974, 1980) has clearly shown that the addition of nic
improves the quality and the age-hardening of Cu-Sn P/M alloys giving a variety
strength properties and elongation values, but the P/M manufacturing cycle for t
alloys is rather intricate. Further literature review indicates that very littl
or no fundamental work has been done in determining the influence of P/M process
variables, especially porosity and sintering parameters (sintering time, temperature, aging time and aging temperature) on the spinodal and discontinuous preci
itation kinetics and the resultant development of mechanical properties in Cu-Sn
P/M alloys. The present work is concerned with the effect of P/M process variab
on the strength properties and microstructure of copper-8tin-8nickel (Cu-8Sn-8Ni
composition by weight percent) alloys.

EXPERIMENTAL

Alloys Cu-10Sn and Cu-8Sn-8Ni were preapred from elemental powders (Kothari, 197
A lubricant, stearic acid 1.0 percent by weight, was added in all mixes. Each
alloy mix was blended for 45 minutes and then pressed into cylindrical specimens
of ~40mm ϕ x 10mm height in a double-acting hardened-steel die at a suitable
pressure to give a sintered density of 7.50 g/cc (~85 percent theoretical densit
Sintering was carried out in a tube furnace in the presence of purified dissocia
ammonia. The sintered specimens were re-pressed and re-sintered to a density of
8.20 g/cc (93 percent theoretical density of Cu-8Sn-8Ni alloy). Both Cu-10Sn an
Cu-8Sn-8Ni alloys were re-pressed and re-sintered at 800°C for 30 minutes (solut
ized), and quenched in a specially designed apparatus attached to the furnace tu
The final re-pressed alloys were annealed at various temperatures and for varyin
times (150°C to 500°C and 10 minutes to 500 minutes respectively).

Mechanical properties were determined for specimens machined from the sintered
blanks using a table-model Instron Universal testing machine at a strain rate of
0.05 cm/cm per minute. The elongation was measured over a 20mm gauge length. T
microstructure was characterised by optical microscopy. The echants for the mic
scopy were 1.5% alcoholic ferric chloride and a freshly prepared mixture of 1 pa
H_2O_2, 1 part NH_4OH, ½ part HNO_3, 2 parts water.

RESULTS AND DISCUSSION

The effect of powder metallurgy process variables (examined on a limited scale)
the mechanical properties of Cu-10Sn and Cu-8Sn-8Ni alloys is shown in Table 1.
Marked differences in strength properties (yield strength, tensile strength and
elongation) can be seen with the addition of nickel in tin bronze.

Both alloys Cu-10Sn and Cu-8Sn-8Ni after re-pressing and re-sintering at 800°C f
30 minutes, showed a slight decrease in density when quenched. These specimens
re-pressed at 138 MPa to bring the final density back to 8.20 g/cc (~93 percent
theoretical density of an alloy).

TABLE I

Effect of processing variables upon the mechanical properties of Cu-10Sn and Cu-8Sn-8Ni P/M alloys (re-sintered, quenched and re-pressed at 138 MPa, density 8.20 g/cc).

Alloy	Processing	Yield Strength (MPa)	Tensile Strength (MPa)	Elongated (%)
Cu-10Sn	Pressed, sintered to 7.50 g/cc density			
	(i) Cooled slowly from sintering temperature	138	168	9.0
	(ii) Quenched from sintering temperature	151	183	4.0
	Re-pressed, re-sintered to 8.20 g/cc density			
	(iii) Cooled slowly from sintering temperature	163	228	11.20
	(iv) Quenched from re-sintering temperature and re-pressed at 138 MPa	186	273	2.6
Cu-8Sn-8Ni	Pressed, sintered to 7.50 g/cc density			
	(i) Cooled slowly from sintering temperature	179	289	4.0
	(ii) Quenched from sintering temperature	217	311	1.5
	Re-pressed, re-sintered to 8.20 g/cc density			
	(iii) Cooled slowly from re-sintering temperature	231	337	6.0
	(iv) Quenched from re-sintering temperature and re-pressed at 138MPa, density 8.20 g/cc.	281	351	2.0

comparison of mechanical properties was made between the specimens of Cu-10Sn and Cu-8Sn-8Ni P/M alloys (Table I). The results show that yield strength and tensile strength of nickel bronze are far greater than those obtained in tin bronze. The effect of annealing temperature on the mechanical properties of these alloys is shown in Figures 1, 2 and 3.

In tin bronze (Cu-10Sn) the strength decreases with increase in the annealing temperature; however, a very large improvement is observed in the elongation. The microstructures clearly show the recrystallized α-grains with annealing twins (Fig. 4). The uniformly distributed pores at the grain boundaries also acted as a sink for dislocation tangles at the annealing temperature.

The temperature-dependence of the strength properties of Cu-8Sn-8Ni alloys is shown in Figure 2. Maximum hardening occurs in the 300 to 400°C interval. The peak is displaced at 375°C for 30 minutes and drops to 350°C when age-hardening time is 200 minutes (Figures 3 and 5). Aging above 400°C results in a softening of the alloys, indicating a change in the sequence of phase transformations. Microscopy indicates that aging beyond 500°C produced the discontinuous phase, (θ phase), in the form of needle-shaped particles (Fig. 6). When quenched, nickel-

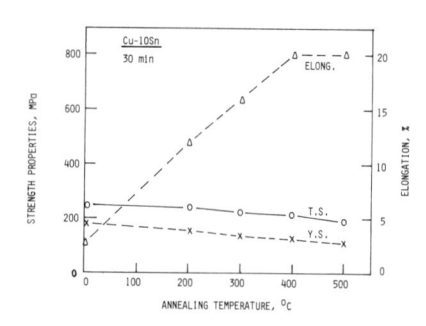

Fig. 1. Effect of annealing temperature upon the mechanical properties of Cu-10Sn P/M alloys (re-sintered, quenched, re-pressed - density 8.20 g/cc) annealed for 30 minutes.

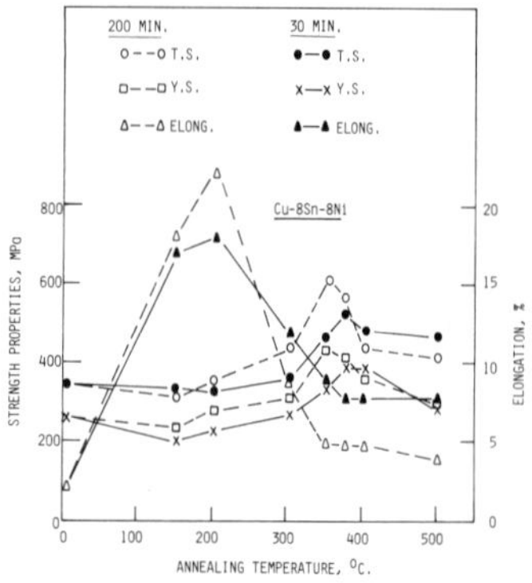

Fig. 2. Mechanical properties of Cu-8Sn-8Ni P/M alloys, (re-sintered, quenched and re-pressed - density 8.20 g/cc) age-hardened for 30 and 200 minutes at various temperatures.

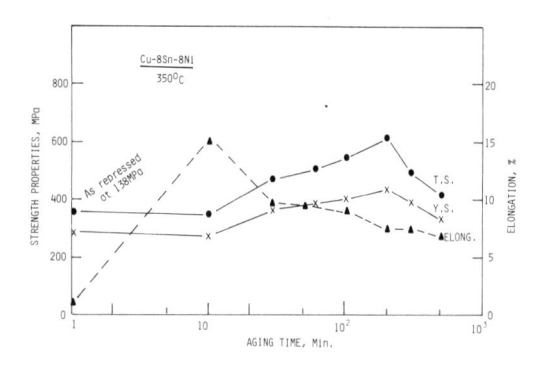

Fig. 3. Effect of aging time on the strength properties of Cu-8Sn-8Ni P/M alloys (re-sintered, quenched and re-pressed at 138 MPa - density 8.20 g/cc) annealed at 350°C.

bronze alloys (re-pressed, re-sintered at 800°C for 30 minutes and quenched) see to contain short-range internal stress fields as a result of the presence of sol atoms in the copper matrix. The glide dislocations appear to overcome these bar iers with the aid of porosity which acts as a dislocation sink. (It must be rem bered that these alloys contain 7 percent porosity which is uniformly distribute mainly at the grain boundaries). These alloys have been further re-pressed and then aged. The short-term aging at low temperature causes a continuous precipitation-spinodal within the α-matrix. The spinodal precipitation grows, generati large internal stresses within the grains. These internal stresses act as a barrier to glide dislocations causing a considerable increase in strength proper ties (Fig. 5). It is possible that super-saturation, grain- and subgrain-bounda free from void surfaces could well account for the increase in the rate of modul ation at low temperature, causing an increase in the yield and tensile strengths of nickel bronze alloys.

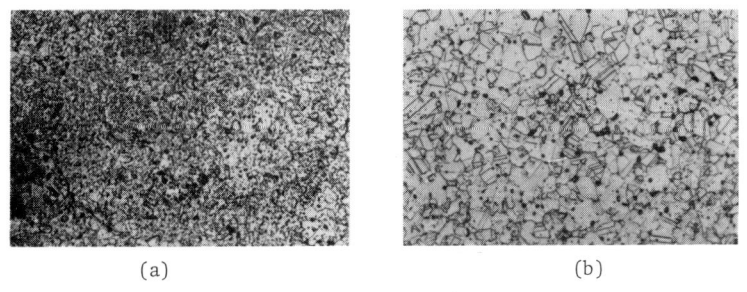

Fig. 4. Microstructure of Cu-10Sn P/M alloys. 650X.
(a) Quenched from re-sintering temperature and re-pressed at 138 MPa - density 8.20 g/cc.
(b) Re-pressed, re-sintered and cooled slowly.

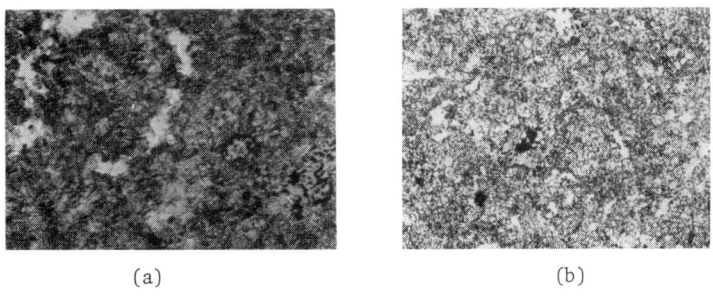

Fig. 5. Microstructure of Cu-8Sn-8Ni P/M alloys, (re-sintered, quenched and re-pressed at 138 MPa - density 8.20 g/cc) aged at 350°C for (a) 30 minutes, and (b) 200 minutes. 1250X.

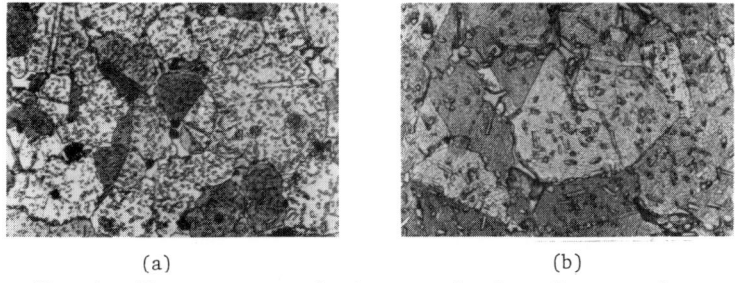

Fig. 6. Microstructure showing growth of needle-type phase (theta) randomly distributed in the matrix. Cu-8Sn-8Ni- P/M alloys - density 8.20 g/cc, aged at 550°C for (a) 30 minutes; (b) 500 minutes. 950X.

MICROSTRUCTURES

e microstructures of Cu-10Sn, re-pressed, re-sintered and cooled slowly in the furce and then quenched, are shown in Fig. 4. The quenched alloys show a very fine-ained texture with porosity distributed uniformly all over (Fig. 4a). The furnace-oled structure showed no evidence of deformation. It contained annealing twins th pores distributed uniformly at the grain boundaries (Fig. 4b).

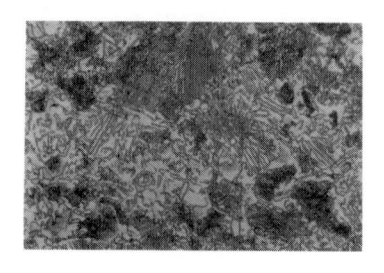

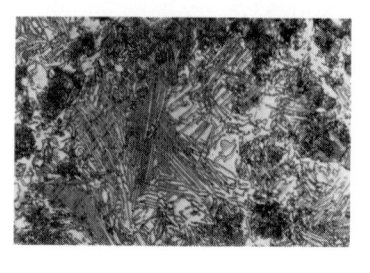

(a) (b)

Fig. 7. Microstructure of Cu-8Sn-8Ni P/M alloys re-sintered and cooled slowly from re-sintering temperature 800°C. 850X. (a) cooled from re-sintering temperature; (b) cooled and aged at 550°C for 30 minutes.

The micrographs of Cu-8Sn-8Ni, age-hardened at 350°C for 30 minutes and 200 minutes, are shown in Fig. 5. The structure shows considerable modulation which increases with annealing time. Furthermore, the modulation in texture indicates fluctuation of composition within the grains. The modulation resembles a possible spinodal decomposition, generating large internal stresses causing a phenomenal increase in the strength properties of Cu-8Sn-8Ni alloys. This spinodal decomposition is stable at low temperature, but with increase in temperature the modulated structure coalesces and transforms into small needle-type precipitants randomly distributed all over the grains (Fig. 6).

With increase in temperature this structure grows, reducing the density of the hardening phase. After extended aging, the needle-shaped particles change into rectangular plate-type particles (Fig. 6). The microstructure of alloys Cu-8Sn-8Ni, density 8.20 g/cc, cooled slowly from the re-sintering temperature showed a lamellar type discontinuous precipitation, similar to the pearlite in steel growing from the grain boundaries and or pores (Fig. 7).

CONCLUSIONS

The results indicate that the addition of nickel in tin bronze increases the strength properties of P/M alloys. The increase in strength is due to a very fine modulated precipitation structure, possibly spinodal decomposition. The modulated structure is not very stable at higher temperatures - above 400°C. With increase in temperature (beyond 400°C) and time, the modulated structure is replaced by needle-type precipitants, θ phase, a discontinuous precipitation, randomly distributed all over the matrix. This causes the strength to drop without increasing elongation.

It must be noted that this is only a preliminary study on the characteristics of Cu-Sn-Ni P/M alloys. The results obtained here are in accordance with the previous work (Kothari, 1974, 1980), namely that the Cu-8Sn-8Ni nickel P/M bronze has much higher strength properties than Cu-10Sn alloys.

REFERENCES

Badia, F.A. (1962). Proc. Amer. Soc. Testing Mats, 62, 665-674.
Bastow, B. and D.H. Kirkwood (1971). J. Inst. Mets, 99, 227-283.
Bell, G.R., F.B. Webb and R. Woolfall (1958). Metallurgia, 233-241.
Berry, D.F. (1972). Powder Met. 15, 30, 147-160.

Ditchek, B. (1978). PhD Thesis, Northwestern University, Evanston, Ill., U.S.A.
Kash, J. and C. Upthegrove (1933). Trans. AIME, 104, 221-41.
Ketz, E., D.R. Hollingbery and R.L. Cavanagh (1974). Proc. 1973 Int. Powder Met.
  Conf. - "Modern Dev. in Powder Metallurgy" ed. H.H. Hausner and W.E. Smith, Vol.
  7, M.P.I.F., Princeton, N.J., 537-583.
Fox, R.L. (1948). Foundry, 76, 6, 103-110.
Kato, M. and L.H. Schwartz (1979). Mat. Sci. Eng., 41, 137-142.
Kato, M., T. Mori and L.M. Schwartz (1980). Acta Met., 28, 285-290.
Kihlgreen, T.E. (1938). Trans. A.F.A., 46, 41-50.
Kothari, N.C. (1974). Proc. Int. Powder Met. Conf. "Modern Dev. in Powder
  Met. ed. H.H. Hausner and W.E. Smith, Vol. 7, MPIF, Princeton, N.J.,
  585-596.
Kothari, N.C. (1980). Proc. Int. Powder Met. Conf., June 20-24, Washington
  D.C., Modern Dev. in Powder Met., Vol. 9, No. 3, MPIF, Princeton, N.J.
Koumani, G. (1972). C.D.A.-A.S.M., Conf. on Copper, Oct. 16-19, Cleveland,
  Ohio, U.S.A.
Lefevre, B.G., A.T. D'Annesa and D. Kalish (1978). TMS-Met. Trans., 9A,
  577-585.
Leo, W. (1957). Metall., 21, 908-912.
Plewes, J.T. (1975). TMS-Met. Trans., 6A, 537-544.
Pound, M.A., A.E.S. Rowley and J.E. Elliot (1960). Powder Met., 6, 129-149.
Price, W.B., C.G. Grant and A. Phillips (1928). Trans AIME, 78, 511-513.
Schwartz, L.H., S. Mahajen and J.T. Plewes (1974). Acta Met., 22, 601-609.
Schwartz, L.H. and J.T. Plewes (1974). Acta Met. 22, 911-921.
Sheppard, T. and A. Greasley (1978). Powder Met., 21, 3, 155-162.
Wise, E.M. (1928). Trans. AIME, 78, 503-514.
Wise, E.M. and J.T. Eash (1934). Trans. AIME, 111, 218-244.

Thermomechanical Treatment of Dual-phase Low Carbon Steels

T. C. Lei, D. Z. Yang and H. P. Shen

Department of Metals and Technology, Harbin Institute of Technology, Harbin, People's Republic of China

ABSTRACT

Three variants of thermomechanical treatment have shown appreciable strengthening effects on plain carbon dual-phase (martensite plus ferrite) steels with nearly 0.2%C. The mechanism of strengthening and microstructural changes are discussed.

KEYWORDS

Thermomechanical treatment (TMT); dual-phase; intercritical quenching; volume fraction of martensite; non-uniformity of strain; microhardness ratio.

INTRODUCTION

The martensite plus ferrite dual-phase steels due to their higher strain-hardening capacity and improved formability (Davis and Magee, 1979; Koo, Young and Thomas, 1980) attract more and more attention of metallurgists and machine designers, but little is known of their deformation strengthening. The aim of the present study was to clarify the effect and mechanism of different thermomechanical treatments, including rolling deformation and intercritical ($\alpha + \gamma$) quenching, and to study microstructural changes of plain carbon dual-phase steels with nearly 0.2%C.

MATERIAL AND METHODS

The chemical composition of the steels used was as follows:

	C	Mn	Si	S	P	Al (%)
Steel No 1	0.21	0.29	0.15	0.030	0.010	0.039
Steel No 2	0.17	0.56	0.38	0.020	0.010	0.056

Hot-rolled rods were forged into strips, normalized at 930 C and machined to specimens with 20 mm width, 150 mm length and different initial thickness in order to obtain a uniform final thickness of 1 mm after different degrees of rolling. Variants of TMT were carried out according to Fig. 1. The quenching medium selected was 5% NaOH aqueous solution and all specimens were tempered for 1 h at 200 C.

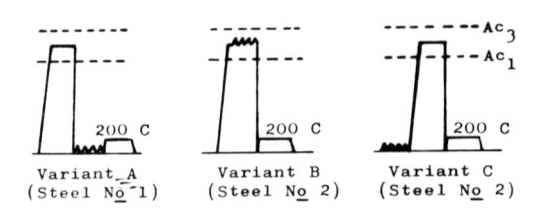

Fig. 1. Variants of thermomechanical treatment.

Tensile tests were carried out with a strain rate of $4.15 \times 10^{-4} s^{-1}$. Microstructures were examined optically and with a graph analyzer, while microhardness was measured by using 5 or 10 g loads.

RESULTS AND DISCUSSION

Variant A (Cold Rolling after Intercritical Quenching)

The volume fraction of martensite (V_M) and its carbon content after intercritical quenching are: 740, 30%, 0.65%C; 770, 40%, 0.48%C; 800, 48%, 0.34%C. With increasing degree of cold rolling the strength increased and the ductility decreased, as shown in Table 1.

TABLE 5 Tensile Strength of Steel No 1, Variant A Treated

Degree of cold rolling (%)	Tensile strength (kg/mm²) 30%M	40%M	48%M	Yield strength (kg/mm²) 30%M	40%M	48%M	Elongation (%) 30%M	40%M	48%M
0	60	75	80	48	53	57	26	21	18
10	96	104	99	88	99	92	10	9	8
23	109	117	110	107	114	108	6	5	5
33	113	122	114	108	118	109	5	5	5
50	121	132	122	111	127	119	4	4	4

The longitudinal elongation of both phases was quantitatively expressed by the aspect ratio (L/A) where L is the length and A the width of grains. It can be seen from Fig. 2, that with increasing degree of cold rolling $(L/A)_F$ increases more rapidly than $(L/A)_M$. The non-uniformity of strain between ferrite and martensite $R=(L/A)_F/(L/A)_M$ increases rapidly at small degrees of rolling and reaches a stabilized value at higher degrees (Fig. 2). This indicates that the deformation of dual-phase steels experiences a transition from a stress-equal model to strain-ratio-equal model as the degree of deformation increases. Microhardness values of the two phases and the microhardness ratio $K=(HM)_M/(HM)_F$ with respect to degree of cold rolling are shown in Fig. 3. The K values decrease rapidly at small degrees

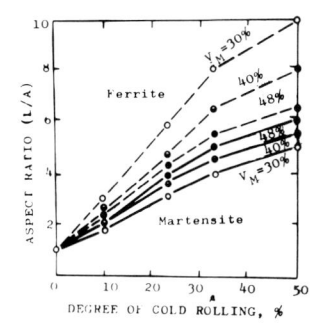

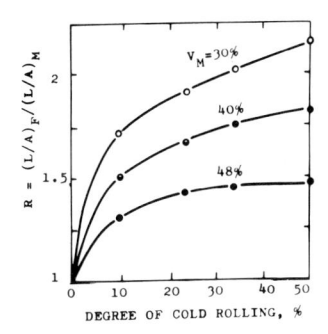

Fig. 2. Aspect ratio (L/A) of the phases and non-uniformity between them (R) vs degree of cold rolling.

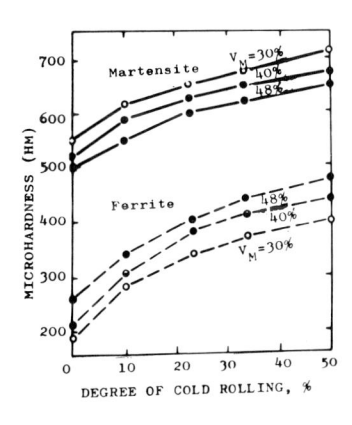

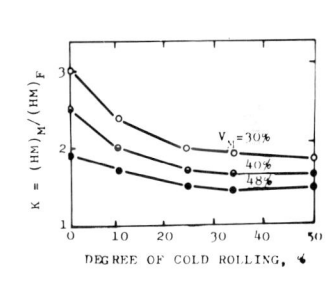

Fig. 3. Microhardness (HM) of the phases and microhardness ratio (K) vs degree of cold rolling.

of rolling and become stabilized at higher degrees. This means that the strengthening of dual-phase steels undergoes a transition from an unequal-strengthening state to a strength-ratio-equal state as the degree of deformation increases.

From data in Fig. 2 and 3 a comprehensive R-K relation can be established as shown in Fig. 4. Here, two independent variables, volume fraction of martensite (V_M) and degree of cold rolling (ε) can be selected by varying the temperature of quenching and the reduction in thickness during rolling, while two dependent ones, non-uniformity of strain (R) and microhardness ratio (K) describe the relative changes in structure and properties between the two phases. For a dual-phase steel with given V_M the values of R and K after a certain degree of cold rolling can be easily found by following along the V_M curve and reading the coordinate values of the points at the intersection with the iso-ε curves.

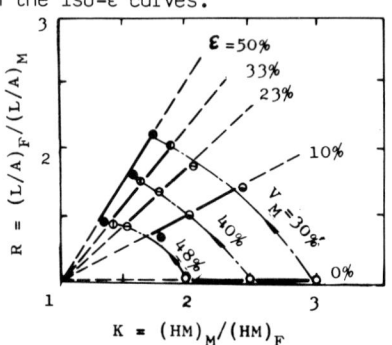

Fig. 4. Comprehensive diagram (R-K relation) of deformation behaviour of steel No. 1.

Variant B (Intercritical Rolling with Direct Quenching)

Two specific features of microstructural changes during rolling at intercritical temperatures have been found, namely: the promotion of ferrite nucleation and the dynamic recrystallization of both martensite and ferrite phases. From data shown in Table 2 it may be seen that with increasing degree of rolling the V_M values gradually decrease, especially at higher temperatures. The decrease in V_M is a direct result of promoted nucleation of ferrite by deformation.

Rolling at intercritical temperatures unavoidably leads to dynamic recrystallization of the ferrite and austenite phases, but the extent of such processes depends on the temperature and degree of rolling. Figure 5 shows that 770 C, 80% rolling gives ferrite obviously recrystallized while the austenite grains remain elongated. The 840 C, 80% rolling gives both phases recrystallized to equiaxed fine grains.

TABLE 2 Effect of Intercritical Rolling on Volume Fraction of Martensite (V_M) for Steel No 2

Rolling temperature, C	Degree of rolling, %			
	0	20	50	80
770	40	38	36	32
800	55	50	48	45
840	90	70	60	55

770 C, 80%   800 C, 80%   840 C, 80%

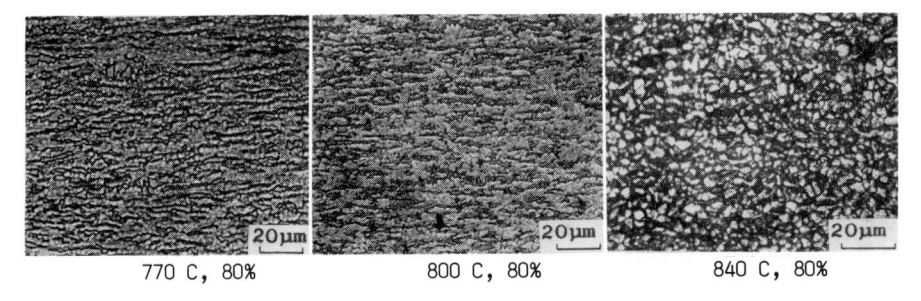

Fig. 5. Optical microstructures of steel No 2 after different treatments by Variant B.

Tensile properties of dual-phase steels (No 2) simply intercritically quenched and treated by variant B are shown Fig. 6. Intercritical quenching without deformation gives a straight line relationship of strength with respect to (V_M), which is consistent with the results of other investigators (Ramos, Matlock and Krauss, 1979). For 770 and 800 C rolling, the tensile and yield strengths increase with degree of rolling, while at 840 C the strength values markedly decrease with increasing degree of rolling which is fully consistent with the decreased V_M as shown in Table 1. A maximum strengthening effect can be obtained by 800 C, 80% rolling which develops a special microstructure with fibrous martensite and very fine-grained ferrite.

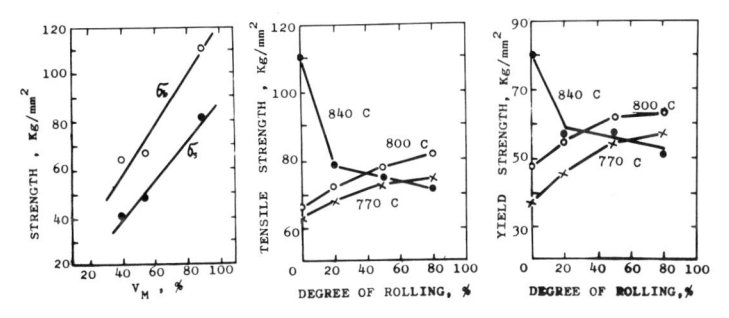

Fig. 6. Strength of steel No 2 intercritically quenched and treated by variant B.

Variant C (Intercritical Quenching after Prior Cold Rolling)

The effect of 70% prior cold rolling on microstructure of intercritically quenched (770 C, 1 min heating) steel No 2 is shown in Fig. 8. Here, a thin fibrous structure has been obtained which means that a short time heating at intercritical temperature may retain the prior deformed structure and hence develops an unidirectionally oriented fibrous structure (similar to commonly used fibre-reinforced composite materials). The tensile strength as also shown in Fig. 7, increases rapidly with degree of prior cold rolling up to 50% and a tensile strength value of nearly 110 kg/mm² was obtained.

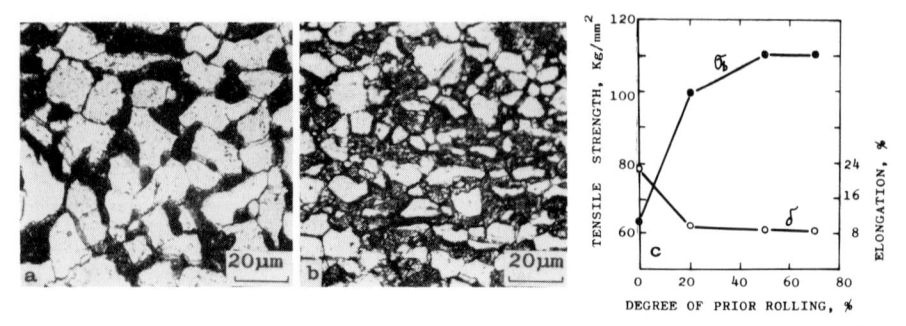

Fig. 7. Microstructure of 770 C quenched steel No 2 without (a) or with (b) 70% prior cold rolling and (c) the relationship of tensile strength to degree of prior cold rolling.

CONCLUSIONS

1. Cold rolling of dual-phase steels can give appreciable strengthening. The deformation process undergoes a transition from a stress-equal to a strain-ratio-equal model while the deformation strengthening moves from an unequal-strengthening state to a strength-ratio-equal state as the degree of rolling increases.

2. Intercritical rolling followed by direct quenching leads to a decrease of the volume fraction of martensite and the dynamic recrystallizaion of ferrite and austenite, but suitable rolling may also give an obvious strengthening effect.

3. Intercritical quenching after prior cold rolling can appreciably increase the strength of these steels by developing fibrous dual-phase structures.

REFERENCES

Davies, R. G. and C. L. Magee. (1979). Physical metallurgy of automotive high-strength steels. J. Metals, 31, 17-23.
Koo, J. Y., M. J. Young and G. Thomas. (1980). On the law of mixtures in dual-phase steels, Met. Trans., 11A, 852-857.
Ramos L. F., D. K. Matlock and G. Krauss. (1979). On the deformati behaviour of dual-phase steels. Met. Trans. 10A, 259-265.

Author Index

A

Abbott, J R	347
Abdellatif, M A	1213
Abel, A	825
Abraimov, V V	63
Aghan, R L	475
Ahmed, Z	255
Akben, M G	499
Alden, T H	741
Allison, J E	1219
Altenpohl, D G	1197
Anglada, M	3
Aono, Y	9, 69
Arsenault, R J	15, 781
Ashida, Y	193
Atkinson, M	481

B

Baburamani, P S	115
Baker, I	487
Balmori, H	983
Bampton, C C	713
Barritte, G S	121
Basinski, S J	819
Basinski, Z S	21, 819
Bassim, N M	839
Bay, B	401
Becker, J	127
Bendeich, D	505
Bendersky, L	595
Bergmann, H W	135, 367
Berveiller, M	443
Birocheau, J	569
Bolling, G F	1045
Bouhafs, M	167
Bouquet, G	275, 281
Bowden, J W	683

Brandon, D G	353
Bronsveld, P M	301
Brown, G G	493
Brown, K R	765

C

Caciorgna, O	927
Carfi, G	223
Carmichael, C	381
Carter, R J	393
Chandhok, V K	185
Chandra, T	235, 499, 505
Chaturvedi, M C	147
Chehimi, C	919
Chen, Y C	333
Cheng, A S	1147
Chopra, P N	735
Christian, J W	319
Chung, Y S	825
Clarebrough, L M	27
Clark, G	773
Clegg, W J	689
Coade, R W	263
Corderoy, D J H	431, 941
Cowling, J M	851
Coyle, R A	535
Crowe, C R	859
Cutler, L R	161

D

Dannemann, K	141
Davies, G J	1121
Deb, P	147
Dehghan, A	757

Delobelle, P	575	**H**		
Demiraj, G	281	Haasen, P	933	
Donovan, P E	361	Hall, E O	393	
Dons, A L	425	Hamel, F	839	
Dover, I R	269	Hansen, N	295, 401	
Doyle, E D	511, 553	Hashimoto, T	173	
Drew, M	867	Hasson, D F	859	
Dubois, B	167, 275, 281	Hatherly, M	381, 541, 523, 1181	
Duerig, T W	263	Hato, H	193	
Duesbery, M S	21	Hazzledine, P M	45, 721	
Dulis, E J	185	Heh, J	333, 895	
Dunlop, G L	665, 695	Herman, J-C	179	
Dunne, D P	235, 505	Hernandez, F	167	
Duquette, D J	141, 879	Hey, A M	223	
		Hirsch, J	965	
E		Hirth, J P	185	
		Hobbs, R M	115	
Eady, J A	269	Honeycombe, R W K	407	
Edwards, L	873	Hornbogen, E	127, 1059	
Edwards, R H	601	Hosomi, K	193	
Embury, J D	1089	Hosson, J Th M De	101, 301	
Esterling, D M	15	Howell, P R	121, 695	
		Hsu, R	781	
F		Hu, C T	333, 895	
		Huang, C K	793	
Falk, L K L	695	Hughes, D S	655	
Feng, X F	1001	Humphreys, F J	625, 747	
Fine, M E	833			
Firrao, D	947	**I**		
Fitz Gerald, J D	735	Inoue, T	787	
Fitzpatrick, N P	387	Ishii, T	141	
Fleck, R G	289			
Forwood, C T	27			
Fourdeux, A	953	**J**		
Franciosi, P	33, 39			
Fraser, H L	313	Jackson, P J	51	
Fukuzato, T	455	Jago, R A	115	
Furukawa, T	1165	Jahn, M T	333, 793, 895	
		Jerath, V	1225	
G		Jonas, J J	499	
		Jones, R D	1225	
Gabrielli, F	607			
Gessinger, G H	263			
Gherardi, F	927	**K**		
Ghosh, A K	713			
Gibeling, J C	613, 741	Kainer, K U	367	
Gifkins, R C	269, 701	Kanert, O	101	
Gil Sevillano, J	547	Kapoor, S	1225	
Gleiter, H	1009	Karashima, S	671	
Golwalker, S	879	Karato, S	753	
Greday, T	179, 211, 1075	Kashyap, B P	707	
Grinberg, A	83	Kassner, M E	581	
Guiu, F	3	Kaufmann, H-J	57, 63	
Gunn, K W	601, 619	Kayali, E S	913	
Gutmanas, E Y	353	Keys, L H	867	
		Kim, H M	1231	
		Kitajima, K	9	

Kitajima, S	95, 413
Knott, J F	799
Kobayashi, S	173
Kolkman, H J	631
Konitzer, D G	313
Kothari, N C	1237
Krishnadev, M R	161
Kubin, L P	953
Kuo, S M	333
Kuramoto, E	9, 69
Kurishita, H	95

L

Laird, C	1147
Lake, J S H	959
Langdon, T G	757, 1105
Lawley, A	1213
Lee, D N	559
Lee, S Y	977
Leffers, T	75
Lei, T-C	307, 1245
Lepinoux, J	953
Liu, T F	895
Löhe, D	199
Lopriore, G J	419
Lorenzo, F	1147
Lubenets, S V	63
Lücke, K	965
Lupinc, V	607

M

MacDonald, G J M	449
Macherauch, E	199, 229
Maeda, K	173
Maki, T	529
Malin, A S	381, 523, 541
Mannan, S L	637
Marchionni, M	927
Marques, F D S	205
Martin, J W	487, 873
Maslov, L	971
Mathy, H	211
McCormick, P G	419
McEvily, A J	845
McGirr, M B	977
McQueen, H J	517
Mermet, A	575
Messien, P	179
Miller, A K	581
Miller, D R	347
Minakawa, K	845
Mintz, B	217
Mitani, Y	983
Molinas, B J	89
Mordike, B L	135, 367, 643
Morgan, J E	643
Morton, M E de	153

Muddle, B C	313
Muir, H	431
Mukherjee, A K	595, 689, 707
Murakami, Y	805

N

Nakamura, H	193
Namba, Y	787
Nes, E	425
Nethercott, R B	535
Netherway, D J	347
Nissenholz, Z	353
Nix, W D	613

O

O'Donnell, R G	437
Obata, M	671
Ocampo, G	281
Ochiai, S	805
Ogle, J C	677
Okaguchi, S	529
Ono, H	431
Ono, N	941
Osamura, K	805
Osborn, C J	467
Oytana, C	575
Ozeri, J	353

P

Pacey, A J	901
Padgett, R A	649
Palmer, G R	1225
Parker, B A	437
Pascual, R	907
Paterson, M S	1137
Paton, N E	713, 727
Payne, T	601, 619
Pedersen, O B	75
Picco, E	927
Piegerova, V	523
Pink, E	83
Plumtree, A	901, 913
Pluvinage, G	919, 995
Povolo, F	89, 589
Pratt, P L	387

R

Rabier, J	953
Ralph, B	295
Ramaswami, B	683
Ranucci, D	927
Retchford, J A	535
Rey, C	443

Ricks, R A	121
Ripley, M I	319
Roberti, R	947
Robin, C	919
Rodriguez, P	637
Rolim, L C	907
Rooum, J A	689
Rosen, A	595
Rubin, K A	581
Rubiolo, G H	589
Ruzzante, J	223
Ryum, N	425

S

Sabayo, M R	989
Sakaki, T	455
Sammis C G	757
Samuel, K G	637
Schneibel, J H	649, 677
Scholtes, B	229
Shaibani, S J	45
Shechtman, D	353
Shek, G K	289
Shen, H P	1245
Shinohara, K	95, 413
Simpson, I D	449
Sinning, H R	933
Smith, G M	407
Smith, R M	235
Snowden, K U	655
Solomon, R G	541
Starke, Jr, E A	1025
Stathers, P A	655
Stobbs, W M	361
Stoloff, N S	141, 879
Sugimoto, K	455
Suzuki, H	327
Swindeman, R W	677

T

Takeuchi, S	173
Tamler, H	101
Tamura, I	529
Tang, N Y	665
Tangri, K	373
Taplin, D M R	665
Thadhani, N N	205
Thölen, A R	107
Thompson, K R L	867
Tormo, J	223
Torrealdea, F J	547
Travitzki, N	353
Tsutsumi, T	69
Tu, M-J	887
Tungatt, P D	747

Turley, D M	511, 553

V

Vale, S H	721
Van Der Wegen, G J L	301
Veyssiere, P	953
Vitullo, G	393
Vöhringer, O	199, 229

W

Wan, C M	333, 793
Wan, C M	895
Wang, D-Z	307
Wang, K-R	995
Wang, T H	333
Wanhill, R J H	631
Watanabe, T	671
Watson, J D	493
Wert, J A	339, 727
Westengen, H	461
White, C L	649
Williams, J C	1219, 1231
Williams, J G	241
Willis, D J	247
Wilms, G R	475
Woodward, R L	467

X

Xu, S-L	1001

Y

Yang, D Z	1245
Yang, M H	895
Yegneswaran, A H	373
Yeldham, R J	1225
Yellup, J M	811
Yeshurun, Y	353
Yoo, M H	677
Yoo, Y C	559

Z

Zaoui, A	39, 443
Zhou, H-J	887
Zhu, J-H	887

Subject Index

A

Abe et al.(1979), 1167
Accommodation slip, 406
Acicular ferrite, 121
Acoustic emission, 839
Activation
  areas, 6
  energy, 195, 202, 656, 667, 701
  enthalpy, 83, 91
  volumes, 73
Adiabatic shear, 814
  bands, 528, 1191
Age-hardening, 193, 264, 645, 1225
Aging, 282, 309
Air, 928
Alkali metals, 22
Alloy additions, 256
Alloy-maraging steel, 194
Alternate-plane model, 782
Aluminium, 102, 108, 353, 401,
  484, 613, 713, 914, 1203
  alloy, 247, 334, 339, 255,
  269, 425, 461
  copper, 102
  extrusions, 1203
  glass composites, 368
  high strength alloys, 765, 1027
  killed steel, 242, 450, 960
  Li-X alloys, 1038
  magnesium, 437, 625
  magnesium-silicon, 419, 873
  nitride, 333, 334, 1122
Amorphous
  metal, 362
  structure, 176

Amplitude dependent, 90
Anelastic deformation, 577
Anelasticity, 721
Anisotrophic expansion, 72
Anisotrophy, 449
Annealing, 351
  twins, 297, 383, 508
Anorthite, 357
Anti whisker, 837
Armour steel, 812
As-cast structures, 1121
As-hot-rolled dual-phase steels,
  1172
Astroloy, 879
Athermal
  component, 741
  martensite, 205
  plateau, 745
Atmosphere at partials, 329
Ausformed steels, 1021
Austenite
  "pancaking", 499
  retained, 1172, 1225
  rolled, 1084
Austenitic
  stainless steel, 529, 901
  steel, 677
Average microstructural
  parameters, 1102

B

Back stress, 229, 233, 830
  reversal, 825

Background damping, 89
Bailey-Orowan relation, 570
Bainite, 895
Bainitic structures, 1087
Bauschinger effect, 77, 229, 431, 494, 781, 826, 956
Bauschinger parameters, 230
b.c.c. lattice, 16, 69
b.c.c. metals, 9, 22
Bend gliding, 413
Biaxial tension, 482
Biaxial test, 438
Bicrystals, 446
Binding potentials, 21
"blocky" cells, 489
Blue brittleness, 199
Blunt notches, 950
Boron steels, 789
Boundary mechanisms, 1106
Brass, 381, 556
Breakaway, 1114
Breakdown, 1107
Bright ground rolls, 960
Brittle-ductile transition, 1137, 1139
Built-up edge, 475
Build-up edge overhang, 479
Bulge test, 438
Burgers - α precipitate, 265
Burgers vector, 28, 954

C

Calcite, 747
Calcium-treated steels, 450
Capped steel, 960
Carbide, 639
    morphology, 186
    precipitation, 137, 153
    thickness, 218
Carbon concentration, 1004
Carbon doped molybdenum, 319
Carbonitride particles, 240
Case depth, 851
Case:core ratio, 852
Cast iron, 199
Casting defects, 1122
Castings, 1121
Catastrophic shear, 1096
Cateclastic component, 1140
Cavitation, 633, 650, 680, 689, 705
Cavity nucleation, 677, 717
Cells, 427, 507, 520, 898
    formation, 591, 1232

"rim", 297
size, 914, 916
structure, 104, 297, 334, 488, 536, 658, 896, 942, 966, 1151, 1158
walls, 742
Cementite morphology, 229
Cementite shape, 231
Charpy test, 214, 789
Chemical
    interaction, 327
    locking, 1014
Chip formation, 511
Chips, 476, 554
Chromium, 257
Chromium steels, 167
Class A (Alloy type), 1110
Class II creep, 595
Class M (Metal type), 1106
Cleavage fracture, 795
Coated specimens, 636
Coating cracking, 248
Coating thickness, 251
Cobalt, 185, 275, 523
    alloys, 275
Coble, 1106
    creep, 667
Coffin-Manson curve, 1157
Coherency strains, 936
Coiling window, 1086
Coincidence boundaries, 672
Cold
    drawing, 985
    rolling, 425
    twisting, 493
    work, 514, 547
Collective direct
    interactions, 1013
Columnar growth, 136
Compocasting, 1131
Composite, 247, 351, 805, 989
    eutetic, 141
    in-situ, 1129
    structures, 1204
    test, 302
    tubes, 1205
    wire, 983
Compressive stresses, 353
Computer model, 15
Computer simulation, 10, 69
Cone angle, 468
Cone source, 1184
Conference organization, 1056
Constant stress, 570
Constant substructure, 615
Constitutive equations, 489
Continuous annealing process, 1167

Continuous yielding, 455
Continuously-cast Strands, 1127
Continuum elasticity, 15
Controlled cooling, 1083
Cooling rate, 215, 1082
Coplanar
  glide, 35
  model, 782
Copper, 34, 95, 108, 295, 431, 517, 535, 541, 557, 689, 819, 907, 965, 1203
  4% Ti, 934
  8tin-8nickel, 1238
  15% zinc, 485
  aluminium, 281, 296, 373, 431, 489, 941
  chromium, 665
  clad aluminium, 983
  neutron-irradiated, 96, 413
  Ni-Fe, 933
  NiZn, 302
  tin, 1237
Corrosion resistance, 256
Cottrell-Bilby theory, 83
c.p.h. metals, 523
Crack, 821, 995
  closure, 847, 870, 923
  extension, 1068
  growth, 845, 846, 1061
  jump, 973
  nucleation, 683
  nucleus, 1061
  opening curves, 921
  opening displacement, 800, 1064
  propagation, 867, 995, 1061, 1147
  retardation, 871
  satellite, 1071
  tip blunting, 870,
  tip damage, 774
  tip ductility, 799
  tip microfracture, 842
  velocity, 356
Creep, 63, 139, 370, 394, 581, 595, 607, 613, 648, 753, 913
  compliance, 652
  exponent, 616
  fatigue mixing, 901
  fatigue superalloy, 631
  fracture, 671
  Harper-Dorn, 759, 1106
  rates, 575, 667
  resistance, 1032
  rupture, 619, 631, 633, 677
  secondary, 668
  strength, 612, 649
Critical

crack size, 836
glide, 35
notch root strain, 950
strain, 201
Cross slip, 45, 51, 54, 99, 536
  spoor, 55
Cube texture, 270, 1194
Cubic δ', 632
Cumulative damage, 632, 636, 819, 1101
Cyclic
  deformation, 3, 1231
  hardening, 685, 875, 1149, 1214
  response, 1214
  softening, 685, 875
  stress-strain curves, 1215
  stress-strain response, 1149

D

Damage accumulation, 837, 1096
  concepts, 906
Damping, high temperature, 89
De-salination plants, 255
Decarburization, 169
Decremental step tests, 1152
Deep-drawability, 1167
Deep-drawing steels, 1168
Defect chemistry, 1138, 1142
Deformation, 118, 240, 735
  and fracture behaviour, 805
  bands, 401
  characteristics, 1165
  configuration, 516
  geometry, 467
  high temperature, 625
  hot, 529
  inhomogeneities, 965
  map, 1119
  markings, 349, 513
  mechanism map, 1143
  texture, 546
  twinning, 967
  twins, 749
Deformed, 118
  polycrystal, 941
  structures, 1232
Degassing, 718
Delay time, 615
Dendritic structure, 315
Denitriding, 169
Density, 689, 1242
Dental amalgam δ phase, 347
Dentritically solidified, 136
Denuded zones, 605

Developing countries, 1197
Dielastic, 1014
Differential strain rate, 708
Differential thermal contraction, 1223
Diffusion, 728
   controlled growth, 982
   controlled screw dislocations, 592
   creep, 703
   flow, 1138
   zone, 985
Diffusive atomic motion, 103
Dimple, 973
   fracture, 795
   mechanism, 887
Dimpling, 862
Dip test, 569, 575
Dipole wall, 550
Directional growth, 1122
Directional solidification, 1128, 1213
Dislocation, 735
   annihilation, 79
   arrays, 801, 802
   arrest time, 437
   climb, 753
   core structure, 24
   core, 69
   damping, 153
   extended, 328
   partial, 50
Dislocation density, 169, 203, 291, 297, 334, 343, 375, 394, 404, 410, 428, 435, 490, 536, 538, 581, 595, 825, 1123
   dipoles, 835
   dynamics, 741
   glide, 3, 753
   helices, 837
   intersections, 105
   labyrinths, 908
   loops, 608, 1151
   motion, 95, 102
   networks, 742
   pile-up, 781, 814, 1011
   pinning, 153
   sheets, 53
   storage, 78
   structure, 3, 401, 944
   substructure, 143, 334, 553,
   surface source, 781
   tangles, 343, 488, 658
   veins, 908
   walls, 908
Dispersions, 1025
   strengthening, 1020

   structure, 223
Dispersoids, 313, 873
   elements, 1078
   particles, 768
Disseminated slip, 396
Domains, 443
Double kink, 16
   nucleation, 322
Double-coplanar model, 782
Double-walled cells, 549
Dual-phase, 1245
   steels, 115, 211, 373, 455, 813, 1084, 1089, 1170
   structure, 127
Ductile
   fracture, 813, 1089
   tearing, 1222
   to brittle transition, 319
Ductility, 150, 161, 248, 449, 649, 704, 805, 1035
   of Ti steels, 183
Dunite, 1142
Duplex
   stainless steel, 505
   structure, 223
Dynamic,
   fracture toughness, 387
   precipitation, 502
   recovery, 439, 505, 517, 529, 544, 615, 639
   recrystallization, 500, 510, 517, 526, 529, 538, 544, 747, 1077
   strain aging, 83, 199, 204, 441, 638

E

E-type change, 708
Ecological movements, 1199
Edge deformations, 269
Edge dislocation, 99
Effect of scavengeing, 1168
Elastic,
   anisotrophy, 24
   interaction, 1018
   limit, 494
   modulus, 1028
   stresses, 361
   waves, 107
Elastoplastic, 39
Electrical resistivity, 986
Electromagnetic stirring, 1127
Electron
   atom diameter model, 503

atom ratio, 327
channelling patterns, 626, 672
diffraction, 317
Electroslag
  remelted, 1126
  steel, 812
Electrostatic locking, 1014
Energy, 1199
Environmental
  attack, 633
  protection, 1199
Equipotential surface, 578
Erbium, 313
Eutectic
  composites, 141
  phases, 766
Explosive-powered fasteners, 467
Exponential stress, 1140
Extended dislocations, 328
Extensometer, 174
Extrusions, 820, 898

F

Faceted regions, 1222
Failure
  analysis, 995
  maps, 1098
  modes, 1100
Fatigue, 683, 819, 851, 867, 907, 919, 1147
  crack, 773, 889
  crack closure, 1223
  crack growth, 845, 887, 1219
  crack propagation, 880
  damage, 1148
  environment, 928
  flexural, 859
  high cycle, 141, 631, 840, 895
  high temperature, 665, 901, 925
  limit, 853
  low cycle, 665, 873, 895, 927, 933
  stage, 954
  strength, 130
  stress intensity, 774
  striations, 862, 887
Fe-Mn steel, 1225
Ferrite
  grain refinement, 235
  martensite interfaces, 118, 456
  martensite steels, 127
  massive, 153
  morphology, 409
  pearlite, 217, 477

pearlitic steels, 223
recrystallized, 148
Ferrous martensites, 1015
Fibre, 805
  -reinforced polyester, 1206
  spheroidization, 333
Fibrous rupture, 799
Filament composition, 859
Final drawing, 308
Fine-grain, 701
  size, 461, 713
Finite element analysis, 393, 921
Flake graphite cast iron, 200
Flexural fatigue, 859
Flow law, 573
Flow stress, 75, 295, 339, 427, 538, 539, 638, 741
Fluid phases, 1139
Formability, 208
Forming limit curve, 559
Forming operations, 1101
Fractional melting, 1130
Fractography, 636, 766, 972, 992
Fracture, 272, 1059
  quasi- cleavage, 813
  mechanics, 1062, 1147
  spall, 811
  strain, 115
  surfaces, 716, 904, 1003
  toughness, 208, 355, 765, 773, 947, 990
  toughness (laminates), 990
  toughness models, 948
Frank-Read sources, 305
Free surface, 362
Freeze moulding, 1128
Friction, 467
  mechanisms, 1014
  stress, 915
  stress strengthening, 1018
Fusion welds, 1122

G

GP zones, 105, 343
Galvanized steel, 247
Gatekeeper, 1054
Geological conditions, 1137, 1143
Geometrical coalescence, 336
Geometrically necessary
  dislocations, 79
Gerber parabola, 856
Glass fibres, 367
Glassy phase, 354
Globurization of sulphides, 179

Godium nitrate, 747
Goodman line, 856
Grain, 1025
   diameter, 1068
   edge effect, 628
   effect, 76, 1010
   emergence model, 702
   growth, 148, 708, 711
   impingement, 236
   misorientation, 29
   nucleation, 381
   refinement, 240, 1026
   size, 218, 292, 339, 396, 462, 533, 608, 637, 959
   size thickness ratio, 271
   stabilization, 461
Grain boundary, 27, 735, 753
   bands, 404
   carbides, 800
   decohesion, 1222
   dislocations, 28, 670, 695, 702
   effects, 1143
   elongated, 707
   enrichment, 653
   junctions, 658
   migration, 628, 672, 702, 751
   precipitates, 645, 695
   "rim", 299
   scalloped, 627
   segregation, 671, 790
   sites, 979
   sliding, 650, 672, 695, 701, 728, 898, 1138
   strengthening, 295, 465
   structure, 673
   voids, 722
Grit blast rolls, 960
Gross deformation bands, 545
Group constraint, 416
Grown-in jogs, 97
Guinier's Law, 679
Gun barrels, 777
Gun steel, 775

H

HSLA steel, 161, 184, 235, 241, 1075
   commercial, 161
Half-value breadth, 887
Hall-Petch
   parameters, 639
   relationship, 558, 1011, 1189
Hard slip system, 322
Hardening, 282, 583
   law, 37
   mechanisms, 195
   models, 481
   precipitates, 877
   processes, 1076
Hardness, 136, 167, 525
   Hart, 571
   hot, 185, 283
Harper-Dorn creep, 759, 1106
Hart's phenomenological theory, 569
Heat treatment, 258, 1002
Heating stage, 955
Heavily cold-rolled, 541
Helium entrapment, 717
Helium temperature, 58, 63
Herring, 1106
High
   conductivity, 665
   temperature damping, 89
Homogeneous
   deformation tests, 1081
   slip, 967
   strain, 1064
Hooke's law, 176
Hot
   drawing, 985
   isostatic pressing, 879
   rolled strip, 241
   tears, 1121
   working, 505, 517, 1077
Household foil, 269
Hydraulic bulge test, 481
Hydrogen
   in steel, 453
   temperature, 63
Hydrolytic weakening, 1141
Hydrostatic
   component, 1139
   pressure, 1093
   tension, 1092
Hysteresis loops, 177, 826

I

ICSMA research, 1045
Image
   forces, 49
   matching, 28
Impact
   notched tensile test, 813
   phenomena, 107
   values, 217
Imposed pressure, 1092
Impurity, 735

"inheriting" heat treatment, 1005
Inclusions, 950, 1090, 1121, 1125
   assessment, 451
   spacing, 802
Incremental unloading, 432
Indentation hardness, 825
Industrialized countries, 1197
Industry, 1047
Inhomogeneous strain, 376
Innovation, 1054
Innovative chain, 1048
Integrated intensity, 887
Intense shear, 515
Intense slip, 824
Intercritical quenching, 1245
Intercritically annealed, 1084
Intercritically rolled, 1084
Intercrystalline
   cleavage, 128
   cracking, 902
Interfacial energy, 1018
Intergranular
   crack propagation, 602
   cracking, 687
   film, 667
   fracture, 789, 1036, 1222
   precipitate, 666
   voids, 666
Interlamellar spacing, 218
Internal
   back stress, 662
   friction, 90, 153, 167, 275, 281
   necking, 799
   stress, 435, 456, 569, 575, 781, 1240
Interparticle spacing, 1031
Interstitials, 168
Intragranular sites, 240
Intrusions, 820, 834, 898
Invisible College, 1051
Iron, 10, 34, 484, 914
   alloys (quenched), 135
   carbide, 978
   silicon, 200
   tin, 671
Irradiated LiF, 742
Irregular migration, 673
Isothermal heat treatment, 265, 408

J

J-ρ plots, 951
Jog-drag, 591

Johnson-Wilson potential, 71

K

Kink interaction model, 60
Knoop hardness, 354
Kossel technique, 36, 443
Koster damping, 153

L

Lamellar spacing, 1068
Laminates, 989
Laser, 314
   glazing, 1132
Latent hardening, 33
Lath martensite, 1005
Lattice
   curvature, 429, 944
   mechanisms, 1106
   rotation, 414, 444
Law of mixtures, 210
Ledges, 377
LI_2 alloys, 301
LiF crystal, 64
LiF, 741
Life prediction, 1147
Limestone, 1140
Lin-Taylor relations, 446
Lip height, 469
Liquid
   gallium, 723
   Hg, 349
   metal embrittlement, 352
Load
   interaction, 1158
   ratio, 845
   relaxation, 742
Loading history, 1098
Local volume fraction, 1101
Localized shear, 549, 767
Localized strain, 1064
Locking mechanisms, 1014
Long-range interactions, 745
Long-range order, 301
Low
   angle boundaries, 294, 404, 1012
   density alloy, 1034
   strain-rate cycle factor, 904
Lubrication, 562
Luders band, 393, 959
Luders strain, 250

M

"m" value, 715
Machinability, 515
Machining, 475
Macroscopic compatibility, 944
Macrosegregation, 1124
Magnesium, 257
Magnesium-yttrium, 643
Managing research, 1047
Manganese, 499
  content, 121
Mantle, 702
Manual metal arc, 121
Maraging steel, 193, 1021, 1225
Marble, 1140
Martensite, 264, 1003, 1082
  athermal, 205
  ferrous, 1015
  island, 455
  laths, 377, 1005
  twinned, 1004
  variants, 206
Martensitic transformation, 275
Massive ferrite, 153
Materials
  constraint, 1200
  conservation, 1201
  high temperature, 643
  research, 1197
  substitution, 1201
Matrix carbides, 680
Maximum plastic work principle, 39
Mean
  jump distance, 102
  loads, 1154
  stress, 1098
Mechanical
  activation, 742
  properties, 212, 247, 255, 283, 646
  twins, 391
Melt spinning, 135, 714
Meniscus freezing, 1127
Metadynamic recrystallization, 1080
Metal
  reinforcement, 367
  surfaces, 107
Metallic glasses, 173, 361
Micro-alloyed steel, 407, 499
Micro-alloying, 235
Micro-crystalline, 1012
Micro-twins, 798
Microarea chemical composition, 1002
Microband, 489, 515, 543, 547, 965, 1183, 1184, 1185, 1186, 1189
Microcracking, 1071, 1138, 1139
Microcrystalline materials, 135
Microhardness, 373
  ratio, 1245
Micromechanical
  concept, 127
  elements, 131
Microprecipitation, 1080
Microscopic compatibility, 944
Microsegregation, 1123
Microstrain, 177
Microstructure, 212, 259, 511, 583, 735, 1059
  average parameters, 1102
  multiphase, 208
  rolling, 523
Microvoid coalescence, 678
Migration recrystallization, 748
Mild steel, 387, 393, 556, 840
Minerals, 1137
Miniaturization, 1199
Misorientation angle, 596
MnS inclusions, 179
Mobility, 95
Mode I crack growth, 845
Mode II crack growth, 846
Modified Sachs model, 76
Modified Taylor model, 76
Modular cast iron, 200
Modulation, 1242
Modulus difference, 17
Moine model, 304
Molybdenum, 10, 58, 188, 319, 619
  alloy, 595
  carbon doped, 319
Moulding methods, 1128
Moving strain field, 388
Multiphase microstructure, 208
Multislip, 34, 41, 944
"multivariance" mechanism, 278

N

n-value, 250
NaCl corrosion, 861
Nabarro, 1066
Nabarro-Herring creep, 758
"necklace" recrystallization, 521
Negative
  creep rate, 575
  transient, 614

"nest" precipitate, 645
Net structure, 223
Neutron-irradiated copper, 96, 413
Ni-Cr superalloys, 607
Ni-CrMo steel, 812
Ni-Fe, 559
Nickel
  aluminium particles, 980
  base composites, 141
  content, 207
  superalloy, 927
Niobium, 3, 10, 90, 222
Niobium-bearing
  carbon-manganese steel, 619
Nitride former, 408
Nitrided steel, 851
Non-coplanar slip, 551
Non-crystallographic slip
  traces, 44
Non-equivalent slip systems, 1138
Non-homogeneous strain, 494
Non-metallic inclusions, 449
Non-uniformity of strain, 1245
Notch, 273, 805
Notch toughness, 793
Notched tensile test, 813
Nuclear magnetic resonance, 103
Nucleation, 1192
  of kink pairs, 7
  on dislocations, 412

O

Obstacle strength, 958
Olivine, 735, 753, 758
Optical microscopy, 519
Order friction effects, 1014
Order locking, 1014
Ordered alloys, 1011
Ordering, 1016
Orowan
  loops, 549, 826, 1020
  mechanism, 421, 610
  model, 195
  relationship, 102, 1026
Ostwald ripening, 977
Overaged, 1220
Overload cycle, 868
Oxide particles, 716
Oxidizing environment, 883
Oxygen content, 121

P

Paraelastic, 1014
Partial fracture mechanical
  properties, 1066
Particle
  coarsening, 421, 977
  dispersion, 295
  hardening, 487
  locking, 704
  radius, 979
  spacing, 1068
  stimulated nucleation, 487
Particular reinforcement, 859
Peak torque, 226
Peak-aged, 875
Pearlite volume, 218
Peierls
  relief, 57
  stress, 16, 21, 70
  stress variation, 23
Penetration
  energy, 468
  resistance, 467
Penetrator
  geometry, 467
  tip truncation, 471
Per capita consumption, 1197
Permeability, 1139
Persistent slip bands, 819, 833,
  907, 934, 1150
Phenomenological theory, 589
Phonon dragging, 1021
Phosphorus content, 130, 789
Pile-up model, 641
Pile-ups, 51, 695
Plain carbon steel, 232
Planar
  faults, 956
  slip, 1220
Plane
  strain ductility, 948
  stress, 993
Plastic, 1203
  analysis, 393
  bending, 417
  deformation, 27, 554
  deformation energy, 971
  deformation hardness (Hart),
    589
  flow, 201, 465, 478, 511
  hole growth, 691
  instability, 128, 512, 805,
    1190
  relaxation, 51
  state equation, 571
  strain accumulation, 1222

strain ratio, 559, 1166
zone profile, 922
zone, 842, 849, 869, 920, 971, 1062
Plate guage, 217
Plugging failure, 470
Polycrystals, 27, 75
Polygonization, 311
Pore coalescence, 947
Porod's Law, 679
Porosity, 1121, 1125, 1139, 1238, 1240
Potassium, 22
Powder, 983
  composites, 981
  metallurgy, 367, 879, 1237
  process, 714
Power law, 582, 704
  behaviour, 1107
  creep, 616, 668, 758, 1140
Pre-aging, 655
Pre-yield, 96
Precipitate
  cutting, 420
  distribution, 194
  growth, 604
  shear strengthing, 1018
  sizes, 194
Precipitated particles, 1078
Precipitates, 1025
Precipitates, 144, 340, 658
Precipitation, 294, 311
  hardening, 101, 1018
  kinetics, 502
Preferentially yielding zones, 458
Pressing, 1237
Pressure
  confining, 1137
  tubes, 289
  vessel steels, 619
Prestrain, 232, 801
Prestress, 432
Primary
  clusters, 416
  materials, 1198
  slip, 414
Prior drawing, 309
Prismatic
  cross slip, 1020
  dislocations, 51
  glide, 27, 30
Process zone, 948
Prominent slip bands, 833
Punching stress, 53
Pyroxenes, 758

Q

Q-spuren, 55
Quadratic activity, 610
Quality control, 1121
Quartz, 1141
Quasi-cleavage fracture, 813
Quenched-in vacancies, 282

R

r-value, 559
Radiation effects, 953
Random boundary, 673
Rapid quenching, 135
Rapid solidification, 1028, 1091, 1122, 1128
Raw materials, 1199
Recovery time, 569
Recovery twins, 1192, 1193
Recovery, 753
Recrystallization, 277, 487, 521, 525, 536, 748, 753, 1080, 1139
  texture, 381, 1193
Recrystallized ferrite, 148
Relaxation curves, 5
Replicas, 402
Research topics, 1056
Resident
  capacity, 1053
  experts, 1053
Residual
  life, 655
  stresses, 363, 494, 851, 871
Retained austenite, 1172, 1225
Rheocasting, 1122, 1130
Rheology, 757
Ripples, 1126
River patterns, 1222
Rocks, 1137
Rolling, 966
  hot, 235
  schedules, 212
  texture, 385, 523, 542, 1191
Rotation recrystallization, 521, 748
Roughness-induced closure, 1223
Rudman model, 304
Rule of mixtures, 248
Rupture ductility, 606

S

S-N curves, 852
S-curve, 1197
SEM channeling contrast, 520
SFE, 1192
Sachs model, 76, 941
  modified, 76
Sandwich structure, 1205
Satellite cracks, 1071
Saturated loop, 1155
Saturation stress, 877, 911, 915
Scale formation, 476
Scaling property, 590
"scalloped" grain boundaries, 236
Scaly surface, 475
Schmid's Law, 40, 48
Screw dislocation, 6, 12, 70, 99
  core, 21
Segregates, 1121
Segregation, 606
Semiconductor crystals, 1142
Sequential mixing, 903
Serrated flow, 83
Serrated yielding, 419, 437
Serrations, 440, 441
Sharp yield, 407
Shear
  bands, 361, 362, 382, 512, 513, 525, 542, 965, 1186, 1187, 1188, 1190, 1192
  bands (cells), 548
  cracking, 478
  facets, 847
  failure, 1096
  fracture, 174, 801, 1139
  mechanism, 394
  plane angle, 512
  strain, 514
Shearable
  particles, 1220
  precipitates, 343, 1153
Sheet metals, 482
Shelf energy, 795
Short Range Order, 653, 1233
Short range dislocation interactions, 42
Shrinkage, 1125
Si flakes, 250
SiC, 860
Silicon, 257
Silicon-oxygen bond, 1138
Similarity criterion, 974
Single
  crystals, 3, 58, 90, 319, 933
  glide, 41
  peak overload, 919

Sintering, 984, 1238
Sintering of voids, 726
Size difference, 17
Size effect, 553
Slip, 320, 833, 1138, 1182, 1187, 1188, 1192
  band etching, 513
  band velocity, 98
  bands, 96, 351, 401, 876, 942
  character, 1233
  distance, 104
  heterogenities, 443
  length, 104
  lines, 52
  non-coplanar, 551
  reversibility, 1222
  secondary, 416, 944
  systems, 447
  transfer, 27
Small subgrains, 558
Small-angle neutron scattering, 678
Snoek damping, 153
Social arena, 1199
Software, 1207
Solid solutions, 1014, 1025
Solid-solution hardening, 45, 101
Solute
  atoms, 17, 437
  diffusion, 437
  drag, 704
  effect, 502
  migration, 203
  weakening, 15
Solution
  hardening, 331
  treatment, 264
Spall fracture, 811
Spatial distribution, 453
Special grain boundaries, 108
Specific energy, 553
Speich, (1981), 1172
Spheroidal γ', 632
Spheroidal pearlite, 1001
Spin-lattice relaxation, 102
Spinel, 758
Spinodal
  decomposition, 933
  precipitation, 1240
Squeeze casting, 1132
Stacking fault, 956
  energy (SFE), 304, 327, 335, 381, 486, 514, 527, 547, 653, 916, 958, 965, 1118, 1182
  strengthening, 1018
Stainless steel, 205, 582, 637, 655

Duplex, 505
Stair-rod dislocations, 279
Static recrystallization, 749, 1078
Steady state, 583
Steady-state flow, 1105
Steel, 493, 919, 959, 1203
   1.9 Mn, 153
   deep-drawing, 1168
   galvanized, 247
   HSLA, 161, 184, 235, 241, 1075
   high strength, low alloyed, 1075
   hot-rolled, 148
   hydrogen in, 453
   low-alloy, 793
   low carbon, 147, 895
   maraging, 193, 1021, 1225
   micro-alloyed, 407, 499
   mild, 387, 393, 556, 840
   Ni-CrMo, 812
   nitrided, 851
   plain-carbon, 232
   stainless, 205, 582, 637, 655
   structural, 799, 867
   ultra-high strength, 787, 1001
Steel line pipe, 407
Stircasting, 1130
Stoichometry, 1017
Stoneley waves, 107
Stored energy, 410, 496, 825
Strain
   ageing behaviour, 1172
   aging, 241, 394, 409
   cycling, 913
   distribution, 949
   free grains, 150
   hardening, 33, 709, 1096
   high, 425, 811
   in-situ, 953
   localization, 949
   non-homogeneous, 494
   non-uniformity, 1245
   rate sensitivity, 4, 439, 462, 685, 708, 730, 741
   softening, 709
   total, 446
   transient, 577, 613
Strength, 161, 582, 735
   differential, 1214
   ductility relationship, 1037
Strengthening mechanisms, 1009
Stress
   back, 229, 233, 825, 830
   constant, 570
   dependence, 656
   drops, 83

   effective, 4, 579
   exponent, 701
   intensity factor, 839, 995
   pulses, 388
   punching, 53
   redistribution, 851
   reduction, 613
   relaxation, 592, 830
   residual, 363, 494, 851, 871
   rupture test, 603
   saturation, 877, 911, 915
   strain curves, 58, 303, 507, 530
   strain relationships, 483
Stretchability, 1167
Striated bands, 701
Structural steels, 799, 867
Subgrain, 149, 333, 401, 530, 625, 750, 1025
   elongated, 555
   misorientation, 428, 595
   "pancake", 427
   size, 402, 581, 595, 596, 616
Subgrains ("pancake"), 427
Submerged arc, 121
Substructure, 213, 426, 519, 903, 913
   constant, 615
   size, 557
Sulphide
   globurization, 179
   nucleation, 181
   shape control, 179
Sulphur partitioning, 184
Superalloys, 649, 879, 1213
Superanelasticity, 721
Superconductivity, 11
Superdislocations, 956
Superficial layers, 1082
Superlattice dislocation, 304
Superplastic flow, 782
Superplasticity, 139, 683, 689, 695, 701, 707, 713, 721, 727, 782, 1141
Superposition, 1009
Surface
   cracks, 143
   defects, 1125
   dislocation, 45
   effect, 353, 953
   geometry, 819
   imperfections, 269
   melting, 314
   removal, 782
   replicas, 834
   solution hardening, 46
   to volume ratio, 714

Suzuki locking, 327
Synergistic
  co-segregation, 791
  effect, 704
  effects, 653

T

TTT curves, 237
Tangled dislocations, 1232
Taylor model, 76, 941
  modified, 76
Tearing, 273
Temper rolling, 245, 959
Temperature dependence, 656
  of crss, 59
Tempered martensite, 788
Tempering, 167, 186, 281
Tensile
  cracking, 478
  prestrain, 242
  properties, 217
  properties, 637
  strength, 276, 308, 561
  tests, 167
Texture, 272, 291, 535, 539, 564, 966
Texture development, 446
Thermal
  activation, 3
  component, 741
  expansion, 370
  mechanical treatment, 333
  stress, 356
Thermally activated quantum-mechanical Models, 63
Thermomechanical treatment (TMT), 147, 205, 307, 793, 1076, 1245
Thickness effects, 270
Thixocasting, 1122, 1131
Thixoforging, 1131
Threshold stress, 702, 709
Ti-Al, 1231
  alloys, 1220
Ti-bearing steels, 179
TiN, 180
Tin segregation, 676
Tin-lead alloy, 990
Titanium, 408, 499
  alloys, 313, 727, 1219
  carbide, 410
Tool steel, 185, 840
Topography, 897

Torsion
  creep, 576
  hot, 224, 518
  tests, 419
Total strain, 446
Toughness, 215, 449, 1005
  austenite grain size, 121
  plate thickness, 766
Trace impurities, 649
Transage 134, 263
Transformation
  dislocations, 157
  induced yielding, 456
  substructure, 158
Transgranular cracking, 884, 930
Transmission electron microscopy, 695
Tube extrusion, 290
Tungsten, 188
Turbulent, 55
Twin traces, 384
Twinned martensite, 1004
Twinning, 12, 351, 519, 536, 965, 1138, 1182, 1186, 1187, 1188, 1190, 1192
  sequence, 384
  shear, 527
Twist tests, 1081
Two-phase
  alloys, 475, 487
  microstructures, 1062
  "two-phase" model (fatigue), 911

U

Ultra-high strength steels, 787, 1001
Uniaxial creep, 570
Uniform plastic strain, 446
Unstable
  flow, 201
  plastic flow, 465
  yielding, 419
Uppermantle, 757

V

Vacancy concentration, 169, 833
Vacuum, 928
  fatigue, 142
  remelted steel, 812

sealed moulding, 1128
technology, 989
Vanadium, 503
Variable
　amplitude fatigue, 1154
　loading, 1147, 1148, 1153
Vermicular cast iron, 200
Viscous glide, 1110
Void, 116, 721, 799, 1090
　coalescence, 119, 691
　density, 118
　growth, 118, 725, 1094
　initiation, 118
　nucleation, 814, 1090
　sizes, 722
Volume
　change, 412
　fraction, 129, 148, 223, 369, 730
　fraction of martensite, 1245

strength, 276, 561
stress, 10
upper, 387
surface, 1100
Yielding, 361, 495, 842
　behaviour, 1166
　unstable, 419
Yttrium, 313

Z

Zener-Holloman parameter, 532
Zinc, 523
　2.5%Nb, 289
　aluminium, 683
Zircaloy, 290
Zone melting, 58

W

Wants List, 1052
Water-rich phases, 1143
Weak-beam microscopy, 328
Weather-resistant steels, 601
Weight function, 995
Welding, 121
Weld-pool solidification, 1132
Weldments, 1121, 1132
Whisker reinforcement, 859
Widmanstatten rods, 508
William's Expansion, 995
Wire drawing, 425, 535
Wires, high strength, 1012
Wohler curves, 935, 1157
Work hardening, 51, 436, 638
　exponent, 462, 1166
　rate, 115, 375
Workpiece surface, 554

X

X-ray diffraction intensity, 887

Y

Yield
　criterion, 364
　point, 154, 303, 320, 959